Understanding the Heavens

Springer

Berlin
Heidelberg
New York
Barcelona
Hong Kong
London
Milan
Paris
Singapore
Tokyo

Physics and Astronomy ONLINE LIBRARY

http://www.springer.de/phys/

Jean-Claude Pecker

Understanding the Heavens

Thirty Centuries of Astronomical Ideas
from Ancient Thinking
to Modern Cosmology

Edited by Susan Kaufman

With 256 Figures
Including Many Historical and Hand-Drawn Illustrations

 Springer

Professor Jean-Claude Pecker
Collège de France
Annexe Ulm
3 rue d'Ulm
75231 Paris Cedex 05
France

Cover picture: Stonehenge from the air. The axis of the monuments is roughly aligned with the Heel Stone, seen on the lower left edge of the picture. The Aubrey Holes, which a modern astronomer could use for eclipse prediction, are seen in the outer ring of white dots. (Cambridge University Collection of Air Photographs: copyright reserved)

Library of Congress Cataloging-in-Publication Data

Pecker, Jean Claude.
   Understanding the heavens : thirty centuries of astronomical ideas from ancient thinking to modern cosmology / Jean-Claude Pecker ; edited by Susan Kaufman.
      p. cm.
   Includes bibliographical references and index.
   ISBN 3540631984 (hardcover : alk. paper)
   1. Astronomy--History. I. Title.

QB15 .P38 2001
520'.9--dc21

                                                                              00-024396

ISBN 3-540-63198-4 Springer-Verlag Berlin Heidelberg New York

Springer-Verlag Berlin Heidelberg New York
a member of BertelsmannSpringer Science+Business Media GmbH

http://www.springer.de

© Springer-Verlag Berlin Heidelberg 2001
Printed in Germany

Typesetting and figure processing by Satz- und Reprotechnik GmbH, 69502 Hemsbach
Cover design: de'blik, Berlin

Printed on acid-free paper   SPIN 10554718   55/3144/mf - 5 4 3 2 1 0

Foreword

This book stems from a series of lectures given to a select group of undergraduate students at Williams College in Williamstown, Massachusetts where I taught at the kind invitation of Professor Jay Pasachoff in the fall of 1989. I want to heartily thank Jay Pasachoff for this wonderful opportunity to resume the activity of teaching the young, a very stimulating one indeed; and I want to thank the students themselves for the many exciting debates I had with them about the questions treated in this book and other interesting subjects.

The book covers almost three millennia of the evolution of ideas about what is now called the "Universe" but which, until the 18th century, was called the "World." It is not always an easy book. It is intended for the learned student, whatever his level, undergraduate or graduate, in astrophysics or in history or related fields. The more difficult passages are either grouped in appendices or indicated by an asterisk ($*$).

I must insist that I am by no means an historian, only an astrophysicist trying to reveal the concepts and predispositions that may be lying behind some of the model universes published in the literature. To advance an understanding of these models, I have, therefore, gone into specific descriptions, sometimes involving mathematical details, with the conviction that this thorough understanding is necessary to follow the evolution of ideas.

I would like to express my thanks to Springer-Verlag and in particular to Professor Wolf Beiglböck, who agreed to publish this book even before it was in a readable form and whose staff helped so much in all the phases of its completion. I want to additionally express my thanks to all of my colleagues and friends who have provided documentation or illustrations, in particular Dr Laurent Nottale, who carefully read the last three chapters. I deeply appreciate the friendly hand given by my two co-writers, Dr Daniel Pecker, for the appendix in Chap. 5, and Dr Simone

Dumont, for Chap. 6, who have both devoted much of their time to developing these two parts of the book from rough text to final version. I am grateful to Dr Suzanne Débarbat who helped with some of the tables and the chronology and to the Chief Librarians of both the Observatoire de Paris and the Institut de France, and their assistants, Mme Pastoureau and Mme Chassagne at the Institut, and Mme Daliès and Mlle Alexandre at the Observatoire whose help was invaluable in finding the needed ancient documents.

English is not my mother tongue. Ms Susan Kaufman in Massachusetts was tremendously helpful in transforming my "Frenglish" into perfect English. We had many exchanges, mostly by fax, over the ocean from Great Barrington to the Ile d'Yeu and vice versa and I must say it was a pleasant and fruitful experience. To Ms Dominique Bidois who converted the text from the original Claris to Word, I also owe many thanks.

At long last, the book is in print with the best possible help from the staff at Springer Verlag. Ms Andrea Kübler and Ms Friedhilde Meyer dealt primarily with the text itself. Ms Brigitte Reichel-Mayer took great care in gathering old documents. Despite my unusual requirements, they kept their good humor and I am grateful to them for their excellent work. Of course, this book is also the product of the efforts of many people behind the scene whose names are unknown to me. Last, but not least, I am grateful to my wife, not only for her material help in the editing process, but still more for her patience and understanding love during the several years of preparation of this book.

This book's appearance coincides with the beginning of a new century. Let us hope that it stimulates more research on the state and evolution of our knowledge and understanding of the heavens.

L'île d'Yeu   Jean-Claude Pecker
May 10th, 2000

Contents

Introduction and Perspectives

Thousands of books have been written about the history of science, notably about the history of astronomy. A selected bibliography is given at the end of this book. Some authors regard the history from an epistemological point of view, trying to find some logic to the history, to discover some trace of the evolution of methodologies and points of view. Some just try to offer examples, presumably typical of the evolution of ideas and techniques. Some defend a theory, some just give the facts. But of course, in merely giving the facts, there are biases which, unavoidably, reveal the attitudes of the writer and his epistemological position. Even more, some historians of astronomy cannot resist the temptation to introduce the scientific achievements in some philosophical frame of reference, often of a religious nature. The bias of the writer is often conveyed subtly, by implication, rather than by any overt statement.

In addition, history of science and epistemology are rarely taught to undergraduate students in physics. It is as if the subject matter were too controversial or too specialized.

On the contrary, and in spite of the potentially controversial nature of the subject, a knowledge of its history is a necessary condition for the understanding of science. Controversial or not, a perspective is badly needed by beginners and by laymen. There is, however, no way for the author of a book, even aimed at a relatively inexperienced group of readers, to hide his preconceptions. So my book, like all the others, has its own idiosyncrasies.

First of all, it must be clear that the book is aimed at the undergraduate student, or the beginner, perhaps even at the layman. Whenever a fact is presented, a demonstration is given, in the most simple, but rigorous, mathematical terms that can be used. Figures are as explicit as possible. Difficult quoted texts, or passages, are indicated in the margin by an asterisk at the beginning, and at the end, of the section ($*$). The style is as direct as possible, and astronomical jargon has been used only in as much as it shortens and simplifies the developments. I hope that in so doing, I have made sometimes difficult subject matter accessible to a wide readership. Moreover, a complete index is given, so that all terms that appear in the text are defined somewhere else in the book. In addition, the birth and death dates of the scientists quoted are explicitly given, whenever possible, at least in the index; the various orthographies of their names (mostly for Arabic scientists) are also given in the index. We hope, therefore, that this book will be easy and pleasant to read.

A second decision I had to make, in view of the fact that the book is based on a one-semester course at Williams College, was to limit the book to the evolution of astronomical ideas, leaving aside most of the instrumental developments (outlined, however, in Chap. 6). As a consequence, one chapter has been added to the book, as compared with the course given at Williams, to cover the classical history of observational astronomy.

Since no book of this sort can be written without some a priori perspective, and mine differs from some of the books previously written on this matter, I must devote some space to explaining the point of view, necessarily subjective, which has guided the writing of this book.

Throughout this book, for the sake of simplicity, Greek and Arabic names habe been rendered in the American style, without accents.

The Theory of Scientific Revolutions

Under the influence of Thomas Kuhn (1957), many modern writers have based their history of astronomical ideas on the emergence, at key periods, of some new paradigm, marking a "scientific revolution." This point of view has the advantage of allowing the writer to unify various disparate developments in the evolution of science. I disagree, at least partly, with this view which I find too extreme.

A paradigm, such as the heliocentric system of the Copernican theory, is composed of many different paradigms of a smaller scope. Perhaps in fact, we should not use the word *paradigm* for the whole set of paradigms, but a word such as *doctrine*. The Copernican doctrine then consists of several paradigms, some of them still of an Aristotelian origin, or even a Eudoxian origin, some of them Ptolemaic, some others still different and more modern. We can consider paradigmatic of the pre-Ptolemaic doctrine the following statements: (1) The very nature of heavenly bodies (stars, planets) differs from that of the sublunary world. The sublunary world is subject to imperfections, fantasies, while the astral world is, on the contrary, perfect, and even eternal (aether-nal, to play on the "aetheral" nature of the astral world); (2) Trajectories of planets cannot be anything but circular; (3) Motions of planets are uniform, i.e. of constant velocity on their trajectories; (4) Only a material body can be located at the center of all these trajectories; (5) This material body is the Earth.

The first paradigm, very dear to Aristotle, was disproved in 1572 by Tycho Brahe on the occasion of the discovery of the famous supernova and confirmed a little later, again by Tycho, with the observation of a comet. In both cases, Tycho demonstrated by accurate measurements that both objects were definitely more distant than the Moon; hence, they did not belong to the sublunary world. Still, they were observed to exhibit variations in brightness and motions of great amplitude. We should add to this that the demonstration is ongoing, with the modern proof that the chemical composition of the Universe is in essence the same as that of our Earth and Moon. The unity of the Universe is one of our modern paradigms, but contemporary scientists are not the first to conceive of this idea. Centuries ago some early thinkers, such as John Philopon, already refused to accept the Aristotelian distinction between astral and sublunary worlds.

The second paradigm was kept alive by Copernicus and by Tycho. Only Kepler showed that it was necessary, in order to properly describe the accurate data obtained by Tycho, to substitute elliptical trajectories. Circular trajectories were no longer adequate, except as a poor approximation. Kepler's first law is one of the bases of Newtonian mechanics, which is still in use.

The third paradigm was refuted as far back as Antiquity! Ptolemy's use of a very artificial construction, that of the *equans* (see p. 105) in about the year 150 of our era forced the planet to move in a uniform way, not around the center of the circular trajectory, but rather, around a point different from this center. The uniformity of motion along the trajectory was, for the most part, lost, though some uniformity was preserved. We should note that uniformity is also preserved by Kepler's second law, which says that a constant area is swept in a given time interval by the vector radius during the motion of a planet on its trajectory (see p. 226–228).

The fourth paradigm was upset very early. Even the Pythagorian Phi-lolaus, in his strange system, envisioned the Earth and the "anti-Earth" turning around some immaterial central fire. In the Ptolemaic system, the planets rotate on a circle, on a uniform path, but the center of this circle is an imaginary point, not a material one. This innovation is the reason why both Averroes and Maimonides, in their faithful admiration of Aristotle and of the Bible (not always in contradiction), rejected the Ptolemaic system.

The fifth paradigm was indeed also overturned in Antiquity with the development of the heliocentric system of Aristarchus of Samos. Of course, this system was not good at describing the planetary motions. Ptolemy was much better at that and this was the basis of the success, the lasting success, of his system. Ptolemy's system was geocentric. He-liocentrism, truly enough, came back (and this will be discussed later in greater detail), notably in the work of Copernicus. It did not convince everyone at once. Even the great Tycho stuck to a form of mitigated geo-centrism. In the 17th century, the detractors of Galileo still fought for geocentrism. But by then it was a losing battle. We certainly no longer consider the geocentric model seriously. Are we heliocentric? In a sense yes: we conceive of planets rotating around the Sun ... but the Sun itself is only in the suburbs of our Galaxy. Our Galaxy is one among thirty (or so) galaxies of the Local Cluster of galaxies and our Cluster is one in mil-lions of clusters.

Can we speak, therefore, of a "Copernican revolution"? Indeed, there was an evolution in our thinking about the world, an evolution which was very rapid and far-reaching from the 14th to the 17th century. In-strumental in this evolution were many factors, some purely scientific, some political, some economic. During this period, translations of the great Greek works, through Arabic intermediate translators, permeated the Western world, where great translators were at work, notably in in-ternational centers such as Toledo. The work of Gerard of Cremona, Ade-lard of Bath (12th century), and Michel Scot (13th century) had a consid-erable importance. Creation of universities everywhere in Europe and even of international centers where different cultures engaged in heated but friendly arguments (the Cordoba school, where Jewish, Arab, and Christian scientists were working together, is a good example, but not unique) added to the intellectual ferment. The discovery of modern printing techniques by Gutenberg allowed ideas to circulate. The great discoveries by travellers such as Columbus, Amerigo Vespucci and many others helped to promote a new global understanding. Dutch opticians improved the technique of making eyeglasses and combinations of lenses were exhibited as a curiosity at fairs. Galileo had the idea of turning them on the sky. The battle between the Roman Church and the Reformed rebels led to the constitution of the Society of Jesus; and the Jesuits, at Rome, had a considerable influence on science at the beginning of the 17th century. This is only a partial list. It is too easy to put the date of the birth of the modern world at 1453, the date of the capture of Constantinople by the Turks. We could just as well select 1492, the discovery of the New World by Columbus, or 1476, the publication of Copernicus's treatise. Why not the apex of the Cordoba school (around the beginning of the 13th century)? Or the creation of the University in Padua in 1222? These are artificial turning points, as I hope I have shown. They represent a turn in History, but History always turns in circles, or more properly in spirals. We are always near some turning point in History; and it is true in science also.

If Kuhn's theory of scientific revolutions harbors a great danger (although I feel certain that Kuhn's prestige itself is not endangered by it), it is to allow people to believe that, at any given time in the history of science, we can forget the past and start again from scratch. This is actually almost true in art, literature, music. Picasso could create without a deep knowledge of Raphael's art; Beethoven could compose almost ignoring Rameau; Joyce could write without having read Cervantes. We cannot deny influences of the past masterworks upon the more recent; but the affiliation is not so clear; and no scale of values is conceivable. In the evolution of scientific ideas, however, to build a statue of Einstein does not mean that we have to destroy that of Newton, and to build that of Copernicus does not mean that we have to destroy that of Ptolemy. Actually, Copernicus's system is largely inspired by the Ptolemaic model, as we shall see (p. 191 ff.). Each generation of scientists builds on what preceding generations have achieved, and improves it.

Interactions Between Politics and Scientific Progress

Another type of analysis of the progress of science is the association of changes with social developments, and the needs, or even the wills, of the decision makers. Jacques Blamont's, in *Le Chiffre et le Songe* (*The*

Cipher and the Dream), for example, documents the history of the determinant relations between political events and the development of science. In particular, with regard to the Copernican system, Blamont notes the primacy of the importance of the historical process. After Saint Thomas Aquinas and the satisfactory merging of Aristotelianism and Christian thought the Church was, at the end of the 14th century, rather firm in its positions with respect to the World. The oriental wars, however, brought to the West many scholars from Constantinople, fleeing from the Turkish conquests. Byzantine science illuminated the intellectual life of Italy. Peripatetic doctrines, Platonic theories flourished in the philosophy of that time. The worlds of finance and of local politics became interested. Cosmo di Medici put his great wealth to use by helping Marsile Ficino (1433–1499) to create the Academy of Careggi. Ficino was a lyric Platonist and tried to reconcile neo-Platonism with Christian doctrine. The Florentine aristocracy, particularly Lorenzo di Medici, Cosmo's grandson, continued to be supportive. Primarily a theologian, but also a very inspired poet, Ficino built a mystical world, centered around a multiple God, simultaneously at the Center and the Circumference. "It is in the Sun that He put His tabernacle." (*de Vita*, 1483) Theology assumed a truly cosmological vision. Light itself became the vehicle of God's thoughts, of God's will towards inferior beings. This constituted a renewal of mystical Platonism, which opened the door to all imagination and all fantasies. It was without doubt an interesting path, but a dangerous one. Blamont noted properly that Italy, soon deeply impregnated by the influence of immigrants from Constantinople, was blinded to a more positivist form of thought. Without following Blamont too closely, we shall describe in the course of this book this intertwining of political movements and ideological debates.

Unlike what happened in Italy, in Northern Europe, particularly in Germany, universities were created one after the other. In this culture of learned princes, skeptical priests, and rich merchants, a new interest developed in science, in knowledge, perhaps to the very extent to which it implied a rebellion against the menace of a tyrannical Roman papacy. These thinkers, priests but also astronomers such as Nicola da Cusa (1401–1464), Georg Peurbach (1423–1461), and Regiomontanus (Johannes Müller, 1436–1476), contributed extensively (see later in this book) to our astronomical knowledge. They, and some others, were the very able predecessors of Copernicus (1473–1533). We mustn't forget, however, that Copernicus travelled in Italy, and must have been influenced also, perhaps in a very determinant way, by the mystical verse of Marsile Ficino such as the poet's work *De Sole*, published in 1492.

On the one hand, we see in Northern Europe, the emergence of Copernicus's almost dogmatic statements about the system of the world, and on the other hand, the Lutheran revolt. The machinery was launched from this active intellectual climate in Germany and Poland, as well as England, Denmark, Bohemia, and France. The momentum returned to Italy only with a sort of self-examination, of introspective criticism which the collapsing and disillusioned Church felt forced to undertake, albeit in a fighting spirit. Under Pope Paul III, the Council of Trent was held (1545–1563) and the Society of Jesus was established under the direction of Ignatius Loyola (1491–1556). At about the same time, the Index Congregation, which had existed since 1215, became a real weapon of the Church and unleashed the full power of the Inquisition now extremely strong and influential in Spain and Italy. Under the energetic leadership of Popes Jules III, Paul IV and their successors in the 16th century, the Counter-Reformation dug in for the long battle. But the Church, preoccupied with the profound social changes of the period, did not bother itself too much as yet with new ideas about the Universe.

The Roman College (the Sacred College) was created by Ignatius under strict military discipline. It was, as Blamont pointed out, bankrolled with New World treasure. "American gold has generated the Roman College." This was the gold of the Spain of Charles V, of Philip II, of the Conquistadors. The Roman College was devoted to Thomas Aquinas's version of Aristotelianism. There, at the turn of the century, we find Cardinal Bellarmin (1542–1621), who was to become the very close friend, and later the not so friendly enemy, of Galileo (1564–1642).

In Italy, the battle raged between the Roman power and the universities of Northern and Central Italy (in Parma, for example, where Galileo taught for several years) especially by the Serenissima Reppublica of Venice. There is little doubt that the Galileo saga was dominated indeed by political rivalries, involving the Roman power and its neighbors and rivals. The Church condemned Galileo after a long history of growing misunderstandings. Galileo, however, was not the only scientist to be involved. Other thinkers and priests were then strongly attacked by the Church on the basis of their ideas. Let us take the case of Giordano Bruno. He was in Rome, in the circle around Pius V, in 1571. Suddenly, in 1576, he fled. He travelled north, had contacts with the Reformers in Geneva, and it is from Germany that his ideas about the Universe started to be disseminated in Latin. Without entering here into the motives for the condemnation, he was condemned by the Inquisition in 1600 and died at the stake. Marc-Antonio de Dominis, an Illyrian scholar and priest, was condemned to exile slightly later. His books, clearly Copernican, were burned. The Index had been slow in fingering Copernicus's treatise after its publication (before the real emergence of Reform as a political power); but it did so quickly, as soon as Galileo was condemned, for reasons which were obviously not merely scientific and theological.

The political and social evolution of the European world was thus a substantial factor in the progress of ideas. However, we hold that many factors escape that logic. It is difficult, it is true, to describe scientific evolution without setting the scientists in their historical context, as means to further the ends of the decision makers, and even as their puppets on occasion. One cannot help, however, to see the chain of intelligences which, transcending political barriers, allowed ideas to jump from Italy to Germany, or Poland, from Denmark or Bohemia to Italy again, and later to France and England. For example, what was more important for Copernicus himself: his travels in Italy, his readings of Ficino, his friendship with Domenico Maria de Novara, a professor at Bologna (both suggested by Blamont); or the political evolution of Europe in this time of great upheaval? We see, at all times and at all latitudes, the perpetuation of an international network of clever minds. They exchanged letters; they read books; they translated Latin into German, or Arabic and Greek into Latin. We cannot escape the idea that a progressive evolution was taking place, each thinker acting from his own motivation and adding his individual gifts to the whole, making use of all the knowledge accumulated before him, as well as all the errors of his predecessors on the way to knowledge.

The intellectual community thus exists indeed as a result of these exchanges. Since the dawn of Antiquity, there has been such a network with its geniuses, its mediocre monks, its plagiarists and its forgers, its strong links and its weak ones.

The Continuous Progress of Astronomical Techniques

Within this intellectual community, the progress accomplished here and there in observational techniques used in astronomy was spreading quickly.

We will come back to the progress of astronomical ideas, but we must give a few examples here of the parallel developments in the technical area. The very fine geographic techniques of Erathosthenes allowed for the determination of the size of the Earth. They were initially conceived more or less for military purposes. The optical alignments permitted by the fine devices of Tycho, a great improvement over more primitive devices such as Levi Ben Gerson's Jacob's stick, allowed for the development of an accurate theory of the motion of Mars and the success of Kepler's ideas. They paved the way for optical astronomy, which began with Galileo's discovery of the satellites of Jupiter. The techniques of time keeping were instrumental in the progress of astronomy in the 17th and 18th centuries. They were developed because of the needs of sailing expeditions. Large telescopes have been as necessary to the progress of modern cosmology as Galileo's very small refracting telescope was for our knowledge of the solar system. We could give many other examples (see Chap. 6).

At this point, we must make note of the one-way direction of this instrumental progression. Once a technical methodology is well used, it is impossible, let us say almost impossible, to revert to a prior methodology. This is the main factor which promoted some illusion of progress in the realm of ideas. Actually, from Antiquity to modern times, the "world-makers," basically theoreticians with a good knowledge of observation, used technical progress and all available tools. Their body of knowledge constantly increased. Their ideas became more and more grounded. Nonetheless, world-makers operate at the borders of the known universe. It is important to understand that for the Ancients, the *World* was essentially the Earth, but the scaling of the solar system by the Greeks led to the use of the word *World* to mean: solar system. In this context, world systems were descriptions of the planetary system which was the only known basis for a consideration of the structure of the Universe. The progress of techniques and knowledge after Galileo expanded the notion of *World* so that it came to mean *Universe*, i.e. everything observable and not observable. In other words, ideas about the *World* at one time fell under the heading of what we now call *cosmogony*, a study limited to our solar system, whereas our more modern understanding of the Universe falls under the heading of *cosmology*. The theories of "world-makers" are based on extrapolations from what we can measure to what we cannot measure. The theories are risky and often "infalsifiable" at the time of their expression. Pushed by technical progress, the border moves further away and one can check predictions based on theories previously advanced. Then, they can be maintained or rejected on a more secure basis, but there remains a border and a great deal of room for extrapolation. The "world-makers," pushed or drawn by new instrumentations, still have much grain to mill.

The Three Streams from Pythagoras, Plato, and Aristotle

The evolution of ideas is marked by a three-way tug-of-war involving instrumentation, observation and theory, by an immersion of this game in the political and economic flow of history, and by regular changes, at shorter or longer intervals, in the succession of paradigms that have characterized systems of the world.

It is the main thesis of this book that, in this continuous evolution, scientists have followed three distinct paths of thought, that of Pythagoras, that of Plato, and that of Aristotle. Before explaining what I mean by that, I would like to make clear that perhaps I should put each of these three names between quotation marks. I will speak of "Pythagoras," of "Plato," of "Aristotle." Certainly, for all three of them, though the body of their thought is broad, sometimes involving contradictory ideas, the coherence in their work is great. We know, however, that medieval Aristotelianism, reviewed and corrected by Thomas Aquinas and many others, is different from that of the Stagirite. We know that the Pythagorian fantasies are more religious than scientific and that Pythagoras would not recognize his own ideas distorted by modern physics. As far as Plato is concerned, what do the astronomical vision of the *Timaeus* and, say, the neo-Platonism of Marsile Ficino have in common? Looked at in this way, the thesis of the three paths appears ridiculous. However, it may be an interesting way of looking at the history of astronomical ideas. At any given time, the three doctrines have indeed coexisted. Sometimes, one of them has been either clearly dominant, or almost extinguished. At other times, they are all very powerful, giving rise to great and fascinating debates, as philosophical as they are purely astronomical. I am close to believing that these debates, generally concluded by the provisional victory of one point of view over the other two, are what Kuhn calls "scientific revolutions."

Let us briefly describe, if not Pythagoras's doctrine, our view of the doctrines of "Pythagoras," of "Plato," and of "Aristotle."

In essence, the Pythagorians, beginning in the 6th century before our era, found in numbers, i.e. in simple combinations of simple numbers, the foundation of everything. The Universe was, in their thinking, dominated by numbers, hence by the very idea of simplicity inherent in some mathematically-minded God. A perfect unity existed between this Universe and this God.

This idea had applications in many fields of human activity, in musical theory for example and in architecture. All kinds of symbolisms were associated with this numerological principle. The "male" principle was associated with the number 3, the "female" one with the number 2. The world was based on the harmony of spheres and matter was organized so that all quantities were expressed by simple combinations of small numbers.

The Pythagorian attitude can be found again in the role assigned to the five perfect polyhedrons by Plato (in the *Timaeus*), and much later by Kepler. One finds similar principles among the Kabbalists, such as the use of "gematria", a numerological system, as a logical principle. Is not the simplicity principle expressed much later by Occam nothing but a reformulation of a *bona fide* Pythagorianism? At the present time, the games with large numbers, the theory of elementary particles, and even the vocabulary invented by some modern scientists to underline their ideas (magic numbers, quark terminology, role of group theories, etc.) seem to me to be strongly influenced by a neo-Pythagorianism. The role of numbers, the importance of geometry, must not be underemphasized; but the trouble with the Pythagorians is that they often demanded that nature adjust to the requirements of numerical simplicity, more than to the observed facts.

Nothing is simple or one-dimensional. Plato, with his Pythagorian ideas, had in addition another type of thinking, of logic, with respect to the sensible world. The famous parable of the cave, described in the *Republic*, expressed the view that the observable world is nothing but an image, a projection on the walls of the cave of the real world external to the cave. The implication is that to acquire some knowledge concerning the external world, the real world, which we cannot really encompass in its totality, one has to start to *save the appearances*, or *save the phenomena*, – in Greek: σώζειν τὰ φαινόμενα.

Although this expression seems to have been used first by Simplicius, it undoubtedly underlies Plato's philosophy of nature. Still, Plato did not imply that saving the appearances was the only requirement of scientific inquiry. For him (under the influence of other thinkers, such as Anaxa-goras), the existence of some *meaning* in the Universe necessarily im-plied the role (as metaphoric as it may be) of the δημιουργός. Plato's spirituality is an essential component of his view of the Universe. The search for the divine is inseparable from the search for causality. There is no science without theology. The role of reason in the organization of the Universe is basic, but this reason is of divine essence. It is clear that neo-Platonism has emerged at several points in the development of science. After all, the Biblical description of Genesis is not that far re-moved and some of the Church Fathers were instrumental in the spread of Platonism. Augustine (354–430) must be included in this group, as well as non-Christian philosophers such as Plotinus (203–270), Jambli-cus, and Proclus, who often went further than Plato. The human mind, partaking of a divine essence, is an essential element of this mode of thought. Much later, John Scot Erigen and Abelard were neo-Platonists. Saint Thomas Aquinas attempted to combine Platonism in its pagan as-pects with an orthodox Catholicism. Is this not the essence of what we now call Judeo-Christian philosophy? Had not Giordano Bruno a hyper-Platonic view of the world? And what about some of the modern devo-tees of the big-bang Universe?

As for Aristotle, his great productivity as a real physicist, based on ob-servation first, made him closer to our modern physicists than any other early thinker. "Saving the appearances" was the essence of his methodol-ogy and one should consider him a real scientist, in the sense we now give to that word. Aristotle believed in an objective Universe; whether it came from an original chaos, or whether it has always been, is almost ir-relevant. His philosophy is much more complex and starts from an elab-orate theory of knowledge. Starting from premises, one has to look for causes based on demonstrations. In other words, there is a logic to the acquisition of knowledge. Of course, there is a great difference between the empiricism of the Aristotelian methods and our modern physics where accurate measurements using adequate instrumentations are the basis of our understanding of the physical world. Nevertheless, there is a continuity there. Aristotelian methodology is after all not so different from that of Bacon or even, in a general sense, from the modern one. The philosophy of Aristotle is a search for causes of phenomena at all costs, but not necessarily for the "primary" causes. It is the noblest task of the human spirit, oriented towards natural philosophy. Aristotle admits some divine existence, some organized mind having brought the

Universe into being from its chaotic origins. Unlike Plato, however, he denies the existence of some conscious project at work *permanently* in nature. Moreover, in essence, whereas Plato had more confidence in reason than in imperfect sensation as a means to knowledge, Aristotle fully rehabilitated empirical observations, which are gathered by our senses.

Medieval Aristotelians pushed the devotion to Aristotle to the extreme, even much further than Aristotle himself would have accepted. They made a gospel out of each of Aristotle's words. The Arabic thinker Averroes and the Jewish philosopher Maimonides are examples of this phenomenon. Out of respect to Aristotle, they rejected the Ptolemaic system. Strangely enough, the emblematic "saving the phenomena" slowly shifted into "saving the Gospels," – whether it be the Aristotelian or the Christian gospels. This shift was present in the minds of many, in their activities and demonstrations. Many of the great medieval thinkers were Aristotelians. Much progress made during this period was the result of the influence of Aristotelianism. Roger Bacon was a good example, among many. The 14th century saw a rebirth of Platonism (Marsile Ficino), but the influence of Aristotle was to decrease only temporarily. Einstein may be considered, for many reasons, a true disciple of Aristotle, as strange at that may seem.

So: simplicity based on the numbers, for "Pythagoras," – reason, and spirituality for "Plato," – and observed phenomena to be saved for "Aristotle." We have seen these three attitudes, coexisting in the works of many scientists, from early history up to the present time. We see them also exemplified simultaneously by different scientists. Most of the time, however, we see a sort of oscillation; periods dominated by Platonism followed by periods of dominant Aristotelianism, interspersed with flashes of Pythagorian fantasy.

The present physics and the present cosmology are clearly at a crossroads. I am personally afraid of a move towards the Pythagorian temptation, which I see as dangerous because of its implicit appeal to faith rather than reason. But the history of science is complex. The unidirectional development of scientific techniques, hence of knowledge, coupled with a tendency towards an erratic return of strange concepts; the appearance of "revolutions" when we only shift from one vision to another, waiting for the pendulum to swing back; the mixed influence of the trends of human history and of intellectual attitudes; the catalytic role of geniuses; the regressive role of religions … all these factors must be taken into account. Seen at a great distance from above, I would prefer to describe the whole with reference to the permanence of the human mind with its three basic attitudes, embedded in the ongoing changes of a developing world where demography, techniques, and economic factors are always pushing, often very strongly, in one direction or another.

We shall see throughout the history of science the countervailing pressures of the three tendencies, a "Pythagorian" one, a sort of mysticism-cabalism, a "Platonic" idealism, tainted by Christianity, and an "Aristotelian" positivism-realism-materialism. Needless to say, such a simplification, and its highly schematicized nomenclature, is subject to criticism. We accept it, since we, ourselves, consider this model no more than a very tenuous Ariane's thread through a very complex history.

CHAPTER 1

Before the Classical Greek Period

1.1
A General Overview of a Rapid Evolution

There are several obvious phenomena which deeply affect everyday life on Earth: night and day (regular behavior), the seasons (regular behavior), meteorology and climate (irregular behavior).

Since the beginning of recorded history, men have been deeply concerned by these phenomena, for obvious practical reasons. They wanted reliable means of predicting their occurrence. Hence, systems of divisions of time were introduced, but also observations of correlated astronomical phenomena, leading to the development of observational astronomy. Hence also, several attempts were made to describe, even to understand the observed phenomena. This led to the construction of *systems of the world* with, to a greater or lesser degree, the intention, first, to *save the phenomena,* in as simple a way as possible and, second, to be able to *predict* them.

The history of astronomy has indeed been a very, very long process, from the shepherds, the travellers and the sailors of early cultures up to the space telescopes and Einstein's theory.

The synthetic Tables 1.1, 1.2, and 1.3, and Fig. 1.1 are a general attempt to represent the earlier steps of this history.

In Table 1.1 (the beginnings of recorded history), one sees the appearance in the Middle East of the calendar year of 365 days and lunar months of 28 days (Egypt, 4241 BC). The Hebrews developed a calendar at a later date, not earlier than 3760 BC, the date given for the beginning of Jewish history, according to the Bible. At about the same time, the Babylonians knew how to "predict" phenomena (astrology), and, throughout that geographical area, numeration, writing techniques and alphabets, and the definition of units of length were developing.

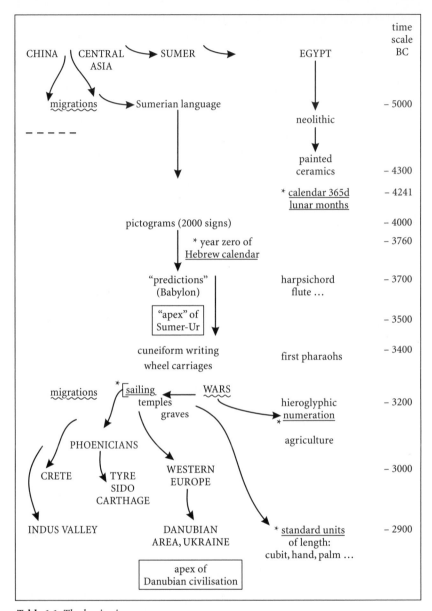

Table 1.1. *The beginnings*

The migrations, as described on the map (Fig. 1.1), were a way of spreading knowledge. The arrows in it are only partially dated. They correspond to arrows in Table 1.1.

The fifth millennium before our era was especially rich in the acquisition of knowledge. During this period in the Middle East, the zodiac was studied, vast areas were mapped for commercial purposes, the first observatories were built, and the first astronomical publications saw the light of day. The spread of culture in the remote north-western part of Europe on the one hand, and in China on the other, was characteristic. This is abstracted in Table 1.2 and Fig. 1.1. After this period, there was rather rapid progress.

Western astronomy was basically born in the Middle East then in Greece (Table 1.3). Although astronomy developed in the Far East, nota-

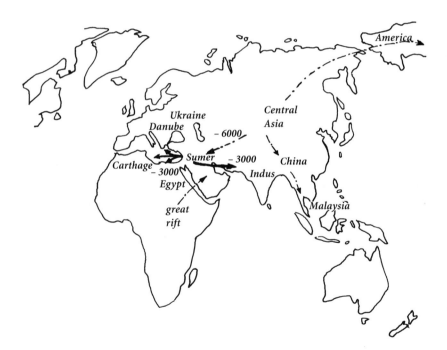

Fig. 1.1. *Great migrations before modern history.* Prehistoric migrations are indicated by broken lines. These are difficult to date precisely. The dark lines mark the migrations from Sumer to the east and west. These are known from historical and archaeological records.

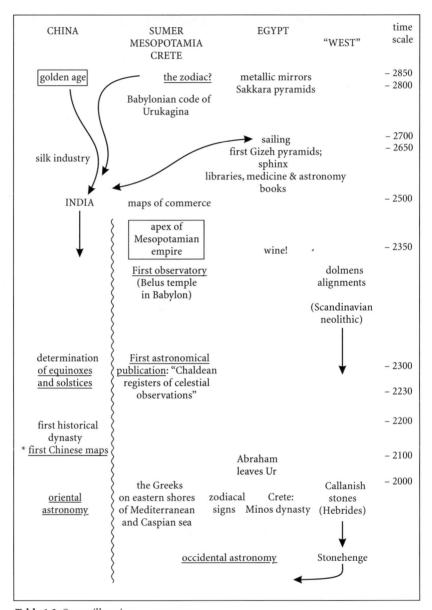

Table 1.2. *One millennium … years ago*

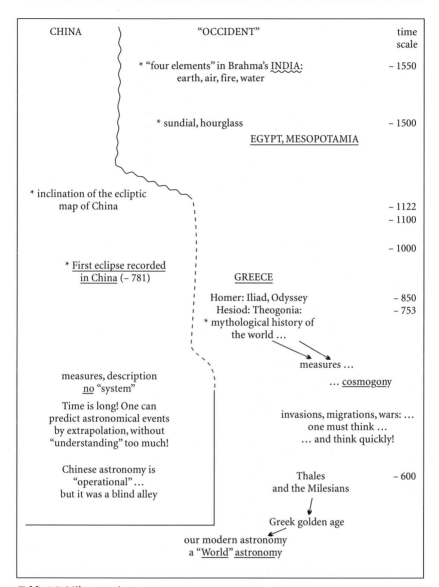

Table 1.3. *Milestones in astronomy*

bly in China, it flourished more efficiently during the "golden age" of Greece, the source of our modern way of thinking.

The reason for this difference from Chinese astronomy (which seemed to be stagnant until further contact with the West in the 17th century of our era) is possibly the relative stability of the Chinese world. The Chinese knew how to measure well, how to properly describe celestial phenomena. However, they did not have any "system" of the World that would permit them to understand by reference to a single model the complicated astronomical appearances. They could predict phenomena on the basis of extrapolation to the future of what happened in the past of their long history. So Chinese astronomy was operational, but it was a blind alley. At the opposite extreme, the Middle East, a melting pot of civilizations, was the locus for invasions and wars, resulting in a long series of forced migrations. Perhaps this explains why people in this part of the world had to learn how to think quickly, resulting in the development of operational models of the World, which would account for past and future astronomical phenomena, acting in a way like a modern computer program.

A first golden age of Western astronomy occurred between 1000 and 100 before our era, – say from Thales to Hipparchus. Then, apart from some important but scattered contributions (Ptolemy, Averroes, Maimonides, etc.), one has to wait until the 13th century in Western Europe to see a comparable development. This was the era of the translations from Greek and Arabic, the development of the printing press, and the great geographic discoveries. Modern astronomy, from Nicolas Oresme to Newton (about 1300–1700), was an outgrowth of this surge of progress (Table 1.4 p. 39). It is difficult to get a feeling for time scales and there is always a tendency to overemphasize the importance of recent history. This may be a reason why the Middle Ages (from 200 AD to 1300 AD) may seem dull, but actually, the struggle over new ideas, for a better understanding of the Universe, has never diminished, as we shall see in the course of this book. In the field of astronomy, at least, one should not speak about the "dark ages" when referring to this period. We must be more modest. Certainly, from our own times, several ideas will stand the test of time, and a few scientists of today will be remembered, as we remember Hipparchus, Copernicus and Newton. Some contemporary ideas, however, now believed to be essential and unassailable, might eventually appear to be minor epiphenomena. In spite of this regular progression of ideas, it is clear that, just as in any other field, demography is the primary determining factor. The fact that the current rate of increase of knowledge appears to be extremely fast can be linked with the fact that the number of active astronomers is much larger than it was at the beginning of our century. The IAU (the International Astronomical Union), which had 207 members in 1922, now has more than 8000 members.

At the time of Plato, there was no world-wide organization of astronomers. Perhaps 100 people in Greece devoted their time to thinking about astronomy, perhaps less, and like Plato, they generally didn't limit themselves to the field of astronomy. The concept of specialization is a modern one. Even at the beginning of the 19th century, there were probably no more than 200 or 300 astronomers and this population hadn't changed substantially since the beginning of our era.

1.2
Elementary Naked-Eye Astronomy

We will briefly discuss the questions asked by early astronomers from the point of view of what is now known about them. We encourage the reader to look into elementary books on astronomy (see bibliography) and to bear in mind that what seems obvious now was very far from obvious at the beginning. The purpose of this short introduction to astronomy is to allow the reader to understand the succeeding chapters in which we will try to explain, in a rather rigorous way, using the tools of geometry, the evolution of our understanding of the solar system, i.e. the known Universe of an earlier time.

1.2.1
The Two Luminaries

The two *luminaries,* the Sun and the Moon, were of course well-known. The importance of the Sun for everyday life was self-evident. Less important as a luminary, the Moon was also studied, as its shape changed periodically. Lunar phases were used for calendar purposes and carefully followed.

1.2.2
The Solar Cycles

The two *solar cycles*[1] and the *lunar cycle* led to the definition of time, and its division into days and years (Sun), and months (Moon).

One should perhaps note that the "non-commensurability" of these three cycles led to many practical problems, both in designing the calendar and in philosophical thinking. In the real world, things are not as simple as some would like them to be. Think how wonderful it would be to have a year of exactly 336 days, divided into 12 months of exactly 28 days, each starting at the exact time of the full Moon? To have a calendar without leap-years, without the Gregorian tricks, and with a perfect reconciliation of months with the Moon's phases would be a marvel! If that were the case, the periodicity of the Moon's phases would have to be exactly 28 days (4 weeks of 7 days) and the period of revolution of the Earth around the Sun exactly 336 times that of the rotation of the Earth around its axis, i.e. one day. That would have been a nice way to justify the 7 days of creation in Genesis – but things are not so simple!

According to modern ideas, the Earth has two distinct motions (the two "solar cycles"), its rotation around an axis, which gives rise to the succession of day and night, and its revolution around the Sun, which gives rise to the succession of seasons. Hence, a vague idea of "day" and "year" emerges from the "model." This looks simple to us and when viewed from the outside, or on a drawing (Fig. 1.2)

From the Earth's surface, however, when one wants to know whether the Earth has made a complete rotation or a complete revolution, it is not so simple, and to uncover the real facts has required a great deal of thought and discussion.

Let us look at this from one point in the northern hemisphere (Fig. 1.3). The Sun is at its highest point in the sky at midday (noon). In the daytime, the Sun rises in the east and sets in the west. After sunset, stars appear, now clearly visible on the dark background of the sky. They follow the same motion as the Sun.

---

[1] The reader is alerted to distinguish between this original use of the term "solar cycle" and what is now called properly the "solar activity cycle," but often abbreviated as "the solar cycle."

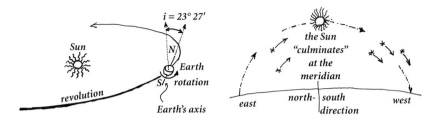

Fig. 1.2. *Rotation and revolution of the Earth.* Note that the axis of the Earth's rotation is at an angle of 23°27' from the line perpendicular to the plane of the Earth's revolution, called the *plane of the ecliptic.*

Fig. 1.3. *Motion of the Sun and stars, seen from the northern hemisphere.* The Sun and stars rise in the east and set in the west, culminating at their passage through the meridian, the plane which contains the polar axis of the Earth, and the vertical of the observing place. The drawing is oriented to the south.

One would expect, if this phenomenon were due solely to the Earth's rotation, to see the same stars rising at any given time after sunset every night.

But this is not the case.

At the beginning of each night, the whole of the starry sky seems to have moved a little bit towards the west with respect to the preceding night, so the stars "moved" slightly more than the Sun during the same time. This is easy to understand with the benefit of the modern "model" we have already alluded to. During the interval of one day, one *rotation,* a "true rotation," each point on Earth is oriented towards the far-away stars in the same way. But the Sun is close by: during the same time, the Earth has moved along its orbit around the Sun. After a "true rotation," a given point on the Earth is not oriented towards the direction of the Sun as it was before. Such effects, called *parallactic* effects, belong to everyone's daily experience (Fig. 1.4).

Hence, one must define two "days": the *sidereal day,* and the *solar day,* the latter being the more important for everyday life.

It is also necessary to distinguish between two independent apparent motions of the Sun. The Sun seems to turn around the Earth, in one solar day, but it also seems to move with respect to the stars. After a complete "year," the Sun will have described a complete turn of the starry sky. The study of the sky (for example an answer to the questions: what star rises just after sunset? or: at what period of the year does a given

Fig. 1.4. *Parallactic effects.* The cyclist sees the nearby trees moving rapidly out of his sight. The very distant line of mountains takes a long time to move out of his view, while the Moon seems to accompany him on his journey.

star, say Sirius, become visible before any other stars?) led to laws which could predict the return of seasons. Aboriginal peoples made use of these observations. The Amazonian Indians planted certain vegetables when the Pleiades became visible (their *heliac* rise) after sunset. In a similar way, one could observe the *heliac set* of a star.

The solar path in the starry sky is called the *zodiac*. Zodiacal groupings of stars, or zodiacal constellations, have been important in the history of astronomy.

As the Earth's axis is not perpendicular to the orbital plane (Fig. 1.2), the apparent path of the Sun (even when we are looking at the sky from the equator) does not go through the "zenith." What then is its apparent path? (Fig. 1.5)

The apparent rotation of the starry vault is a reflection of the Earth's axial rotation (Fig. 1.6). The Earth's axis, therefore, is also the sky's polar axis, not far from the "Polar star" and the plane of the equator is also the *celestial equator.*

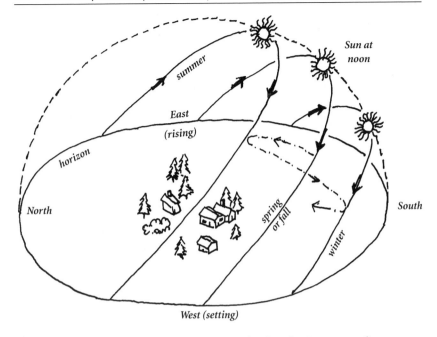

Fig. 1.5. *Apparent motions of the Sun.* During the day, the Sun moves from east to west (unbroken line). During the course of the year, its apparent orbit moves (broken line) from the great circle corresponding to the spring equinox (which goes through the zenith if the observer is located on the equator) to a more northern trajectory in summer, then back to the great circle at the fall equinox, and finally to a more southern trajectory. The two extreme daily trajectories are symmetrical with respect to the equinoctial daily trajectory.

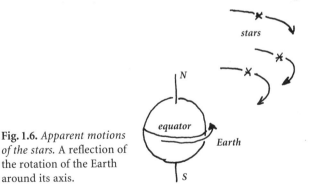

Fig. 1.6. *Apparent motions of the stars.* A reflection of the rotation of the Earth around its axis.

We shall see how important the reflections, or images on the sky, of the Earth's revolution around the Sun have been in the history of astronomical thought. The drift of the Sun during the whole year, with respect to the starry background, is a reflection of the Earth's revolution. Hence, the Sun's apparent motion is linked to the orbital plane. It stays within it. The apparent trajectory of the Sun stays in a plane with an inclination of 23°27′ with respect to the equator (see Fig. 1.2). This is the *plane of the ecliptic* (Fig. 1.7).

This very brief description of the main aspects associated with the double motion of the Earth has many consequences. In general, the Sun is not in the equatorial plane. Its apparent motion in the sky during the day changes from day to day. Its height above the horizon at midday (noon) is a function of the date. Twice in the year, the Sun is in the equatorial plane. Then, it stays on the equator, rises exactly in the east and sets exactly in the west. At noon, if seen from the equator, the Sun is then exactly at the zenith. Otherwise, the Sun culminates at noon at a different point, in the meridian plane. When the Sun is in the equatorial plane, the length of the day is equal to the length of night. These two instants define the "equinoxes." At this time (2000), they fall on the 20th of March at 7 h 30 m, in Universal Time (spring) and the 22d of September at 17 h 28 m in Universal Time (autumn), as seen in Fig. 1.5.

The year may be defined as the interval between two equinoxes of the same kind, for example, two vernal (i.e. spring) equinoxes.

It is necessary, however, to be careful with this definition! The year is the interval between two vernal equinoxes (the *tropical year*), but also the interval between two successive passages of the Sun in front of a given point of the stellar vault (*sidereal year*). These would lead to the same definition if the axis of the Earth did not change in direction over

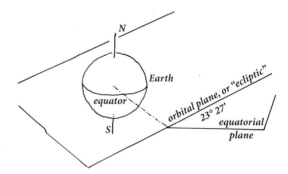

Fig. 1.7. *The plane of the ecliptic and the equatorial plane (see also Fig. 1.2.).*

Fig. 1.8. *The precession of equinoxes.* The polar axis of the Earth does not always point in the same direction N in the sky. It moves around a circle over a period of 26000 years. It is now pointed to a location near the polar star.

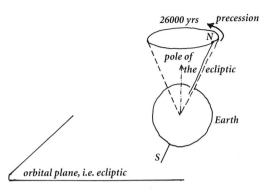

time, but, again, this is not the case. The two "years" do not have the same length. The axis of the Earth actually rotates in a conical motion with a period of about 26000 years (Fig. 1.8). With that period, the direction of the sky where the Sun is at the vernal equinox (the *vernal point* or *γ-point*) drifts with respect to the stars and describes the complete zodiac. This phenomenon is known as the *precession of the equinoxes*. Because of this precession, a tropical year will not be equivalent to a sidereal year. The tropical year is indeed the one which determines the seasons and which is, therefore, the more important for ordinary life on Earth.

This fact, known since the time of Hipparchus, means that our Polar star is near the pole now, but was not so millennia ago and will not be so millennia from now. It means also that the zodiacal constellations are regularly moving with respect to the vernal point. The astrological signs were named about 2500 years ago, but because of the drift described above, each sign is now located in a constellation different from the one for which it was originally named. For example, the sign Aries (the Ram) coincides with the constellation Pisces (the Fishes) (Fig. 1.9) and so forth.

1.2.3
The Moon's Cycle

The Moon accompanies the Earth in its revolution. This is known now. The Moon also moves (as does the Sun) with respect to the stars. It will appear at the same location on the starry background after a certain time, called the *sidereal month*. As the Sun also moves with respect to

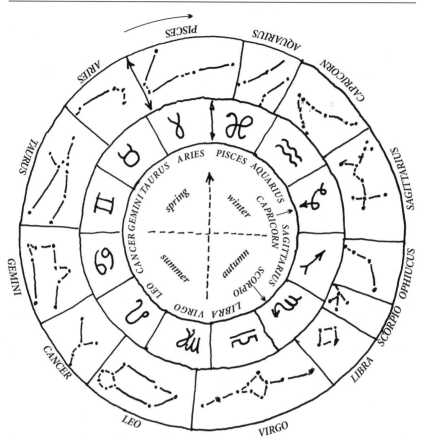

Fig. 1.9. *The zodiac.* The signs of the zodiac (inner crown), named for constellations, are located with respect to the equinoxes and solstices according to the locations of these constellations in Antiquity when the signs were designated. In our time, the true constellations have shifted with respect to the seasonal divisions. Note that they are not crossed by the Sun in one month. In some cases, that time is shorter, in some cases longer. Note also that there are thirteen signs, not twelve. The thirteenth is Ophiucus, the Man who carries the Serpent.

the stars, the interval between two identical phases (for example, two successive full Moons) is different. This is the *synodic month,* a natural unit, easy to observe, and slightly longer than the sidereal month.

Theoretically, the full Moon occurs whenever the Earth is just between the Sun and the Moon, the new Moon whenever the Moon is between the Sun and the Earth. Those two situations should result in, respectively, solar and lunar eclipses. However, the two orbits are not actually in the

Fig. 1.10. *The Moon's orbital plane.* Because the Moon's orbital plane differs from the plane of the ecliptic, the full Moon rarely coincides with a solar eclipse and the new Moon rarely coincides with a lunar eclipse.

same plane. It is only when the Moon is in the plane of the ecliptic, at full Moon or new Moon, that one can observe an eclipse. This is the origin of the word "ecliptic" (Fig. 1.10).

1.2.4
Calendars and Problems

The natural unit of time is the solar day, but the orbit of the Earth around the Sun is not strictly circular. It is elliptical, as we have known since Kepler's time. The velocity of its motion in its orbit is, therefore, not constant. This results in the fact that the duration of the solar day, in terms of sidereal time, is not constant. We have, therefore, defined a *mean solar day* divided into 24 hours.

Time, expressed in fractions of the solar day, is measured by sundials, or gnomons. We can watch the shadow move over the course of the day. At noon, it has its smallest length on a horizontal plane. In order, however, to arrive at *mean solar time,* the time as determined by sundials must be corrected by a term (the "equation of time") which may reach 15 minutes, plus or minus, and which varies from one day to another with a periodicity of a year.

Note that:
1 sidereal year = 365.2465 mean solar days
1 tropical year = 365.2422 mean solar days.

For the same reason, i.e. the ellipticity of the Earth's orbit around the Sun and the non-uniform velocity of the Earth on its orbit which results from its ellipticity (as shown by Kepler), each season (roughly one quarter, geometrically, of the orbit) has a different duration (Fig. 1.11).

Fig. 1.11. *The duration of the seasons.* The length of the seasons is not equal as the Earth is not at the center of the apparent orbit of the Sun. Assuming the motion of the Sun on its apparent orbit to be uniform, Hipparchus determined the *eccentricity* of the Sun's apparent orbit.

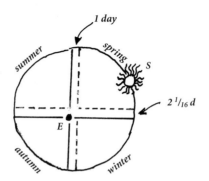

Spring lasts 92.76 days; summer 93.66 days; autumn 89.84 days, and winter 88.98 days.

The sidereal month (defined from the apparent phases of the Moon) varies from 27.03 to 27.61 days from month to month. The four quarters of the Moon's motion have unequal duration, because the Moon's orbit around the Earth is also an ellipse, not a circle centered at the center of the Earth. Neither the month, nor the year contains an integer number of days. Nor does the year contain an exact number of months.

In addition, we know that many gravitational forces, tides and so forth, affect all these values, which evolve slowly, very slowly, over the millennia. Neither cycle has a constant duration, but if these latter effects are very small, the above-mentioned ones are really overwhelming, and make the construction of any satisfactory calendar a terrible and almost impossible task.

Two conditions must be fulfilled to design a reasonable calendar: the mathematical ability to manipulate *fractions* (which permit the determination of "large" cycles, such as those governing the return of eclipses) and the permanent, careful, quantitative measurement of positions of the Moon, the Sun, and the stars in the sky at well-defined times.

The Babylonians acquired a higher level of mastery of fractions than the Egyptians, hence their dominant influence on later Greek science. A system of numeration based on divisions by 60 (degrees into minutes, minutes into seconds) also originated in this period.

Among the old instruments devised to measure astronomical time intervals were the *gnomons* (a type of sundial) (Fig. 1.12). At night, one used the *merkhet,* and the stars themselves (Fig. 1.13).

Hourglasses and clepsydras also provided measurements of physical time. Strangely enough, however, the Egyptians, who used them, did not

Fig. 1.12. *The gnomon.*
During the day, a vertical
stylus has a shadow. The
movement and location of
the shadow allows us to
determine and measure
time. The gnomon is a
particular form of sundial.

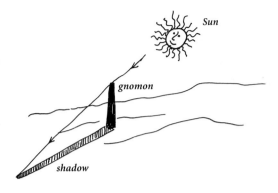

Fig. 1.13. *The merkhet.* At
night, this instrument
enables the observer to
determine the apparent
height of a star above the
horizon.

employ them even to try to study the evolution of astronomical phenomena (Fig. 1.14).

There is actually no real a priori reason for astronomical phenomena to define the same time as physical phenomena. In other words, there is no reason a priori to believe that astronomical motions are uniform at all scales, but so strong was the "evidence" that there can be only one form of time, that not much thought was given to this question until only very recently.

Fig. 1.14. *Hourglasses and
clepsydras.* These instruments, which became with
time much more elaborate
than in this simplified figure, measure *physical time,*
which may differ from *astronomical time,* however
we choose to define it, and
irrespective of which astronomical phenomenon we
may select as a basis.

1.2.5
Philosophers in Egypt and Mesopotamia

In Egypt as in China, a stable society encouraged a long series of recorded observations. The ancient world was characterized by a feeling of
periodicity, of the "eternal return" of everything. This explains the care
given to burial, as well as the use of astronomical observations to predict
the return of celestial bodies. The sky was a map, a guide, but there was
no interest in trying to understand why it is as it is. For this reason, there
was a lack of interest in irregularities, be they apparent, such as planetary motions, or even real, such as meteorological phenomena.

Like Chinese astronomy, Egyptian astronomy was therefore condemned to immobility – a dead end!

In Mesopotamia, in Babylonia, the situation was somewhat different.
The sky was seen as a model of the human realm. A correspondence was
thought to exist between the fate of empires (remember: there, they often
fell) and the fate of heavenly bodies. The prediction of "strange" phenomena (motions of planets, comets, eclipses) was of great concern to
them, although they did not really master any of these predictions. This
explains the birth, in that part of the world, of both astronomy, as we understand it today, and its extension to astrology which the Babylonians
regarded as a natural application of their observations of the sky. These
ideas from the Middle East spread quickly (through wars, invasions, battles, migration) into the eastern part of the Mediterranean, i.e. to Greece.

The next step, which led the Greeks to modern astronomy, was therefore a systematic effort to describe precisely, to "model," hence to predict, the motion of planets.

1.2.6
Eclipses

Among the regular phenomena that were known were the solar and lunar
eclipses. The regularity of eclipses was, of course, not immediately apparent as they appeared at significant intervals and locations that were
distant from one another. A solar eclipse occurs whenever the Moon is
seen between the observer and the Sun. The observer is then in the shadow projected by the Moon on the Earth. The observer can see the solar
disc occulted by the disc of the Moon, in successive steps, until complete
occultation, before a progressive return of the solar disc. A lunar eclipse

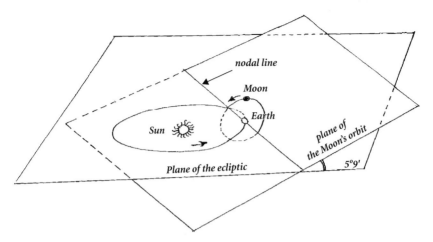

Fig. 1.15. *Circumstances of eclipses.* The planes of the Moon's orbit around the Earth and of the Earth's orbit around the Sun are different. Their intersection is the *nodal* line. In general, there is no correspondence between a solar eclipse and a full Moon, and no correspondence between a lunar eclipse and a new Moon. The eclipses occur only when, at the time of a full or new Moon, the Moon is also on the nodal line.

occurs whenever the Moon is passing in the shadow projected in space by the Earth illuminated by the Sun. Then, the bright Moon becomes progressively darker, before emerging from the Earth's cone of shadow.

It is easy to understand why new Moons do not always produce solar eclipses and why full Moons are not always eclipsed. This is because, as already noted above, the plane of the orbit of the Moon around the Earth does not coincide with the plane of the ecliptic, as shown in Figs. 1.10 (p. 29) and 1.15.

Figure 1.16 shows the mechanism of lunar eclipses. They are visible from all points from which one can see the Moon at that time. They last several hours, four hours and some minutes for the longest total eclipse, perhaps half-an-hour, or even less, for a partial eclipse. All observers are overwhelmed by the magnificent colors of the full Moon when it is plunged in the Earth's shadow.

In Fig. 1.17, we see the mechanism of solar eclipses. As the shadow of the Sun occupies only a limited area on Earth, the total eclipse is visible only along the trajectory of this shadow, as the Moon moves. From the shadow (umbra), one sees a total eclipse. From the partial shadow (penumbra), one sees a solar crescent. A total eclipse, therefore, does not

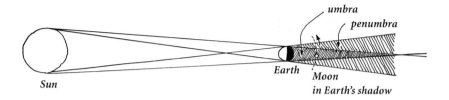

Fig. 1.16. *The mechanism of lunar eclipses.* Because the Sun is not a point-source but has a diameter, the shadow projected in space by the Earth, illuminated by the Sun, contains two different regions. In the penumbra, we still see some part, large or small, of the Sun. Only a part of it is hidden by the Earth. In the umbra, the Earth completely occults the Sun and suppresses the sunlight. However, even in the umbra, the Moon can assume dark colors, reds, browns, and yellows, caused by the refraction of solar rays in the Earth's atmosphere. When the Moon crosses the umbra, we have a *total* eclipse; the eclipse is *partial* when the Moon is crossing the penumbra.

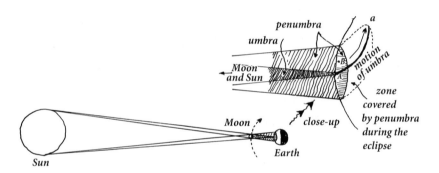

Fig. 1.17. *Mechanism of a solar eclipse.* The fact that neither the Sun nor the Moon are point-sources (each has a diameter) and the fact that their apparent diameters are almost identical results in a division of the Moon's umbra, as projected on the Earth, into two very different zones. In the penumbra, which is rather large (several thousand km in size), one can see the solar disc, only partially eclipsed. The sunlight is still quite visible. In the umbra, the diameter of which rarely exceeds 200 km, the Sun is completely hidden by the lunar disc. During the course of a solar eclipse, the zone of the umbra moves on the ground until the cone of shadow moves out from the Earth.

last long. Since the extent of the shadow on Earth is a few hundred kilometers at most, its duration is at the most on the order of seven minutes.

We have seen earlier that the Moon has an elliptical orbit around the Earth. Hence, its distance to Earth varies. When this distance is large,

Fig. 1.18. *Mechanism of an annular solar eclipse.* It may happen that the apparent size of the Moon's disc is smaller than that of the Sun. Then there is no point on Earth where one can see the Sun totally eclipsed. In an annular eclipse, one observes a ring of Sun around the disc of the Moon.

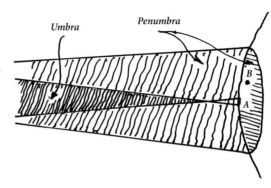

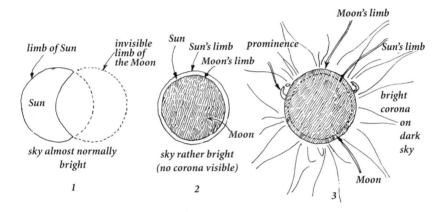

Fig. 1.19. *The appearances of the eclipsed Sun.* At left (1), a partial eclipse; one sees a crescent of the Sun; the Sun is only partly hidden by the Moon; the light is almost like that in full daylight. In (2), an annular eclipse; the Sun is limited to a thin ring around the disc of the Moon. In (3), the Sun is totally eclipsed. One can see, on the starry sky, the bright features of the solar corona (which gave rise, in the ancient Egyptian paintings, to "winged suns.")

the totality of the solar disc cannot be hidden completely by the Moon, which then appears smaller than at times when the distance is shorter. Then, at its maximum, the eclipse is said to be "annular." We can see a ring, the Sun's limb, around the Moon (Fig. 1.18).

The appearances of solar eclipses can have, therefore, very different characteristics (Fig. 1.19).

1.2.7
Early Knowledge of Planets

The planets (or "wandering stars") were known to the Babylonians. They were of two kinds. The motions of Mercury and Venus seemed to precede or follow closely that of the Sun. Venus, which appeared either as an evening star or a morning star, was not really identified as one single object but as two different ones for a very long time. Mars, Jupiter, and Saturn, on the other hand, appeared to move all across the sky, in an irregular way, but always in the vicinity of the ecliptic.

Their motions were noted and even measured with whatever accuracy (i.e. inaccuracy) was then available, but they were not described in any systematic way. This will be the task of Greek astronomers, from Thales to Ptolemy. Figure 1.20 shows that the complex appearances are disconcerting, but the models are very simple indeed (Fig. 1.21 and 1.22).

The right parts of Figs. 1.21 and 1.22 show the Ptolemaic conception (2nd century of our era). The left parts show the Copernican

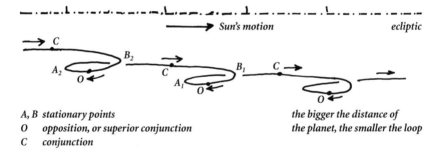

A, B stationary points   the bigger the distance of
O   opposition, or superior conjunction   the planet, the smaller the loop
C   conjunction

Fig. 1.20. *The apparent motion of planets.* In this figure, one can see that planets move close to the ecliptic, and follow strange but periodic paths. Some points on these apparent trajectories are *stationary*. The planet seems to stop there for a short time. On points like C (*conjunction*), the planet is moving in the same direction as the Sun on the nearby ecliptic. On points like O (*opposition or superior conjunction*), the planet and the Sun are moving in opposite directions. Note that in this figure, on the whole, the motion of the planets is about the same as that of the Sun. This is due to the fact that the figure represents only the outer planets, Mars, Jupiter, and Saturn. The apparent trajectories of planets present retrogradations and loops, different from planet to planet. The greater the distance of the planet, the smaller the loops and the intervals between them appear to be. This is a concept that has only been understood by modern astronomy.

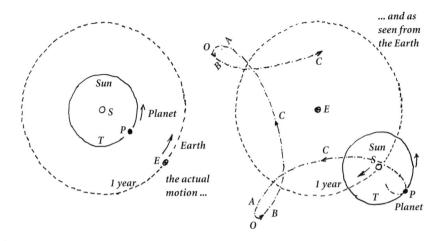

Fig. 1.21. *The planetary model for an inner planet (Mercury or Venus).* In this figure, the actual motion is represented, in a simplified form, as being circular and uniform centered on the Sun. The motion, as seen from the Earth, reflects a combination of the (apparent) motion of the Sun around Earth and the motion of the planet around the Sun. It dispays loops. We see the location of oppositions (or superior conjunctions) O and conjunctions (inferior conjunctions) C and of the stationary points A and B, as in Fig. 1.20.

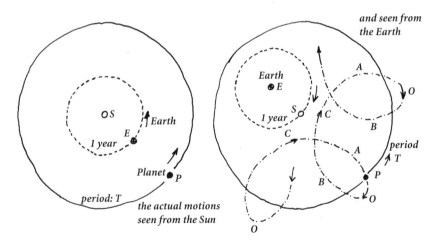

Fig. 1.22. *The planetary model for outer planets (Mars, Jupiter, or Saturn).* For the explanation of this figure, see the caption for Fig. 1.21. Note that the one-year period (that of the Earth's motion around the Sun) is that of the interval between two successive loops for the outer planets. For the inner planets, that interval is the period of the planetary revolution around the Sun (same notations as in Fig. 1.21).

conception (14th century). Even Ptolemy's model appears rather simple! It is not obvious how to proceed from the appearances to the models.

1.3
Pre-Socratic Greek Astronomy and Cosmology

1.3.1
A Few Milestones

Table 1.4 gives, in a condensed and rough way, a few important milestones in the history of astronomy in ancient Greece and later.

Note that not only do we know only vaguely the birth and death dates of the philosophers and thinkers of the period, we hardly know the period of their *acme*, i.e. of their most powerful intellectual activity, generally around the age of forty. What is worse, we know very few of the works of these men and women. The first scientists for whom substantial texts remain are Plato and Aristotle. Even then, a text which presents typically Platonic-Aristotelian principles (for instance the famous: σώζειν τὰ φαινόμενα) turns out to have been written in the 6th century of our era by Simplicius! Those who contributed to the progress of knowledge and understanding before Plato and Aristotle are known either because Plato or Aristotle quoted them (rarely verbatim), or because later historians of science, or doxographers as they are called, devoted commentaries to their works. In this manner for example, the thinking of Aristotle's student Theophrastes was later transmitted, and of course transformed, abbreviated and even completely distorted, by various compilers. This often leaves the impression of a very chaotic development and it is a difficult task to suggest reasonably logical affiliations among the early philosophers. Their various intuitions, therefore, influenced medieval scientists in a very strange and illogical way. Some followed Aetius (1st century AD), some the "pseudo-Plutarch," some Hermeias, some even Saint Augustine. Needless to say, many introduced their own ideas, sometimes impregnated with Christianity, to their descriptions of the Ancients.

I do not claim to have acted, in the following pages, in any original way, as a true historian of science, who would, necessarily, go back to the original sources. Being primarily an astronomer and a teacher, I have, rather, tried to give some coherent view of what I have learned in the

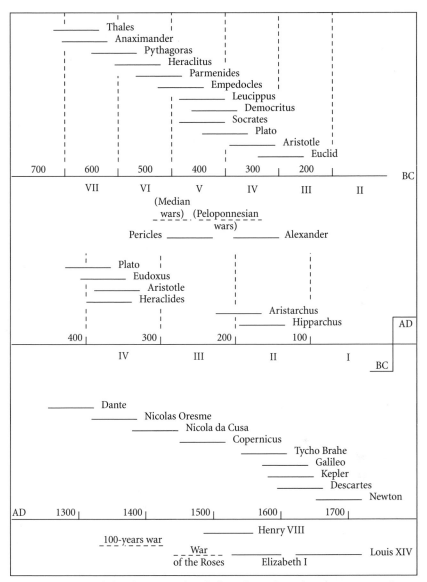

Table 1.4. *Getting a feeling for time scales. The lines drawn for each author are approximate.*

reading of *bona fide* historians of Greek science, such as Tannery, Duhem, and Neugebauer, whose main references are given in the bibliography of this book.

1.3.2
Three Permanent Tendencies

During this period, and in spite of historic uncertainties, we can see the emergence of the three different approaches which we shall see crisscrossing each other all the way throughout the history of science and into our modern cosmology, in a sort of permanent ballet. They coincide, in the Greek world, with a certain diaspora of the Greek scientists (Fig. 1.23).

Although this is an oversimplified view, we can isolate three affiliations, three families, often mixed and interacting with one another as follows:

Pythagoras and the Pythagorians:
They were firm believers in the perfection of the Universe and in the existence of simple mathematical or geometrical relations which govern it, so that no matter what observations are made, they must comply with numerological properties. If some property is observed that is difficult to fit into a nice numerical or geometrical pattern, it must not be relevant to our understanding of the Universe. In essence, the Universe is governed by a necessary harmony, the harmony of spheres.

Plato and the Platonists:
Their thought was based on the belief that nature exists but is not fully accessible to observation. A global, idealistic vision (a Creator-God) is superimposed on the observation of phenomena. One must save the observed phenomena, but only in the context of this overreaching principle which promotes coherence and which belongs to the realm of speculative metaphysics.

Aristotle and the Peripatetic School (the Aristotelians):
For the Aristotelians, observation came first. "Save the phenomena" became the main directive, provided one limited oneself to the simplest possible hypotheses (circular or uniform motions, spherical geometry, infinite duration of the Universe). It did not exclude the belief in some

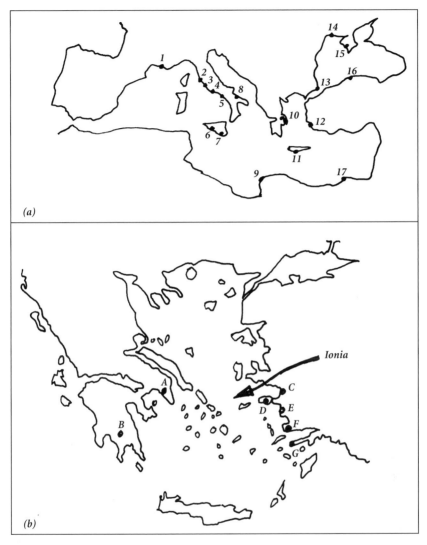

Fig. 1.23a,b. *The Greek diaspora.* From the 8th century BC, migrations resulted in the colonization of the borders of the Mediterranean sea (**a**). Ionian, Dorian, Achean, and Greek colonies prospered. On this map, we have plotted only some of them, typical of those flourishing at the apex of Athenian power: 1. Massilia (Marseilles); 2. Rome; 3. Naples; 4. Cumae; 5. Croton; 6. Gela; 7. Syracuse; 8. Taranto; 9. Cyrene; 10. Athens; 11. Crete; 12. Miletus; 13. Byzantium; 14. Olbia; 15. Theodosia; 16. Sinope; 17. Naucratis. The Aegean sea itself (**b**) was surrounded by intellectual and political centers such as: A. Athens.; B. Sparta; C. Ephesus; D. Samos; E. Miletus; F. Halicarnassus; G. Cnidos, and several others.

God, but it did not explicitly include it in the description of the heavens. God's role is quite different from that of Plato's δημιουργός (analogous to a Creator).

1.3.3
Thales (of Milet) and His Followers, the Milesians

The dates of birth and death of Thales are quite uncertain, but fall somewhere from 640–635 to 548–545 BC.

Thales, exiled from Phoenicia to Milet, possibly for political reasons, later became one of the seven "wise men" of the Greek tradition. Unfortunately, no written text remains from him. He was the head of the Ionian school. We have to rely upon later authors and compilers, or even fabulists, to know about his accomplishments.

Thales is known (at least Herodotus so reports) to have correctly predicted the solar eclipse of May 28th, 585 BC and, as it has been claimed, the eclipse of September 30th, 618 BC. This is certainly an essential achievement, even if partly (or completely) legendary.

It is not clear how he could have predicted a total solar eclipse in complete ignorance of everything that we know about the mechanisms involved and the periodicities of eclipses. It was not until two centuries later, in the time of Eudoxus, that the theory of eclipses began to take shape. What is more likely is that it was known by experience that lunar eclipses occurred at the time of a full Moon. It was possible to predict an eclipse, with a small but not negligible chance of success, at any full Moon for which the Moon's height above the horizon was about that of the Sun, at their respective culminations. A similar, though less likely, prediction can be made for the occurrence of solar eclipses. Once in a while, a prediction made with a small probability can nevertheless become true, somewhere in the eastern basin of the Mediterranean. That is possibly what happened. The Chaldeans, before Thales, knew the period of 223 lunar months (the *Saros*) which works well for lunar eclipses and can be used as a way to predict solar eclipses as well, though with less certainty (see Figs. 1.16, 1.17, and 1.18). It is possible that Thales, who came, remember, from the eastern shores of the Mediterranean, knew of the works of Chaldean astrologers.

Thales is also, possibly, the one who described the Little Bear (Ursa Minor) as a constellation (according to Callimachus). He knew that the Moon does not produce its own light and that it is illuminated by the

Sun. He knew of the existence of solstices and equinoxes and was able to predict them. He knew the seasons and their inequality. He knew of the heliacal set of the Pleiades. He knew the apparent solar trajectory from solstice to solstice on the starry vault. He measured the solar and lunar diameters as 1/720 of their apparent orbits. Since the orbits have an angular extent of 360°, this results in a diameter of 1/2°. The modern values are practically the same.

Thales was of course a philosopher and, as such, interested in everything. He had ideas about the nature of the Universe. For him, water was the essential component of everything. Earth floats on water, like a piece of wood, more or less a flat raft. This idea, quite primitive (Fig. 1.24), resembles that of the Hebrews or of the Egyptian papyri, but it has nothing to do with astronomical knowledge. Primitive ideas and more advanced astronomical ideas coexisted in Thales' thought.

There is a famous anecdote about Thales. He is the one who is said to have fallen in a well (he seems even to have died that way) because he was looking at the sky, but this might as well have been some kind of a metaphorical invention. We should note also that it is recorded that

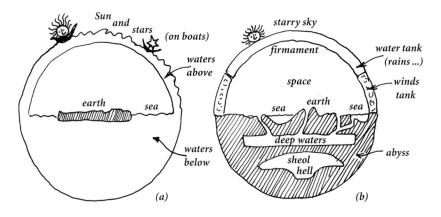

Fig. 1.24a,b. *Some primitive views of the world.* On the left (**a**), the Egyptian vision of the Universe, as described in old papyri. Thales' vision is quite similar to this one. This fact, incidently, contradicts the mythological notion that he could actually have predicted an eclipse. If he did have this ability, his logic must have emerged from empirical consideration, not from any model of the kind represented here. On the right (**b**), the Hebrew vision of the Universe, according to the Bible and ancient Jewish commentators.

Thales acquired a monopoly in the olive industry because he made good meteorological predictions. This may be a reflection of the superiority of theoretical knowledge in the eyes of his commentators.

Anaximander (Milet) (620–547 BC) was a friend and follower of Thales. He is certainly the man who first used the word οὐρανός (ovranos) to designate the sky and perhaps the one who suggested κόσμος (cosmos) to designate the world as a whole. His philosophy is quite similar in its emphasis to that of Thales. He formulated the concept that everything is made of a combination of "principles" and "elements" (a concept much elaborated by 20th century physicists with their notion of particles and interactions). This gives rise to an unlimited number of combinations (air, water, and other elements). The parts are changing, but the whole is without change.

Although it may be legendary, Anaximander is known as the inventor of gnomons, as a maker of time-keepers, and as the one who first drew a map of the Earth and sea. He discovered the inclination of the zodiac with respect to the equator. According to Aristotle, Anaximander also discovered the intervals and orders of the "wandering stars," the planets. He also used the eclipses (that of May 585 BC?) to approximately determine the proportions within the Sun-Earth-Moon system. Anaximander was no doubt an experienced geographer and astronomer. He seems to have invented (or borrowed from the Babylonians?) the πόλος (polos), a spherical sundial (Fig. 1.25).

Possibly, this spherical instrument gave Anaximander the idea that the Earth, at the center of the "whole," was a thick and flat disc; its thickness 1/3 of its horizontal diameter. It is suspended and immobile at the center of three large rings, one for the Sun, one for the Moon, one for the Milky Way. The Moon is, like the Sun, a very "pure" fire and the Sun is "not smaller" than the Earth. It seems that these three rings had a strange structure, full of fire, but with a round hole letting the light go through, a moving hole of course (Fig. 1.26). There was much elaboration on this system, which would perhaps have explained all the apparent motions of the Sun and Moon, but perhaps we are asking too much from the few passages concerning Anaximander in the works of Theophrastes and his successors, the pseudo-Plutarch, Hermeias, and Aetius.

Anaximenes (Milet) (550?–480 BC) seems to have been the last of the true Milesians, a student of Anaximander, but also a follower of Thales. We think that he furthered the development of the time-keeping instrumentation of Anaximander. For him, the boundless ἀπειρών (apeiron) is

Fig. 1.25. *The polos*
(πόλος). Said to have been
invented by Anaximander,
the *polos* is a spherical
sundial. One end of the
stylus is at the center A of
a sphere. Its shadow B is
projected on a hollow
spherical surface. During
the solar motion on the
sky, its end describes a circle, parallel to the equator
of the sphere, to the equator of the sky. The drawing
represents a polos limited
by its equator. It only
works all day, therefore, in
summer and spring. A
horizontal plane can be
used instead of an inclined
one. Then, the equator is a
circle only one half of
which is drawn inside the
hollow sphere.

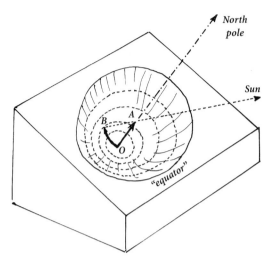

not the "whole," as it is for Anaximander, but only part of it. Air is now
the main principle (remember: for Thales, it was water). Anaximenes
asked the question: Is the celestial sphere the limit of the boundless?

His cosmogony seems to have been quite sophisticated and strange.
The Earth is flat, sustained by air (as are the Moon, Sun, and other stars,
which are, in essence, fire). The celestial vault is basically solid, carrying
the celestial bodies around the Earth. It turns around the Earth like a hat
can turn around a head. High mountains may hide the Sun, which would
explain the night. He certainly did not go as far as Anaximander's model.
This puts our chronology in some doubt as well as the notion that he
was a student of Anaximander.

The air, the main principle of nature, is God. This is a step along the
path of constant interplay between astronomy (physics) and metaphysics. Very contradictory statements are given by commentators about
the achievements of Anaximenes.

The Milesians were heavily criticized by Aristotle and the Peripatetics.
We shall see later what they had to say.

Fig. 1.26. *The World, according to Anaximander.* The Earth is a cylinder with a diameter of unity (1) and a height of 1/3. The solar ring, on which a hole represents the Sun, has an internal diameter of 27 and an external one of 28. It is located in the plane of the ecliptic, along the zodiac. The lunar ring, on which a hole represents the Moon, has diameters of 19 and 18 times the Earth's. It is located in the plane of the orbit of the Moon (hence it has to move over time, with respect to that of the Sun). The stellar ring is associated on the sky with the Milky Way. Its diameters are 10 (outer side) and 9 (inner side). This ring should be transparent to the intense light coming from the Sun. The rings rotate, so that the hole-Sun and the hole-Moon move at their observed rates around the Earth.

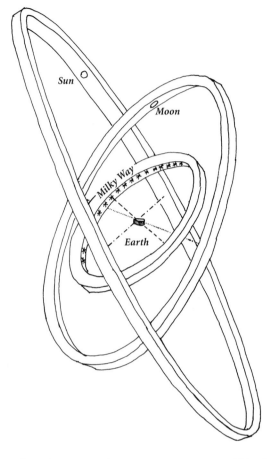

The size of the rings (1, 9, 18, 27) can be expressed as: 1, 1 × 9, 2 × 9, 3 × 9. This obviously indicates a strong Pythagorian influence. This interpretation is based on the work of Tannery and several others.

1.3.4
Pythagoras and the Pythagorians

Pythagoras, probably from the island of Samos (his approximate dates are 580?–500? BC), appears more as a symbol than as an historical figure. There is actually no trace of Pythagoras's opus. We have thousands of testimonies, but few are reliable, being mostly second-hand. We know him as the founder of a religious sect, which searched for a harmonious way of living and for some spiritual-political influence, more than as an astronomer or a scientist. Actually, Aristotle never spoke of Pythagoras, only of the Pythagorians.

He is the one (more likely than Anaximander) who gave the envelope of the Universe the name κόσμος (cosmos) which means beauty, order, and Universe as an organized whole. It was his belief that perfection and organization are concepts more important than truth. He seems also to be the one who coined the word philosophy (φιλοσοφία). The essence of Pythagorian philosophy is this word κόσμος, as it was used by Pythagoras's followers who indeed conveyed, much more than he did, his views about the Universe. Pythagoras was also a mathematician, expert in arithmetic and geometry. This had almost metaphysical consequences in his thought. The number is the "One." It rules the material world. In order to actualize the basic concept of order, it is said that Pythagoras recognized a duality of principles, ἀπειρών (apeiron) the continuum, the unlimited, the fire, the space, the fluid, the light ... and πέρας (peras), the principle of individuality, of discontinuity, the solid, the obscure, the Earth. This duality is the simplest possible system, but the Pythagorians added some dimensions to it.

Take the example of the description of the world according to Petron, an obscure Pythagorian: "There are 183 worlds, arranged along an equilateral triangle of 60 worlds on each side. Each of the 3 remaining worlds constitutes an angle, whereas the 180 others, in an impeccable order, describe the sides of the triangle, turn around it, like the round of a chorus." (*quoted from Plutarch*). Plutarch attributes this strange view to Petron's Dorian origin (Fig. 1.27).

The Pythagorian tradition of mystical numerology has existed for centuries. Alcmaeon, according to Aristotle, said that the number of principles is indeed 10. They are organized in two parallel series according to the two Pythagorian principles, as follows:

Fig. 1.27. *The Universe according to Petron.* This Pythagorian held the obscure but typical view that there are in existence 183 worlds (3 + 3(5 × 12)) arranged in a triangle. A triangle is the image of the letter Δ (delta), initial of the name of the God Zeus (Διός), and representing in Greek numerology, the number five, which, like three, was regarded as a sacred number.

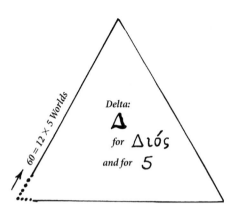

limited	vs	unlimited, or illimited
odd		even
many		one
left		right
male		female
quiet		moving
straight		curved
light		darkness
good		bad
square		oblong

If we look at each pair individually, there is nothing unnatural in these dualities. The arbitrary thing is to limit their number to exactly 10. Other philosophers have proposed many other organizations. Even Aristotle was guided by similar ideas.

Alcmaeon again, according to Aetius, noted that "the lunar eclipse is caused by the fact that its shell is lying on its side." The Moon is seen as an unstable oscillating shell. In this tradition, the most reasonable ideas coexist with strange systems, of a mystical nature.

Much later, Kepler's *Mysterium Cosmographicum* (see p. 217) goes very far along the strangest lines and even now, many naive people, but also cranks and crooks, generally in unpublished papers, are tempted by exaggerations of Pythagorian aesthetic principles. Even such great figures as Eddington and Dirac, in the thirties and forties of our own century, (see p. 515–516) were influenced by this type of thinking. This also applies to the model of the set of quarks and gluons introduced a few decades ago. A construction of this sort takes its value only (hence rarely) when fully confirmed by some degree of direct observation.

So far, so good, and so long to the Pythagorians. We could quote many more, but they drifted too far from observations, unlike Anaximander. Still the Pythagorian Church (as it was actually called) lasted for many centuries.

1.3.5
Between Pythagoras and Plato

The evolution of the philosophical scenery may be deduced from Table 1.4. The early schools of the Milesians and the Pythagorians developed and new schools formed in their turn.

The Milesians discussed the nature of God, necessarily associated with man and with nature. They are the precursors of the Platonic-Hebrew tradition. Their influence is felt in the ideas of such people as Xenophanes, who gave root to the school of Elea (the Eleates); but also Heraclitus, who was more a philosopher than an astronomer.

Heraclitus of Ephesus (540–480 BC) was called the "obscure." His book, which Aristotle was very familiar with, was devoid of any punctuation. For him, *the* principle was fire, not water, as in Thales, or air, as in Anaximenes. Fire was seen as the primitive and final state of the Universe. It was in essence identical with λόγος (logos) (a similarity with the Hebraic Genesis). Fire is the "One," but it is in motion.

Xenophanes of Collophon (540–440 BC?), a follower of Anaximander, and probably the founder of the Elea school, was concerned with the nature of God, an essence not born from anything, one and immaterial. Hence, Xenophanes, and after him the Eleates, slowly came to the idea of a world that was spherical, eternal, and without motion. Clearly, the idea emerges that we must make a definite distinction between "truth" and "opinion." This is a necessary condition for the progress of knowledge. Xenophanes had many strange ideas about the sky, such as the existence of several successive suns and the mention of an eclipse which had lasted one full month!

Parmenides followed Xenophanes' tradition and can be considered the true founder of the Eleates (540–460 BC). He continued to pursue the idea, similar to the Pythagorian Alcmaeon, of the "binary" nature of all

things and developed a systematic description of the Pythagorian duality. There is a further treatment of this idea in Aristotle's metaphysics.

In the field of astronomy, Parmenides claimed that the Earth was
spherical and occupied the center of the world. The Earth, he thought, is
divided into 5 zones (1 tropical, 2 habitable, 2 glacial). The basic duality
is between fire, associated in essence with the Creator (the δημιουργός),
and earth, essentially the material realm of creation where created beings
are located. The birth of the idea of the sphericity of the Earth, stemming
from the general unity of the Universe, is the beginning of the great astronomical period of the Greek geometers.

Parmenides' influence was enormous. It is not surprising that there
are some differences of opinion among his commentators regarding his
cosmology. Major Eleates were Zeno and Melissus who shared similar
points of view. They were followed, but with some major changes in
orientation, by the Sicilian Empedocles.

Zeno (490–430 BC) expressed the idea that there are many worlds and
that emptiness, or vacuum, does not exist (an idea which will influence
many astronomers in the Middle Ages). Four qualities, assimilated to elements – hot, cold, dry, wet – interact and are combined to form everything. Zeno is also well-known as the author of the famous paradox of
Achilles and the tortoise (in which he takes a stand against Pythagorianism), and of that famous arrow (noted by Archytas), "qui vibre, vole, et
qui ne vole pas" (Paul Valéry, in *Le Cimetière marin*). Both argue for the
continuity of the Universe, for the unity of "being" (i.e. of the real
world), using convergent series of fractions.

Empedocles of Agrigento (?–550 BC) spoke of four elements, air, fire,
and water, of course, to which he added earth, as basic to material creation. The idea will become expanded in Aristotle's *Physics* (but is also
present in Plato's *Timaeus*). Apart from the four elements, there are two
principles, friendship and hatred, which are forces, or interactions. So
Empedocles' vision is of a dynamic Universe, not a static one. This is another vague idea that will prove very far-reaching. In spite of this refinement in his thought about the nature of matter, the astronomical ideas of
Empedocles are quite naive. For him, the Moon was an "iced fire," while
the Sun was a reflection of fire, as in water.

While the Pythagorians were continuing in their strange, mystical tradition, Anaxagoras (500–428 BC), whose influence on Socratic ideas is
thought to have been great, developed some of the Milesian ideas. He
thought that everything was made out of *homeomorphic* corpuscles (of

similar essential shape and geometrical properties) and that the intellect governed the principle of motion. This is very close to the ideas of Empedocles. Anaxagoras's astronomy is again simple-minded. The Sun, now, is a burning stone, larger than the Peloponnesis (enormous to the ancient mind!). On the Moon, there are houses, mountains, valleys. The Milky Way is the solar light reflected by dark stars. Anaxagoras is known to have interpreted the fall of the meteorite of Aegos Potamos as a stone falling from the Sun. Interesting is his remark that the pole is not at the zenith. Anaxagoras explained this anomaly with the hypothesis that, in the beginning, the pole was indeed at the zenith, but that later on the stellar vault inclined. Note, incidentally, that like Socrates and Galileo after him, Anaxagoras was accused of impiety because of his astronomical views. Pericles saved him, in spite of the accusation put forth by the historian Thucydides.

Archelaus (5th century BC), continued the thinking of Anaxagoras, developed it, and realized (perhaps the first one to do so in history?) that the Sun was the largest of all celestial bodies (including Earth? How could this be?). The Universe of Archelaus is illimited.

Diogenes, again in the Milesian tradition, thought of the Moon as a burning stone, as Anaxagoras saw the Sun. He compared it to a volcanic stone. Reasonable ideas coexisted with the wildest views about heavenly bodies.

In the meanwhile, the Pythagorian school developed independently along the lines already visible in Petron and Alcmaeon. Sometimes, this development was marked by a real breakthrough, such as that of Oenopides, who seems to have discovered the inclination of the zodiac on the ecliptic. According to the compiler Jamblicus (250–325 AD), the Pythagorians (even in his time) constituted a long-lived and important school. They could even be considered a Church or a sect. There were then 218 Pythagorians listed by name, including many women. This precise accounting is an indication of a group, Church, sect, or party, which implied formal membership. Oenopides, whom we have already mentioned, also determined the length of the tropical year. We must also mention Meton (5th century BC), who made the important step of discovering the "Great Year" of 19 years duration. Actually, he noted that 19 tropical years are equal to 235 synodic lunar months. Meton's cycle allows for the prediction of eclipses.

If we study this computation, using modern figures and methods, we can see that it is typical of the Pythagorian orientation of mind (who else would have thought at that time to look for these numerical relations?).

In the example below, a refers to the lunar month, b to the tropical year; $a = 29.5306$ days; $b = 365.2422$ days. The ratio b/a can be achieved relatively well by a rational fractional number, at different approximations:

$b/a = 12/1$ with an error of 11 days in a year
         99/8   with an error of 1.5 day for 8 years
         235/19   with an error of 0.1 day for 19 years
         (Meton's ratio)

Actually, 235 lunar months are equal to 6939.7 days and 19 years are equal to 6939.6 days. So Meton's cycle is correct to a rather high degree of precision, but the two figures are not completely equal. After 3000 years, the error is only about 15 days.

Philolaus of Taranto, another famous Pythagorian (5th century BC) also had great confidence in numerical and geometrical constructions. He thought that there was fire at the center of the Universe. The spherical Earth turns around it in one day keeping its inhabited side facing the fire. The Sun, the Moon, and the planets also turn around that central fire. The Sun, a glassy volume, reflects the light of the central fire. Stars, located on a stationary sphere, close the system. Beside us and the central fire, there is some "necessary" anti-Earth. In that way, we have 10 bodies in the Universe, quite a nice number, since $10 = 1 + 2 + 3 + 4$. Still playing, quite arbitrarily, with the numbers, he goes so far as to give estimates of distances to the central fire. The radius of the Earth's orbit being unity, the anti-Earth is at a distance 3 from the central fire, the Earth at a distance $3^2 = 9$, the Moon at a distance $3^3 = 27$, Mercury at a distance $3^4 = 81$, Venus at a distance $3^5 = 241$, and the Sun at a distance $3^6 = 729$, a particularly nice number since it is both a cube of 9 and a square of 27! Although such a theory has no basis, except intuition, it contains at least one intuition which later proved to be correct. It is not a geocentric theory, however crazy it may otherwise look!

Other Pythagorian and mythological oddities are Oenopides' notion that the Milky Way was initially the path of the Sun's motion. The Sun, horrified by Thyeste's banquet (during which the Atrides more or less ate each other, according to cannibalistic practice), started to turn the other way, along the zodiacal circle. Certainly a crazy and theological idea, but associated with the already noted true discovery, by the same Oenopides, of the inclination of the zodiac.

According to Hippocrates (460–377 BC), the famous doctor, indeed the creator of medicine, the tail of a comet is due to light refraction and belongs to the *sublunary* world, whereas the head is made of "quintessence" (which we will encounter again when discussing Plato and Aristotle). This idea is a premonition of the precise distance determinations of comets made much later by Tycho Brahe in the 16th century.

We could fill pages and pages with this mixture of observations, intuitions, and religious or mystical thinking during this period when a direct commerce with the gods was part of everyday life and the knowledge of the World itself was limited to the Mediterranean area and to the Middle East.

One unique idea emerges from so many contradictory strains of thought. Behind the discussions about unity, duality, plurality, the accepted or hidden interventions of God or gods, and the models, geometrical or numerical, aimed at describing and predicting, each school believed that the World was, if not yet explained, at least explicable. Each school built geometrical or numerical models, even some "chemical" (the four elements) descriptions to account for it. This effort, although hardly worthy of the designation "scientific", may be the real common basis for the explosion of Greek science.

CHAPTER 2

Classical Greek Astronomy

2.1
Plato's World

2.1.1
The Myth of the Cave

The known works of Plato are many. They are available mostly because Plato, having created the Academy in Athens, had many students and many colleagues who made a point of transmitting the teachings of the old master. Plato's works were actually transcribed and translated many times and into many languages. Obviously, we shall not even attempt to present here a detailed discussion of Plato's life or his philosophy and its tremendous and lasting influence.

His astronomical work is mostly described in his dialogue *Timaeus*, but in *The Republic* there is also an important text (Sect. 7.1), commonly called the "cave myth," which pertains to his astronomical views.

In this metaphorical story, Plato describes the World as it would be seen by people enclosed in a dark cave. The cave is open to the daylight at one large entrance, but the men are chained, forced to stay inside. They can only see images on the walls of the cave. These are shadows of distant objects projected by a powerful light, a fire, which burns behind them. The knowledge that the men inside can attain concerning these objects is only what can be learned from looking at their shadows. Quite probably, they would assume that these shadows constitute the unique truth. Contact with the World outside would be almost unbearable to them. This metaphor suggests that scientists and philosophers should perhaps be more open to the unobserved or even to the unobservable. For Plato, the "light" is also here the symbol of philosophy.

This rich allegory, in essence, claims that observation alone cannot yield a complete knowledge of all things. It creates, in addition, although without really saying so, the vague impression of the basic need for a three-dimensional view of the Universe.

For astronomy, the implication is that the "world of appearances," the stars seen against the night sky, is only an imperfect projection of the ideal World. Employing reason and holding to a belief system are more important than observing. The gods, or God, are there, and that is enough, or almost enough.

The Academy created by Plato conveyed this philosophy for many centuries and, in a way, prevented many astronomers and philosophers from observing the world around them. The act of observation was considered a priori to be inadequate and biased, possibly even unnecessary as it would show only a wrong image of the true World. This is a view which will find many advocates in the Christian schools of thought, as we shall see.

In *The Republic*, after the cave myth, Socrates explains to his student Glaucus that the best areas of study to which a man could devote his life are the sciences of numbers and geometry. This reflects the influence of Pythagorianism. The observation of nature was not held in particularly high esteem. Strangely, the third science mentioned by Socrates is astronomy. This is only so, however, because astronomy provides a detailed knowledge of the seasons and was useful, therefore, in agriculture, navigation, and the military arts.

Some other dialogues mention cosmology or astronomy, often in a different context. For example, in the much quoted *Phaedon*, which describes at length the death of Socrates and perhaps his last message to his pupils, there is a long passage in which the Earth is described. Contrary to Anaxagoras, quoted by Plato (*Ph. 97*), Socrates considers the Earth to be a sphere (*Ph. 108*) located at the center of the World. It is a large corrugated sphere, full of cavities, each of them comparable to the cave of the myth. One of them, in essence the Mediterranean basin, is inhabited by men. They have, of course, a very distorted vision of the sky, just like the limited vision of the external world of people imprisoned in the cave. These cavities are filled with air and water, which makes it impossible for people inside to see properly what is outside, i.e., in the ethereal universe, the astronomical one. The main purpose of Socrates's description is, however, not only to discourse about the shape of the Earth, but to introduce a very complex mystical system (Fig. 2.1). In this system, all rivers, including not only rivers of water, but also rivers of fire,

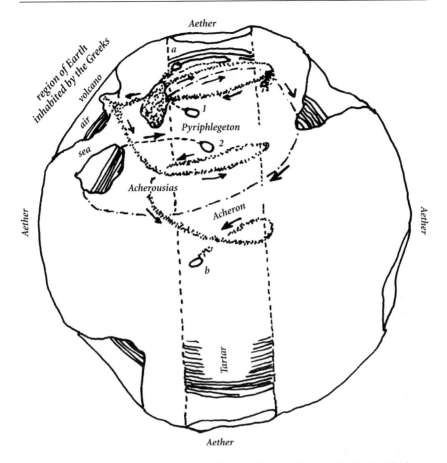

Fig. 2.1. *Plato's Earth.* The Earth is, roughly speaking, spherical. In the details, however, the sphere is corrugated. In this figure, following the interpretation of the *Phaedon* by O. Baensch, we see the interior of the Earth. This is the part which is affected by life and death. Men and other living things live in cavities, from which the real World is hidden by the air and the sea. They are separated from the aether. The Earth is axially pierced by the Tartar which might be a great river or a hole. It is, in any event, a large volume where water is oscillating, provoking flux and reflux in the water channels or rivers that cross the interior of the Earth. These rivers (the Acheron, Cocytes, Styx and Pyriphlegeton, of which only the first and last are represented here) linked the realm of the living to the realm of the dead (from a to b, and from 1 to 2 respectively). The Pyriphlegeton is a river of fire which flows through lake Acherousias. At some points, it travels very close to the Earth's surface where it is visible as volcanic lava. From far away, the Earth appears in its ideal form, as a perfect sphere embedded in aether.

come from the Tartar and end in it. The Tartar is an axial hole, dug through the Earth from one side to the other. The motion of waters in it successively pumps out and nourishes the rivers. This explains the observed circulation of water. In addition, the rivers of fire may come close to the surface, causing volcanoes. The main purpose of this system is to explain how the souls of the dead evolve in the interior inferno of the Earth. The Hebraic view (see p. 43) was not so different in this respect.

2.1.2
The *Timaeus*

Plato arrived at his maturity at a time that was already the beginning of the decline of Greek thought. Half a century after Pericles' death, one can see the decomposition manifested in the conflicts among the armies of Athens, Sparta and Thebes, and the menace of the growing power of Macedonia. Plato collected and synthesized the work of the thinkers who preceded him, often twisting the old teachings into a new form of his own. He named his various writings after the earlier philosophers and set up dialogues alleged to have taken place between them and the old master, Socrates. Therefore, Plato's works, in particular his cosmology, described at length in the *Timaeus*, may be considered a compendium, a "text-book" of the period.

As Plato was not primarily interested in astronomy as such, but in the philosophical aspects of it, emphasis is often put on the metaphysical approach, which is a logical follow-up to the cave myth in *The Republic*, and the description given in the *Phaedon*. Plato was interested in the reason for the existence of the World. He believed that Creation came about "because the Creator (δημιουργός), was in essence a good being." He constructed everything in his image, good in essence, perfect, ideal. The World so created is by nature the most beautiful World possible, and even better, it is really a living entity, gifted with a soul and an intellect. It is this perfect World that man seeks to apprehend fully. However, (see above) neither in the cave myth in *The Republic* nor in the description of the Earth given in the *Phaedon* can man know reality in its totality. He can only infer it from two main principles. What is observed has to be taken into account and properly described. What is not observed must be reconstructed, according to a *model* of the World, created in the image of the δημιουργός, which is to say, perfect. Plato, like our modern cosmologists, built models of the Universe and of the nature of matter at the microscopic level. In his modelling, Plato was often guided by Pythagorian ideas; but contrary to them, he put a great emphasis on *saving the phenomena* (σώζειν τὰ φαινόμενα), which from then on became a guiding principle for both Platonists and Aristotelians. In modelling the World, Plato devoted a great deal of attention to the various sciences and branches of knowledge. Meteorology, for example, was taken into account. He was, however, more interested in man himself, than in the cosmos. His investigations took him into such areas as medicine and pharmacology.

The following is a discussion of the elements involved in Plato's overall modelling of the Universe, beginning with what constitutes the essence of the universal soul, the Soul of the World, primeval to every other construction.

2.1.3
The Four Elements

As we have said, God, who is essentially good, created the World in his image. Therefore, the World is "animated;" it has an *anima*, a soul (ζῳον), in the same way that an animal has a soul. As God is unique (the absolute animated being), so is the World. It is corporal (σωματοειδές), hence visible, and tangible. The idea of it being visible suggests that fire (f) is the first essential element, while the idea of tangibility refers to earth (e), as the the second essential element. The Universe must be beautiful. A link must exist, therefore, between earth and fire. The link is naturally (and this takes into account Plato's extensive knowledge of the discussions of previous astronomers and previous schools), through water (w) and air (a). A very strange linkage is symbolized by ratios, which do not seem to come from the Pythagorians, and have been (perhaps daringly?) extrapolated from the *Timaeus* by its commentators, such as Pierre Duhem whose analysis we have followed.

One could write (Fig. 2.2):

$$\frac{a}{w} = \frac{f}{a} \quad \text{or} \quad \frac{air}{water} = \frac{fire}{air} \quad \text{or again} \quad a^2 = fw \,.$$

Similarly, one can write:

$$\frac{w}{e} = \frac{a}{w} \quad \text{or} \quad \frac{water}{earth} = \frac{air}{water} \quad \text{or again} \quad w^2 = ae \,.$$

Fig. 2.2. The ratios *f*:*a*:*w*:*e*, according to Plato, where *f* is fire, *a* is air, *w* is water and *e* is earth – *ffe* equals *aaa* and *fee* equals *www*. Expressed in two dimensions, this means that air is to fire as water is to air and earth is to water as water is to air.

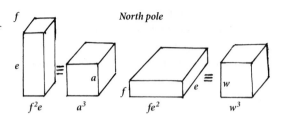

One can generalize in three dimensions (Fig. 2.2). Combining the above two equations by the use of elementary algebra leads to:

$$a^3 = f^2 e \quad \text{and} \quad w^3 = fe^2$$

These symbolic equations are interesting in that they imply the use of irrational numbers (ex: $w = \sqrt[3]{fe^2}$,) contrary to the practice of the Pythagorians, who rejected them.

The four elements are constantly evolving into one another. Water turns into stones and earth, or vaporizes into air. Air, in the process of combustion, becomes fire; and fire, in its turn, when extinguished, becomes air again. When air is condensed and compressed, it becomes clouds and fog, which, under further compression revert to water. In other words, matter can have several forms, changing each into the other. The continuous apparent change in the nature of concrete things, susceptible, therefore, to corruption and generation (the "sublunary world," or the world at the bottom of the cavities in the *Phaedon*, or inside the cave) implies the existence of another reality. This other reality is space, which enables us to locate elements and matter.

2.1.4
Space

Some atomists (such as Democritus or Leucippus) prefigured Hamlet by setting up an opposition between "non-being" or vacuum (τὸ μὴ ὄν) and "being" (τὸ ὄν). Being consists of bodies (human beings, animals) and matter (αἴσθησις, η δεύσις), but also elements, i.e. "ideas" (νόησις) of the model of being. According Plato, one has to distinguish "space" (ἡ ξώρα) from the two other categories of reality, the noblest (νόησις), that of imperishable ideas, and the lowest, that of perishable bodies. The latter are known to us by sense perception (αἴσθησις), to the birth of which Plato gave the name *genesis* (η γένεσις). Space, as defined by Plato, seems to be a notion which differs somewhat from empty geometrical space. It is linked closely to the existence of matter. This idea was expressed before Plato by the Pythagorian Archytas of Taranto and crops up again in the concept of General Relativity introduced by Einstein (20th century).

The idea that God is a master of geometry is primary in the hierarchy of Platonic thought, inherited in part from the Pythagorians. Since God operates according to geometrical principles, he has to describe the four elements in a geometrical frame of reference. In the *Timaeus*, Plato describes at length the shapes of the elements, reduced to their elementary forms (a sort of atomic theory). He first describes the three regular polyhedrons of which each side (or face) is an equilateral triangle. The fourth polyhedron, the cube, has a square face. These four polyhedrons are then equated with the four elements (Fig. 2.3). Fire is the most mobile and aggressive figure, the tetrahedron, with its four pointed corners. Air is represented by the octahedron and water by the icosahedron. The cube, with squares as sides, represents earth. The *Timaeus* contains a very lengthy analysis of the physical properties of

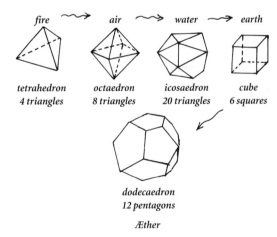

Fig. 2.3. *The five Platonic elements and volumes.* There are only five regular polyhedrons. They correspond to the four elements plus the *quintessence* or *aether.*

these four elements and of their mutual changes, presented as a description of the true sensible world. It is fascinating to read and it gives a rather good idea of how matter was understood before modern chemistry.

Actually, we know, as did Plato, that there are five (and five only) regular polyhedrons. This can be easily demonstrated by classical geometry. The fifth is a dodecahedron, the sides of which are pentagons. Plato does not discuss it in the *Timaeus*, but he does mention it in *Epimenides,* another of his dialogues. It is not certain whether Plato considered the sphere a sixth regular polyhedron (which it is, as a limit case when one considers the sphere as a polyhedron with an "infinite" number of sides). In any case, it was necessary to assign something to the fifth polyhedron. Plato associated it with the aether (αἰθήρ). Out of the aether, the World soul formed immaterial beings, sort of ectoplasms, along with air, fire, water, and earth. Clearly, the view expressed in the *Phaedon* is different. There the aether is only involved in the ideal World, not the material World, in other words the astronomical World, not the sublunary one (obscured by the dense atmosphere of the cavities). Soon after, Aristotle followed that last line of thought. For him, the intermediate concept was replaced by the "quintessence," or fifth essence, called again αἰθήρ by him, although in Aristotle the aether is associated with both the dodecahedron and the sphere. It is important to remember that there is still room for a great deal of interpretation. Commentators, since Greek times, have had many opportunities to read Plato and Aristotle according to their own biases. Only the original texts can reveal the authors' meanings, but they are often extremely difficult to translate, understand, and interpret.

Are these geometrical figures only symbols of the macrocosmic World organization on a microcosmic scale? Or are they real "particles," which matter is made of, as Democritus, in his truly atomic theory, believed? It seems that Plato adopted the atomic point of view, but later neo-Platonists (such as the Christian John Philopon) went much further in the symbolic phantasm: "As the dodecahedron has twelve faces, God made the world as twelve globes incorporated each in the other." Earth, water, air, fire (four elements, four spheres contained in one sphere) were nested and followed by the orbs of the Sun, the Moon, and the five planets (seven spheres), and finally by the "starry vault," making altogether twelve concentric spheres (see Fig. 3.3, p. 144). A similar model is also described by Plato in the *Phaedrus*, another dialogue (see Fig. 2.6, p. 65).

Although Aristotle was, as a skeptic and rationalist, quite critical of some of these constructions and their mystical implications, it must be mentioned that he credited Plato by noting that: "Everyone says that location is something to consider; only Plato tried to say what it was."

2.1.5
Platonic Astronomy

Because it is the perfect object of Creation, the form of the World has to be spherical, an exactly polished sphere, without any irregularities. This is indeed the sixth perfect figure we have alluded to, an "infinitehedron," to coin a bad neologism.

Among the seven possible motions designed by the Creator, the one which best fits the sphere is rotation (Fig. 2.4).

In this model, there are two concentric spheres, one being the *essence of the identical*, the other the *essence of the different.* They rotate (Fig. 2.5) around two axes.

The essence of the identical rotates around the polar axis, from east to west. The essence of the different around the ecliptic axis, from west to east. The identical is the dominant power (κράτος). Its rotation, the *diurnal* motion, gives rise to the succession of days and nights. It is more di-

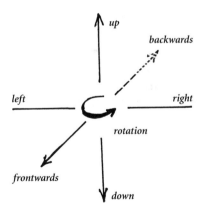

Fig. 2.4. *The possible motions in space.* There are seven possible motions. The only one which could be attributed to the World is rotation.

Fig. 2.5. *The essence of the identical (I) and the essence of the different (D).* The structure of the World, according to Plato, opposes the *identical*, which imposes its behavior on all astronomical bodies in an identical way, to the *different*, which is the germ of variety, of diversity. The two motions are opposite to one another. The essence of the identical turns around the polar axis in the diurnal rotation. The essence of the different, which will characterize each planet, turns in the opposite direction around the axis of the ecliptic.

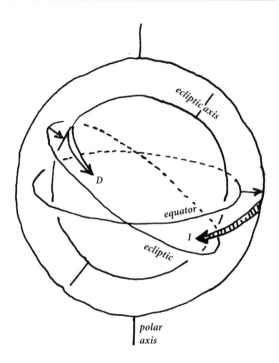

vine in essence, and this divine property is transmitted to the whole World inside it (or under it, as often expressed). The distinction between the two essences is not an easy thing to understand. According to the Pythagorians (see p. 47), God starts with two complementary essences, from which everything else emerges. The essence of identical, the indivisible, is basic, a sort of divine self-reference. The essence of the different, the divisible, is diversity. Mixing the two essences, God created a third, intermediary one and with the three essences constructed a unique species (ἰδέα = species in Greek). The sense of *idea* here is not the same as in modern English. This species, or idea, is the Soul of the World, the principle of all life, of all motion, in the World. The system described by Fig. 2.5 results from a bisecting of this unique species. One half becomes the circle of the equator, the other the circle of the ecliptic. They intersect one another to form the Greek letter χ. Each circle will define the motion, respectively, of the outer and inner orbs of Fig. 2.6.

One must admire the ingenuity of Plato in constructing a system which does a rather good job of describing diurnal motion and the seasonal changes (ecliptic motion) and is based almost entirely on principles of a metaphysical nature.

In Plato's system, the wandering bodies, or planets, are linked to the internal sphere (the different) but the essence of the different takes multiple forms. In place of the internal sphere, a more extended version of the system introduces several spheres. There are a total of seven different spheres plus one identical sphere, the latter reflecting diurnal motion, which affects all the stars equally, both the luminaries and wandering stars. We shall come back to this figure, which in some sense was completed much later by John Philopon (p. 132 and Fig. 3.3) with the addition of the four spheres of the elements in the center. We shall also discuss the problem of the distances of these spheres to the central one, which has a radius, by definition, of one.

What about the Moon? It is borne by the sphere of radius 1 (Fig. 2.6), which surrounds the elements of the sublunary World. The rotational velocity of the internal spheres is much smaller than that of the supreme orb. This accounts for seasonal motions, for diurnal vs. annual changes in the Sun's location (a), and could be associated with some kind of spiralling motion. Venus and Mercury, however, (on spheres b and c) are sometimes behind, sometimes in advance of the

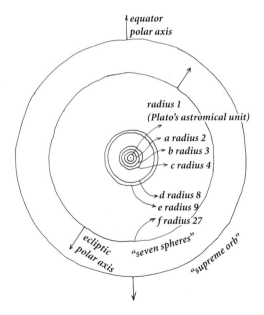

Fig. 2.6. *The Platonic system of homocentric spheres.* The Earth is at the center of the system. The sphere of radius 1 (unity of the distances within the system) is that of the Moon. The spheres a, b, c, d, e, and f bear the Sun, Venus, Mercury, Mars, Jupiter, and Saturn, respectively. Outside the seven spheres, in motion around the ecliptic axis, is the *supreme orb*, which produces the diurnal motion and is identical to that of Fig. 2.5.

Sun (from Ἑσπέρα, the evening star, to Ἑωσφόρος, the morning star in the case of Venus). In a way, a power oppositional to that of the supreme orb drags spheres b and c towards a. This is, in part, a heliocentric concept, which Plato was forced to take into account by the observation of the motions of these two planets. Plato was not the only one to have this in mind. Venus and Mercury were often called satellites of the Sun. The planets Mars, Jupiter, and Saturn occupy the spheres d, e, f. The supreme orb accounts for the diurnal motion and drags the seven other spheres. In addition, it carries with it the fixed stars. It is interesting to note that the double motion of the wandering stars gives them an apparently spiralling motion which is often mentioned in the Platonic dialogues.

A very interesting statement in the *Timaeus* is that the internal radius of the surface of each of the seven spheres (nothing is said about the diameter of the supreme orb) relates to one another like the numbers 1, 2, 3, 4, 8, 9, and 27 (noting that $27 = 1 + 2 + 3 + 4 + 8 + 9$). This was an important aspect of the Pythagorian-inspired arithmetic of Plato's constructions, but some commentators have other theories. Even among Plato's immediate followers, namely Adrastes, Clearchus of Soli, Crantor, the great Eudoxus, the famous Eratosthenes, Theodorus of Ascania, Thrasyllus, and Plutarch the historian, other interpretations have been suggested. Some were based on a combination of seven figures, which are between them like the terms of two geometrical progressions of ratio 2 (1, 2, 4, 8) and ratio 3 (1, 3, 9, 27). The successive ratios of each radius to the preceding ones are also rather simple: $2 = 2 \times 1$; $3 = 3 \times 1 = 1.5 \times 2$; $4 = 2 \times 2$; $8 = 2 \times 4$; $9 = 3 \times 3$, and $27 = 3 \times 9$. Some have found similarities between musical intervals and those used in describing the planetary spheres. In fact, the musical analogy is very frequent in Pythagorian-inspired astronomy, from Philolaus to Plato and much later to Kepler.

In the famous allegory of Em, Armenius's son and the spindle of necessity (in *The Republic*, X), Plato seems to describe this system of eight (7 + 1) spheres again. However in this case, he also expresses the idea that some planets may be wandering in spirals, around the ecliptic, a rather new idea.

Spiralling motions, particularly of the Sun, suggested that a true periodicity of the planetary system is neither one day, nor one year. Actually, Plato knew of the retrogradations, of the complexity of the planetary motions. He also knew of the inequality of the seasons but he didn't incorporate this knowledge into his theory as it would have impaired the beautiful organization of concentric spheres. In any case, all these irregularities moved Plato to consider that there must be a Great Year, which would allow for all planets to return to the same mutual positions. Meton's 19-year long Great Year (which took only the Sun and Moon into consideration) was known to Plato. In order to take all the planets into account however, the Platonic Great Year had to be much longer. This implies the periodicity of behavior of the Universe, an idea that we shall revisit in modern cosmology, sometimes even more in theology (the myth of the "eternal return"), with generally no more success than Plato. The Platonic Great Year was estimated to be on the order of 10000 years, but there is little sound argument for that. Some authors, starting from the vague indications given by Plato, arrive at a Platonic year of 760000 years.

Similar ideas emerged from religious myths elsewhere, notably in India. They were described by the Arabic scientists Massoudi (around 913) and Al Birûni (around 1031). Their *Kalpa* is equal to 4320000000 years. It is possible that these ideas, which belonged then to the Indian tradition, were brought to India by Alexander the Great and were indeed adapted from Plato's works. On the other hand, the Chaldeans (such as Berosus, about 260 BC) knew them also, and even before Plato, philosophers like Anaximander and Oenopides expressed similar ideas.

Exactly as in the case of Meton's cycle, we are confronted with the problem of finding rational fractions, approximately representative of the *least common multiple* (or L.C.M.) of all the periodicities involved, as observed in Plato's time. This is a very interesting discussion and one that interested philosophers over a long period of time. Oenopides and Pythagoras considered Great Years of 8 y, 19 y (the "luni-solar" one), and even 59 years, according to Philolaus. Heraclitus considered a cycle of 18000 years (or 10800 years?). Diogenes used a cycle of 18000 (from the planets?) × 365 (from the Sun?) which equals 6570000 y, without any apparent logic to it. More reasonably, Democritus suggested a cycle of 82 y and Hipparchus one of 304 y. According to Aristotle, Aristarchus used 2484 y, Aretus 5552 y, Dion 10884 y, Orpheus 120000 y, and Cassandra 600000 y. Of course, the skeptical Aristotle did not propose a number! Some even suggested an infinite length for the Great Year, which in a sense is quite reasonable and follows from the incommensurability of the periods. This would rule out most of the theological implications.

We see, at the time of Plato, several alternative lines of thought emerging. On the one hand, there was the notion of the eternal return, man by man, event by event. This is a position of extreme determinism and in apparent contradiction with what we now know as the *Second Law of Thermodynamics* (see p. 369 and p. 517). Plato did not posit such an exact recapitulation of events, directly inspired by religious views and implying the immortality of the soul and metempsychosis. A belief in a modified or partial return was held by some people. According to this view, there would be a "similar," not identical, return of people and events. Finally, there were also skeptical people who did not believe in an interconnection between human events and the realm of planetary motions.

For Plato, the *primary* cause of all motions, all phenomena, was the motion of the Soul of the World, the motion of the Supreme Orb, created and governed by the Deity.

One important aspect of Plato's cosmology was the Earth's immobility. However, there are contradictions within the *Timaeus* on this subject. According to Timaeus, "God has made Earth, our nurse. It is rolled up around the axis of the World. It is the guardian of day and night, and it produces them. Among all gods who are under the sky, it is the oldest." Aristotle later interpreted the first part of this statement of Plato's as an indication of the rotation of the Earth around the World polar axis. However, elsewhere in the *Timaeus*, diurnal motion is attributed to the spherical starry vault, which is indeed the Supreme Orb. In the *Phaedon* and *The Republic,* the Earth is said to be "without motion." It is possible that Aristotle did not properly express Plato's views. At the very least, it is evident that the quoted sentence is ambiguous. It is likely that, in his turn, Aristotle was inadequately interpreted by Simplicius and others, and that his views regarding the *Timaeus* were quite different from what has come down to us.

In modern language, we could say that, according to Plato's views, the Earth is suspended and immobile for "reasons of symmetry." It is a principle of equilibrium, clearly expressed in the *Timaeus*. That Earth is at rest seems to be Plato's final conclusion, the commentator Simplicius notwithstanding.

According to Plutarch (but never stated explicity in any of Plato's works), Plato in his later years expressed regret for having put the Earth at the center. Like the Pythagorians, who put fire at the center, he seemed to think that the center should be attributed to some being more powerful than the Earth, in other words the Soul of the World, which animates the Supreme Orb. Plato's views about this controversial problem are perhaps better expressed in *The Republic, II.* There Socrates, in his discussion with Glaucus, outlines, as we have described previously, what should be taught to the young. Numbers and geometry, then astronomy, but only for rather mercantile purposes, if we follow Glaucus. Socrates, however, looks at it with a nobler mind: "I cannot," he says, "accept that a study directs the soul upwards, unless it has for its very object what is and cannot be seen." True astronomy (not merely utilitarian observational astronomy) keeps the young from turning their minds to the heavenly gods with false or sacrilegious judgment, which would be harmful to the City. In other words, the true astronomer must save the appearances and using geometry and arithmetic properly, uncover the true nature of motions, which is to say, abstract the observed. We are back to the myth of the cave. This abstraction goes so far as to invest celestial bodies with souls. They are gods, or at least god-made bodies. They are powers. The astronomer has to model the astronomical observations, even at the expense of simplification. We know the Earth is not a sphere (*Phaedon*), yet we still prefer to model it as a sphere. We recognize the inequality of the seasons, but we have no way, in our concentric spheres model, absolutely no way, to save this observation.

2.1.6
Concluding Our Visit to Plato's Dialogues

Plato was primarily a philosopher, not an astronomer. His astronomical knowledge was, for the most part, indirect, the result of hear-say. However, Plato, like his number-enchanted predecessors, Pythagoras and Philolaus, wanted to model the observed parts of the world, to abstract them. For the remainder of his vision, he relied on the mathematical perfection of numbers and geometrical figures and on some concept of the Creator's design.

Both schools were perfectly honest, deeply impregnated with metaphysical beliefs, and quite impressed by the perfection of mathematics, but also willing to attempt to save the phenomena. It is only their descendants who have followed different paths.

The Pythagorians, even before Plato, but certainly after him, operated from the premise that the harmony of the spheres and the perfection of numerical combinations had more truth in them than necessarily imperfect human-made observations. More and more, they abandoned any attempt to save the phenomena.

The Platonists maintained their desire to save the phenomena, including phenomena that were not known to their predecessors. However, they also maintained that the World has a Soul and that the Universe was created by some benevolent God, well aware of the best mathematics, obviously his own creation as well. It is very hard to distinguish, in the Platonic school, the purely physical foreground from the religious background, from the metaphysical implications. Often these thinkers had to choose between saving their idea of God's creation and saving the phenomena and often, it was the latter that they, sometimes reluctantly, sacrificed.

2.2
Plato's Contemporaries: Eudoxus, Callippus

2.2.1
Eudoxus and his System of Homocentric Spheres

Geometrical astronomy, which tries to save the phenomena through well built models, was the essence of Plato's cosmology.

Eudoxus was the one who put these precepts in writing and followed them through with a considerable ability. The basic text was copied from Eudoxus by Eudemes, from Eudemes by Sosigenes and finally by Simplicius, who introduced it in his comments on Aristotle. It is there that we find Plato's affirmation of his basic need to save the phenomena, i.e. the apparent motions of the planets, using circular and perfectly regular motions. This was indeed the essence of Plato's cosmological philosophy: a combination of perfect geometry and the correct representation of the best observations.

Plato's model required a combination of circular motions sharing the same and unique center. At this center, he put (note the contradiction with the comment by Plutarch quoted above) the Earth's center itself. The problem is then quite straightforward, at least to our modern eyes: conceive of a number of spheres, concentric (or *homocentric*) around the same center, which is the Earth's center, and reproduce, with different choices of axes, direction and velocity of rotation, the apparent motions of the planets. The followers of Plato assumed, moreover, that the circular motions were always circular in the same direction and at a constant velocity. They all added to these requirements the impossible dictum expressed by Simplicius, repeated in Plato's oral teaching, and later inherent in Aristotelian philosophy: *Save the phenomena* (or save the appearances: σώζειν τὰ φαινόμενα). This philosophy was expressed for example in the textbooks, typical of the Greco-Roman commentators, of Aratus (ca 300 BC), Geminus (ca 50 AD) and still better of Simplicius (VIth century).

These principles, with minor changes, guided scientific/philosophical thought from the earliest Antiquity. Certainly not, however, up to the time of Copernicus. By that time, the idea of the Ptolemaic eccentric suggested a very different way of looking at things. Indeed, many ways of saving the appearances have been devised.

Unfortunately, homocentric spheres did not save some of the well-known appearances. Meton and Euctemon knew the different durations of the seasons, as did Democritus. Plato himself knew about the retrogradations of the planetary motions. The first one to really make a serious attempt to respond to the Platonic challenge was Eudoxus of Cnidos (408 BC–355 BC). He was a student of the Pythagorian Archytas, but also of Plato himself. He spent sixteen months in Egypt, working with priests, and there devoted much time to observing the planets with a rather high level of accuracy, as can be seen in Table 2.1. He had his observatory in Heliopolis. Later he founded a school in Cyzicus, where Polemarchus was a student, a school which he later removed to Athens. He finally ended his life in Cnidos, where an astronomical observatory was built for him.

Eudoxus built a rather elaborate system (Fig. 2.7) of concentric, homocentric spheres, of various axial orientations. In essence, the number of spheres necessary to describe the motion of any given wandering body (planet or sublunary) can be as large as needed. The external sphere (Plato's supreme orb) is the one which is responsible for diurnal motion and turns from east to west, uniformly. The second sphere turns from west to east, uniformly, with a period equal to the *zodiacal revolution* of the planet. Its axis is perpendicular to the ecliptic.

Table 2.1. *Eudoxus' data*

Planet	Zodiacal revolution		Synodic revolution	
	Eudoxos	Now	Eudoxos	Now
Venus	1 y	1 y	570 d	584 d
Mercury	1 y	1 y	110 d	116 d
Mars	2 y	2 y 322 d	260 d	780 d
Jupiter	11 y	12 y 315 d	390 d	399 d
Saturn	30 y	29 y 166 d	390 d	378 d

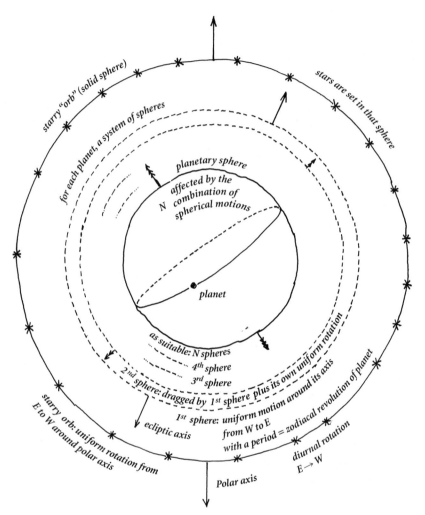

Fig. 2.7. *The general scheme of Eudoxus's homocentric spheres, for one planet.* For any given planet, Eudoxus's system requires the starry orb (Plato's supreme orb) for diurnal motion and the first sphere for ecliptic motion of a period equal to the zodiacal revolution of the planet. Then, it requires as many spheres as deemed necessary to account for all of the observed motions of the planet. The planetary sphere, the last one, bears the planet and is affected by the combination of the rotations of all of the spheres exterior to it. It is desirable to keep the number of spheres as small as possible, but sufficiently large to save the phenomena. The successive spheres have neither the same axis, nor the same period of revolution around it.

The center of the planet under study is located or implanted, to use a more physical metaphor, like a diamond on a ring, on the equator of the internal sphere, irrespective of the number of spheres between that internal sphere and the diurnal sphere of fixed stars. All spheres are animated by motions successively more complicated, when moving from the outside in. The first sphere has its rotation, a simple one, the more obvious one. The second one is dragged by the first sphere's rotation which is combined with its own, resulting in a certain composite motion. In its turn, this composite motion is combined with the proper rotation of the third sphere to give the composite motion of that third sphere, and so on down to the internal sphere, the one which bears the planet.

In the general framework of this model, one can describe as follows the motions of the seven wandering bodies, the two luminaries and the five planets.

The lunar theory is the first and the simplest. It was known from observation that the Moon's latitude varies from 5° N to 5° S of the plane of the ecliptic. This explains why solar eclipses do not occur at every new moon, or lunar eclipses at every full moon. The particular spheres system for the Moon is thus described in Fig. 2.8, and accounts for this observed fact, as well as for the length of the lunar months, using only three spheres. The third sphere, the lunar sphere, has a polar axis inclined by 5° with respect to the axis of the ecliptic.

Here we must make a digression. How can lunar eclipses be predicted? The points at which the projected apparent orbits of the Moon and the Sun on the spherical starry vault cross are the *nodes*. An eclipse occurs when the full or new moon coincides with the Moon's location at one of the two nodes, and this rules the prediction of eclipses. The lines of the nodes were called the "Dragon," symbolizing the mythical animal which was thought to devour the Moon. This designation gave its name to the *draconic* month, of (as we now know) 27 d 5 h 5 mn 36 s. This is the length of time between two successive passages of the Moon by the same node. The second sphere, the ecliptic one, turns in 223 lunar months (or about 16.5 y or again about 6030 d). This motion properly accounts for the movement of the nodes along the ecliptic circle.

The solar theory is constructed along the same lines as the lunar theory, but it does not work out as well at all. The precession of the equinoxes (see p. 92, 123 ff.) was not yet well known, but it had to be taken into account in some way. This confirms that Eudoxus was really committed to saving the phenomena. However, the ellipticity of solar appar-

Fig. 2.8. *The Moon's motion according to Eudoxus.* Two spheres (in addition to the supreme orb) are necessary. This allowed Eudoxus to account for the inclination of the plane of rotation of the Moon around the Earth with respect to the plane of the ecliptic (about 5°).

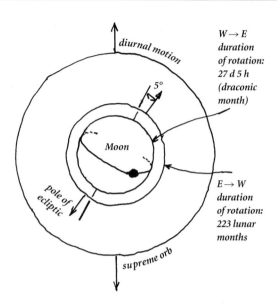

ent motion (using our modern terminology) which gives rise to the seasons, was not taken into account and the Sun was displaced from the ecliptic in order to reestablish the true appearances of the third sphere, the solar orb. The solar orb turns in one year (not surprising). The ecliptic sphere turns extremely slowly, possibly due to a vague idea that Eudoxus might have had about the very slow precession of the equinoxes. This is unduly complicated and, in addition, did not reproduce the seasonal variations. Clearly, the Sun's true motion, unlike that of the Moon, is not well adapted to a system of homocentric spheres. It is simpler, in a way, than Eudoxus understood it to be. Note that the equator of the ecliptic sphere (the second one) is not assumed a priori to be the Sun's orb. This means that, in the mind of Eudoxus and his contemporaries, the ecliptic was not essentially defined by the apparent motion of the Sun around the Earth, but was rather a circle, common to all planets. In fact, it was the zodiacal circle, the road on the sky for the Sun and planets to follow.

The planetary theory introduced a new refinement – the motion on the planetary sphere (Fig. 2.9) of the *true position B* of the planet around a *mean position A* of the planet. As B turns around A, the true curve described on the sphere bearing A by B is an interesting shape. It looks like a *lemniscate*, an "∞-shaped" curve called a *hippoped* because it describes the motion of a horse trained to walk dragged either by one side of the mouth or the other at a constant distance from the dragging man. Schiaparelli, in 1874, computed the exact shape of the hippoped (Fig. 2.10).

In this schema, several orbs are necessary. There is the external one (responsible for the diurnal motion), the ecliptic one, and the last one,

Fig. 2.9. *The planetary motions according to Eudoxus.* In addition to the supreme orb and to the ecliptic motion, a combination of two spheres with almost exactly compensating motions allows for a representation of the motion of the true planet B. The mean position of the planet A follows only the ecliptic motion. The two additional spheres and the rotation of the true planet around the mean position are tricks that provide an explanation of the excursions of the planet out from the plane of the ecliptic, its retrogradations and loops.

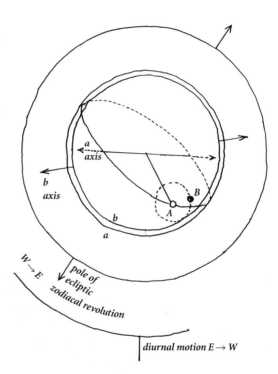

Fig. 2.10. *The hippoped.* In Fig. 2.10, the true planet B describes a complicated curve around the mean position A, when projected on the ecliptic sphere. This curve describes the excursion of the planet out from the ecliptic. It is called a *hippoped*, or, in less zoological terms, a *lemniscate*.

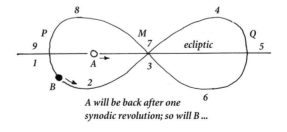

which bears the planet, both in its true and mean positions. This last orb is made out of two sub-spheres, *a* and *b*, one bearing *A*, the other bearing *B*. They essentially coincide, but with different axes. The rotation of the ecliptic sphere from west to east is accomplished in a time called the *zodiacal revolution of the planet*. The true planet *B* accomplishes its revolution around its mean position *A* in one *synodic revolution*. Note that this circular motion is quite similar to the epicycle of the Ptolemaic system a few centuries later (see p. 95ff.), and that the synodic revolution is never very different from one year. This time is rather well determined by the observations of retrogradations and changes in latitudes, phenomena Eudoxus would not have been able to explain or describe by considering only the true planet. The axis of the sub-sphere *a* is oriented with respect to the axis of sub-sphere *b* with a small angle α, which varies of course from planet to planet.

The true position *B* of the planet on the sub-sphere will describe on the sub-sphere *a* a motion which accounts for the appearances (retrogradations or loops and changes of latitude of the planet during its synodic revolution). The organization of the sub-spheres is then as follows (Fig. 2.9): In the plane of the ecliptic, Eudoxus traces the diameter perpendicular to the radius of the circle which will meet the mean position *A* of the planet. This diameter will be the rotational axis of the *a* sphere, which bears the mean position of the planet. Its revolution will be accomplished in a time equal to the synodic revolution. The axis of the sub-sphere *b* will form with the axis of *a* an angle smaller than 90°, and the sub-sphere *b* will accomplish its revolution around the axis of sub-sphere *b* in a time equal also to the synodic revolution. These two revolutions, almost compensating each other, will account for the strange, but observed motion of the true position of planet *B*, not around the mean position *A*, but projected on the ecliptic sphere, allowing *B* to intersect the ecliptic four times during the synodic revolution, at P, Q, and twice at M (Fig. 2.10). Of course this strange motion has to be added to the diurnal motion (of the first sphere) and to the ecliptic motion (of the second sphere). It only accounts for the irregularities of the zodiacal path of the true planet. The mean planet will continue to describe the ecliptic. During the course of one zodiacal revolution, the hippoped will occur in successive different places around the plane of the ecliptic.

2.2.2
The Reform of Calippus

Calippus (370–300 BC) was a student of Polemarchus, hence a second-degree pupil of Eudoxus. According to the later comments of Simplicius, Calippus was always a good observer. In the course of his observations, he found that he was not entirely satisfied, and decided to attempt to correct some inadequacies in the Eudoxus constructions. We have mentioned the seasonal problem. Calippus was even more concerned by the fact that the mean position of Venus could not be aligned with the Sun, which, because of its third sphere, was outside of the ecliptic. Calippus noted several other oddities and worked at correcting the Eudoxus system to account for them.

His first move was to note the different durations of the seasons (Table 2.2). He observed the changes in the apparent diameter of the Sun, which we now know are caused by the ellipticity of the apparent motion of the Sun around the Earth. Calippus added two spheres to the motions of the Moon and the Sun, and one each to Mars, Mercury and Venus, leaving Jupiter and Saturn with the four spheres assigned to them by Eudoxus. Using 40 spheres (instead of the 33 homocentric spheres of Eudoxus), he managed to eliminate most of the difficulties. But, in particular, did he account well for the seasons? Simplicius is far from clear about that. If the spheres remain homocentric, it is not at all clear how the changing apparent diameter of the Sun could be explained. As far as the duration of the seasons is concerned, a certain manipulation of the five solar spheres might do the trick. It was a beginning. Calippus and Eudoxus certainly improved the homocentric system of Plato, but they were still far from a satisfactory solution. Would Aristotle, this protean genius, who was a pupil of Plato and who had worked with Eudoxus and Calippus, find a better way so save the phenomena? All the phenomena?

Table 2.2. *Calippus and the seasons*		Duration according to Calippus	Modern values
	spring	94 d	94.17 d
	summer	92 d	92.08 d
	fall	89 d	88.57 d
	winter	90 d	90.44 d

2.2.3
Astronomers at the Time of Aristotle

The cases of Eudoxus and Calippus are quite interesting in that they really were observers and their geometrical constructions were based upon a very good knowledge of the sky. They were not the only ones engaged in astronomical research during this period. Some astronomers left written texts. Others were quoted by the commentators of the following centuries.

For example, Autolycus of Pitane (around 360–290 BC) is known for his attempts to improve the Eudoxus system and for his famous controversy with Aristotherus. He was, so to speak, a "modern." Before him, there were some purists, for example Polemarchus of Cyzicus, a pupil of Eudoxus and teacher of Callipus, who would only teach the Eudoxian concepts. Autolycus, on the other hand, in his *Moving Sphere* and *Heliac risings and settings*, worked very thoroughly on the geometry of the motions of points located on the homocentric spheres. His work helped scientists of a later period to better master the geometry and kinematics of astronomy.

Rather than going into detail regarding the work of these astronomers, however, we will look instead at the emergence of Aristotle and Aristotelianism, which exerted such a powerful influence up to the Renaissance. Even now, by virtue of his methodical rationalism, his reasonable skepticism, and his sense of the "physics" of things, not only of their geometry, or their kinematics, the influence of Aristotle is still pervasive.

2.3
Aristotle's World

It is interesting to note that the name of Aristotle has very contradictory implications in the literature. For some commentators, the very mention of his name throws them into a frenzy and promotes a violent attack on medieval dogmatism. For others, he symbolizes skepticism and rationalism. He could even be considered a physicist, in the way we now understand the word. Actually, there are many contradictions in Aristotle's thinking, from our modern vantage point, and one could not expect it to be otherwise, given the breadth and extent of the man's thought.

Aristotle was born in 384 BC at Stagira, in Macedonia, and is often referred to as "the Stagirite." He was a pupil of Plato, about 30 years younger, and he adhered closely to Plato's teaching in his early works. However, he diverged rather early from the old master and created the Lyceum around 335 BC as a response to Plato's Academy, which was founded fifty years previously. This is the birthplace of the so-called *peripatetic* school (so named because participants discussed philosophy while walking along a promenade).

Aristotle left an immense body of work, dealing with all aspects of knowledge, having the character of an encyclopaedia, in which his cosmological views are only a relatively small part.

2.3.1
Aristotle's Cosmology of Homocentric Spheres

Aristotle readily accepted the idea of homocentric spheres, rooted in the Platonic general description of the world. However, his ideas contrasted in an important way from those of Eudoxus and Calippus, who were good at geometry, but were not really philosophers or physicists. They described the system planet by planet, without (unlike Plato actually) paying much attention to the global mechanism of motions. Aristotle, on the other hand, wanted to design a system coherent in itself, which would not describe the planetary motions separately, but in the context of the whole World, while at the same time preserving the independance of each motion. Moreover, the system had to have some physical reality. Unlike Plato, he did not regard a geometrical model as sufficient. The Eudoxian system was full of empty spaces, especially with the large distances between the successive systems of planetary orbs, as given by Plato. Aristotle, in his physics, denied the possibility of the existence of a total vacuum. Therefore, he needed to put something betweeen the different systems of spheres. They could not be lined up side by side, because then they would no longer be independent. In order to achieve this, Aristotle intercalates between planets M and P some spheres rotating in the opposite direction, exactly compensating, for planet M (inside), the rotation of the external systems of spheres associated with planet P (outside). Each sphere transmits the diurnal motion to the ones interior to it. The diurnal motion which affects them all is given to the whole system by the first orb, the Supreme orb, or starry vault. This restoration of the diurnal motion at each step is one of the reasons for the compensating spheres (Fig. 2.11).

Note the materiality of the successive orbs in which each planet is en-shrined, like a precious stone in a glass sphere. They are not immaterial geometrical spheres. Each of the successive spheres, whatever its role, is associated with the one just inside it by a material axis of rotation. One can characterize the mechanism by the number of the spheres it needs: for Saturn, three spheres, then three compensating spheres, then three Jupiter spheres, three compensating spheres again, etc. Altogether, there are 55 homocentric spheres. As noted later, according to Sosigenes, the system could be simplified to 49 by suppressing six of the compensating spheres. Still, it appears to be a complex system, but not a purely ab-stract, geometrical one. It is, in fact, a model that one could consider building in a workshop.

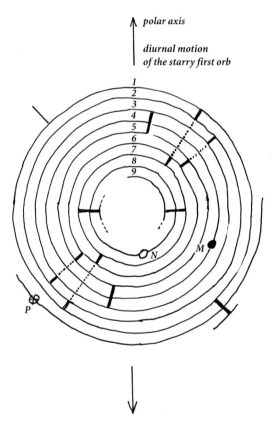

Fig. 2.11. *Aristotle's compensating spheres.* Planets in this figure (inspired by Simplicius's account of Aristotle's cosmology) are labelled P, M, and N. Spheres 1, 2 and 3 rotate around their axes, but transmit the combination of their motions to sphere 4 (bearing the planet M). Spheres 5, 6 and 7 are there as *compensating spheres* to cancel out the motions imposed upon M and to leave only the diur-nal motion acting on it. Diurnal motion is com-mon to all planets. This system allows the planets to be physically associated with each other, but to maintain independent mo-tions.

polar axis

diurnal motion
of the starry first orb

The main contribution of Aristotle is not, however, this complex combination of nested solid spheres, but in the physical ideas that lie behind
it; ideas which belong more to physics than metaphysics.

2.3.2
Aristotle's Physics

Like Plato, particularly in his consideration of the Earth's structure in
the *Phaedon,* Aristotle noted as an obvious fact that there was not yet a
real science of the accidental or the irregular. There was a need, however,
to model the observed world. Unlike Plato, Aristotle came to this conclusion for operational, not metaphysical reasons, and the two arrived at
quite different models. Aristotle's abstraction is made out of solid
spheres, not ideal geometrical figures without materiality as in Plato.
Moreover, whereas Plato regarded living beings as creatures of God,
Aristotle did not rule out the scientific investigation of entities which
could be born and later die. Therefore, Aristotle claimed that behind
generation and corruption there are causes and principles, which in their
turn may have been born and may die, and which can be the object of
scientific study. The object of science then is to study what evolves, what
moves, what changes in the material world. The connection to any God
or Soul (the *primary cause*) is very far indeed from Aristotle's scientific
approach. In this spirit, Aristotle left scientific treatises on natural
science, on everything that can be observed: falling bodies, vapors and
clouds, their formation and their motions, animals, plants, and so forth.

In other words, mathematical theory and metaphysical considerations
on the one hand and sense perception on the other have opposite values
and roles in Plato and Aristotle. Again, we shall find these opposed states
of mind represented in all eras of the history of science. In modern cosmology, for example, Einstein has a typically Aristotelian attitude, while
others, such as Weinberg and Hawking, maintain an attitude closer to
Plato's. It must be understood, however, that this is only an approximation, since the meaning of the words has changed over the centuries.
Moreover, in every scientist, there is a cohabitation, or a succession, of
states of mind, sometimes Pythagorian, sometimes Aristotelian, and
sometimes Platonic.

Aristotle, in his rationalist approach, discussed several problems already touched upon by Plato, notably those concerning space, time, and
location. Within the enormous body of Aristotle's work, we shall limit

ourselves to a discussion of some points that we have already mentioned in connection with Platonic philosophy.

Aristotle readily accepted the notion of the Platonic Great Year which returns again and again, *ad infinitum*. This meant that the local motion was the primary motion and the principle behind all other change. The World, what we would understand as the Universe, does not undergo corruption or generation. This statement had to be reconciled with the observation that the sublunar world is, on the contrary, characteristically subject to variations, deaths, and births.

This discussion led Aristotle to state that the vacuum does not exist since an immaterial sphere could not drag another sphere. It also led him to the conclusion that the World was finite. Beyond the first sphere, which is to say beyond the sky, there is nothing. There is neither a vacuum nor matter of any kind. There is Nothing. There is no "place," no room for either the vacuum or matter.

A first corollary that emerges from this principle is that outside of the World there is no motion, as there is no space. A second is that no true straight line can be infinite, nor any longer than the World's diameter. Aristotle opposed the physical idea of a straight line to the geometrical Euclidian straight line, prefiguring some sort of Riemannian geometry, or curved space physics, by more than two millennia.

What then constituted "local motion?" It had to be a rotation. No rectilinear motion was possible, because of the necessary finitude of the path. The first motion had to be a rotation around an axis (not unlike Plato's concept of the first motion).

The substance which turns with a constant velocity of rotation around an axis is the aether (αἰθήρ), so named because it returns ceaselessly, for eternity (ἀεί). Is this aether Plato's aether or the quintessence? In the Aristotelian context, it is quite different from the four elements, unmixible (unlike in Plato) with any of them, untransformable into any of them, and unformable from any combination of them. Aether is characteristic of the astronomical World. It is not subject to the generation and corruption of the sublunary world.

This idea (which we know now to be false and misleading) was the basis of all ideas about the nature of the Universe from Antiquity through the Middle Ages, and up to the time of Tycho Brahe's discovery (1572) of a distant supernova.

Nothing in this theory is incompatible with the Eudoxian system of spheres. Rather, it is complementary to it. Aristotle was, in fact, very interested in the Eudoxian system which, as we have seen, he himself worked on and improved.

For Aristotle, the Earth was spherical. The idea was not new. It was expressed by Plato in the *Phaedon* and elsewhere, but Aristotle devised a proof, which was far more exacting than Plato's. Actually, the very idea of "proof," of testing the quality of a model, is an Aristotelian one. Aristotle argued that the Earth's shadow was always circular during lunar eclipses. The curvature of the Earth is also demonstrated by an observer who sees different constellations when moving from north to south. These observations will later lead to the measurement of the size of the Earth by Erathosthenes.

Another observation that derives from the Earth's sphericity is the fact that falling bodies tend towards the Earth's center and fall along the vertical, but the vertical direction varies from point to point. In addition, the spherical shape of the water surface on the large seas and oceans was noted. Archimedes' later account of his demonstration of this phenomenon was much more convincing than Aristotle's, but it was completely forgotten until it was translated into Latin (1269) by William de Moerbecke and later plagiarized by Nicolo Tartaglia (1543).

2.4
The Legacy of Plato and Aristotle

The Platonic and Aristotelian legacy has left its mark on science up to modern times and will perhaps continue to do so for a long time into the future. As we shall see, the idealistic point of view of Plato and the materialist point of view of Aristotle always had their respective defenders among scientists. Nevertheless, many details of their systems have been forgotten. The two methodologies and perspectives remain, but in the light of the explosion of data, it is quite clear, in the Aristotelian spirit, that the modeling had to evolve.

For example, the theory of the four elements (plus one, aether) lasted up to the time of Tycho Brahe, and even up to the modern chemists, Lavoisier and others. There were new concepts about the aether, which emerged in the 19th century and had varying degrees of success and failure (see p. 382). The difference between the nature of the sublunar and astral worlds was also erased by Tycho's observations. The idea of circular motions was entrenched until Kepler introduced elliptical motions to explain observations of Mars. Aristotle's understanding of the rise and fall of heavy bodies lasted up to the emergence of Galilean kinematics and Aristotelian dynamics lasted until the development and acceptance of Newtonian dynamics. The homocentric spheres did not even last that long. Neither Heraclides, Hipparchus, Ptolemy, nor later Copernicus, ever entertained such a concept.

During the "dark ages," medieval times, a peripatetic school fought continuously for the letter, not only the spirit, of most of the Aristotle's views. Many Christian scholars were strongly influenced by Platonic idealism. Some of the models which resulted were crude, some were quite refined. The developing tradition was often a mix of the principles of Plato and Aristotle and their descriptions of the World. Their two names became the lasting symbols of Greek wisdom.

2.5
The Heliocentric Systems

The axes of the Eudoxian spheres are all oriented in almost the same direction. Since the trajectories of the Sun, Moon, and planets are always in the plane of their respective equators, they are almost in the same plane, i.e. "coplanar." However, the phenomena are not properly saved. The Eudoxian system does not explain the inequalities of the seasons or the changes over time in the brightness of Mercury, Venus, and Mars, which we now know explain their relative distances to Earth. It does not account for the difference in appearance between total and annular solar eclipses. Is it possible that a plane representation, that is projecting all the trajectories on a single plane, would make up for these inadequacies in the system? How would such a representation account for the fact that the trajectories are not perfectly coplanar, but involve excursions out of the ecliptic which are quite essential to the Eudoxian system? Several astronomers worked with the plane model to try to account for unexplained phenomena such as the variation of the distance to the Earth of all wandering bodies. Heraclides, Aristarchus, and Seleucus are the best known of these. Their contributions, especially that of Aristarchus, have been of paramount importance.

2.5.1
Heraclides (from the Pontus)

Heraclides was born at Heraclea, on the Pontus (what we now call the Black Sea), and lived primarily in Athens during the 4th century BC. He was a follower of Aristotle, but while at the Academy, also involved himself with the ideas of the successors of Plato. He was such a daring and prolific writer that he received the nickname of *Paradoxologos* (paradoxical discourser).

Heraclides suggested that Venus and probably Mercury are turning around the Sun (Fig. 2.12). This easily explains the change in brightness over time of both planets. It also explains why Venus never departs from the Sun by more than about 47°, and Mercury never by more than 28°. It is possible, but this hypothesis is quite controversial, that Heraclides was defending strictly heliocentric ideas, similar to those later presented by Aristarchus and Copernicus. Did Heraclides ever suggest that the Earth was turning around the Sun and not the reverse? In spite of many later testimonies that he did, it is quite difficult to answer that question. It is even likely that in his day the question did not appear to have any scientific significance.

In the following centuries, the influence of Heraclides is apparent. It seems likely that some later Pythagorians defended similar views, perhaps Hiketas and Elephantus. They seem to have abandoned the ideas expressed earlier by Philolaus (see p. 52) about the central fire and the

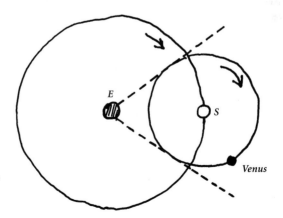

Fig. 2.12. *The Heraclides system.* See detailed explanations in the text for Figs. 2.12–14.

anti-Earth, but they kept the Earth in motion. Heraclides' theories were adopted by several astronomers. One could say that the system of Tycho Brahe in the 16th century was a modern revival of the Heraclidean system.

2.5.2
Aristarchus (300–250 BC?)

Aristarchus of Samos lived after Euclid and before Archimedes, who reported at length about the Aristarchian system. Although born in Samos, it is certain that he spent much of his life in Alexandria, where he observed the solstice in 280 BC.

Aristarchus must surely be considered one of the greatest astronomers of all time. He is certainly a precursor of Copernicus, who was not ignorant of his works. He invented, or at least developed, several instruments of precise observation of directions and of time. He also made precise determinations of various dimensions in the solar system. As we shall see, these determinations were later basic to our understanding of the solar system.

According to Aristarchus, the Sun is at the center of the system. The Earth and planets turn around it. The Earth also turns around its axis. The sphere of fixed stars is stationary and at a considerable distance (Fig. 2.13).

A corollary of the Aristarchus system is that the radius of the sphere of fixed stars is considerably larger than that of the Earth-orbit radius. If that were not the case, each star would appear on the sky to have a small annual motion, a result of the motion of the Earth around the Sun. Such a motion is the indication of a *parallax* (see p. 23). However, no such motion was actually detected at that time (it was actually only detected in the 19th century, see p. 347). This argument was convincingly used later by Kepler, in defending the idea of an extremely great distance to the fixed stars.

The system of Aristarchus did not account for the seasonal inequalities, nor for the different "sizes" of solar eclipses. It did, however, account for the variation in brightness of planets over time.

In addition to this hypothesis, reported by Archimedes, we know of another essential work of Aristarchus, which came down to us directly. This is his treatise *Concerning the size and distances of the Sun and Moon.* His determinations were precise; his geometrical demonstrations

Fig. 2.13. *The Aristarchus system.*

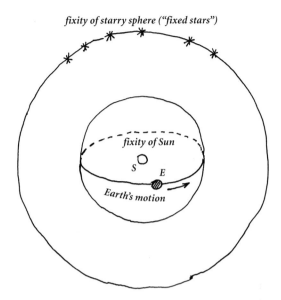

were rigorous, in the Euclidian spirit, and we can consider these measurements a real basis for an estimate of distances in the Universe. In a later chapter, we shall describe Aristarchus's methods more fully.

2.5.3
Seleucus (150–100 BC)

Aristarchus's ideas about the structure of the Universe were presented by Archimedes as hypotheses. Seleucus defended the same model as Aristarchus did, but with much more conviction, almost as gospel truth.

Seleucus was born in Chaldea, at Seleucia, a town located on the river Tigris.

In addition to being a committed Aristarchian, he suggested that tides were connected to the Moon's revolution around the Earth. Actually, it was quite a naive explanation in its details, implying air and water in motion between the Earth and the Moon. A similar idea was proposed around 1650 by one of the founders of modern dynamics, René Descartes.

These early heliocentric theories were soon abandoned.

2.5.4
The Provisional End of Heliocentrism

Since the 5th century BC, some philosophers had argued that the Earth
did not belong at the center of the Universe, but they were subjected to
persuasive criticism. The Greeks regarded the Earth as a sacred object
and geocentrism was a way for Greek culture to assert its theologically
driven conception of the Earth. Aristarchus was severely criticized by
the Stoic Cleanthes, in the name of commonly accepted religious beliefs,
because he wanted to move the very focus of the Universe. Cleanthes
wanted to have Aristarchus judged for impiety. This would have meant
the death penalty (remember Socrates!). Fortunately, Aristarchus was
never condemned or even judged, but this illustrates the fact that, at that
time, many scientists felt that it was prudent to keep their knowledge a
very close secret.

The main logical reason for the rejection of heliocentrism in scientific
circles was a typically Aristotelian one. It was not believed that heliocentrism could properly describe the observed phenomena, or "save"
them. The heliocentric theories of the day were only qualitative speculations. Rules, precise rules, allowing for computations, i.e. predictions, of
the apparent motions of wandering stars, were still lacking. At this stage,
as long as the stars were thought to be at a "very great" distance (a rather
vague idea), there was no essential difference between the Eudoxian
point of view and that of Aristarchus. Whether the Sun is turning around
the Earth, or vice versa, is essentially irrelevant. What matters is the ability of a model to properly represent the apparent motions of planets and
their apparent behavior.

In addition, it seems that Apollonius of Perga, the great master of geometry, and Hipparchus as well, entered into the controversy, against heliocentrism. Several bona fide arguments were actually given.

The first argument noted that heavy bodies fall towards the center of
the Earth. When they reach the location where they remain without motion, a place of rest so to speak, they stay there. This suggests, by extention to the Earth itself, that the Earth has reached its place of rest. Not
only is it at the center of the Universe, but it does not move in any way.

The second argument suggested that if the Earth were moving, even
only turning around its own axis, objects present in the air (clouds, objects launched from Earth) should be affected by this motion. They
could never overtake the necessarily fast motion of the Earth. They

would be left behind. This argument is, however, specious. After all, planets, the Sun, and the Moon move without falling. It was this observation which led Aristotle to defend the idea that the astral world was essentially made of aether, of a different nature from objects known on the Earth. Much discussion took place regarding the strange properties of this aether which was said to prevent the heavenly bodies from having the fate of falling stones.

The absence of stellar parallaxes was used by Aristarchus to defend the extremely great distance of the sphere of fixed stars. This third argument was, however, also used against heliocentrism. Naturally, the starry vault had to be larger than the Earth, but, said his opponents, Aristarchus described a much larger than necessary distance in order to allow for the sphere defined by the terrestrial orbit around the Sun.

None of these arguments was actually powerful enough. Nonetheless, the heliocentric hypothesis, which did not account for all observations, was quietly abandoned. This may have been primarily a function of the enormous prestige of Hipparchus who was opposed to heliocentrism. There was a tendency for all of his ideas and achievements to be accepted in a sort of package deal by his successors, notably Claudius Ptolemy, who formalized geocentric astronomy, so completely and so well, as an efficient working hypothesis. The move of the center of intellectual life from the Aegean sea and Athens to Alexandria was certainly operational in this new tendency.

2.6
Hipparchus and His Successors up to Ptolemy

2.6.1
Alexandria as an Intellectual Center

After the division of Alexander's Empire, Alexandria, a new city of the future, was founded by a successful and rich general of Alexander's, Ptolemy Sôter (i.e. the Saviour) (350?–285 BC). He and his successors wanted to make Alexandria a commercial and maritime power, hence a great political power, but also a center of intellectual activity. It is likely that Ptolemy Sôter was strongly influenced by his spiritual father, Alexander, who was himself a pupil of Aristotle, the Stagirite (Stagira is in Macedonia!). For him, science and knowledge were necessary ingredients in the building of a modern empire. Many people of talent came, therefore, to

Alexandria from all parts of the world. They came from Greece, of
course, but also from India and the Arabic realm. Out of this melting pot
emerged, almost in the sense we understand these words today, a real
"scientific community," around two institutions of considerable impact,
the Museum and the Library.

Among the first stars of the Alexandrian intellectual sky, we must
mention Euclid (323–283 BC). He certainly spent some time in Alexandria around 300 BC and his *Elements*, the basis indeed of all mathematics
up to modern times, must have been in circulation there. Straton
(335–268 BC), an Aristotelian, worked at many treatises in physics,
which are now lost. With the encouragement of Ptolemy Sôter, a great effort was made in geography and cartography. For this purpose, instruments were built, already quite elaborate, such as the famed *armillae*.
They were also used for astronomical determinations, which were necessary for engaging in geographical work. Aristyllus and Timochares,
around 300–280 BC, constructed stellar catalogues which were used by
their successors. During this period, as we have said, Aristarchus jour-
neyed from Samos to Alexandria and out of Alexandria came the works
of Aratus, a poet more than a philosopher. He did much to popularize
the astronomical knowledge of his time, especially Aristotle's astronomical treatises.

It is no doubt impossible, in such a short book, to pay adequate trib-
ute to the many thinkers, philosophers, and scientists who moved from
Athens to Alexandria, often for economic reasons, and who ultimately
shaped our vision of the world. Many of them are completely unknown,
even to the educated public. Only the specialists knew of them. We must,
therefore, keep in mind that the names we remember are only the most
famous in a huge and continuous flow of intelligence over the eastern
Mediterranean sea.

Astronomy benefited greatly from several of these thinkers. We must
mention Erathosthenes, born in Cyrene (273–150 BC), later a student in
Athens, but one of the brightest astronomers and geographers in the
Alexandrian empire. He was called there by Callimachus, the learned Director of the Library, whom he succeeded. We shall discuss his astronomical achievements later in this book.

Archimedes (287–212 BC?), a Sicilian, was born and died in Syracuse,
but spent a large part of his life in Alexandria. He was a friend of
Erathosthenes, the leader of a school of engineers, and one of the greatest physicists of all time.

Another scientist whom we cannot omit was Apollonius of Perga (262–190 BC). It seems that he was attracted to Alexandria by Aristarchus's presence there, but he stayed on much longer. In Alexandria, he developed his theories of conical sections, a monumental contribution to geometry, and developed theorems that were indeed the basis for the epicycle-eccentric theory of the planetary motions. There is no doubt that Hipparchus made great use of Apollonius's geometrical demonstrations.

2.6.2
Hipparchus

The greatest astronomer of this period, who probably spent little time himself in Alexandria, was Hipparchus.

Hipparchus is regarded as the founder of modern astronomy, i.e. of Ptolemaic astronomy, which was certainly remarkable for its predictive power and the accuracy of its description of celestial motions. This astronomy remained in place, sometimes criticized, sometimes applauded, until the fifteenth century of our era.

We know little about Hipparchus. Born in Nicea (Bithynia), he lived in the 2nd century BC. We know that he observed the sky in Rhodes in 128 and 127 BC. Those were years of great political turmoil in Alexandria. Hipparchus probably stayed in Rhodes to escape from it. Many of his books are known to us only by their titles. Only one is still available to us and it is not one of the most interesting. We rely on Claudius Ptolemy who made Hipparchus's views known three centuries later in his famous work *Almagest*, which has since been translated, published and republished a great many times. This circumstance makes it unfortunately quite difficult to distinguish between the accomplishments of these two remarkable scientists.

In the following section, therefore, we will discuss the concepts of the epicycle and eccentric circles as they developed in the work of these two authors.

Before that, however, we must describe the progress made by Hipparchus himself in arriving at a knowledge of the various observable characteristics of the solar system, the necessary basis for any model.

Hipparchus determined that the Earth is not at the center of the circle described by the Sun and he determined the location of the Earth in this circle by measuring the unequal length of the seasons (Fig. 1.11, p. 30).

From Ptolemy, we learn that Hipparchus's calculations yielded 94.5 days between the vernal equinox and the summer solstice and 92.5 from the summer solstice to the fall equinox. Using only these appearances, Hipparchus demonstrated that the distance between the center of the Sun's circle (the "center of the World") and the Earth is about 1/24 of the radius of the eccentric circle, and that apogee precedes the summer solstice by about 24 1/2°, the ecliptic containing 360 of these "degrees." We shall come back to this discovery of Hipparchus. He also showed the orbit of the Moon around the Earth to be eccentric.

Hipparchus knew the motions of the nodal line of the Moon (see Fig. 2.35, p. 125). He accurately measured the apparent motions of planets on the sky, their retrogradations, and their stations. He discovered the *precession of the equinoxes*, an extremely important accomplishment, which we shall return to later.

Hipparchus devoted a great deal of attention to the mechanisms by which one could reproduce the apparent motions of the planets. He was not the only one. Apollonius of Perga, as we have said, applied his knowledge of geometry extensively to this problem. In essence, they introduced the eccentric circles and the epicycle circles. Still, the motion of these circles was over simplified and did not yet manage to save the phenomena.

2.6.3
From Hipparchus to Ptolemy

Hipparchus and Apollonius had successors. Adrastes of Aphrodisias and his pupil Theon of Smyrna demonstrated the equivalence of the descriptions of the epicycle and the eccentric. The famous Latin writer Pliny the Elder was also strongly influenced by Hipparchus, whom he quotes with much grace. Unfortunately, Pliny's knowledge of geometry was limited, so his explanation of Hipparchus's theory is seriously flawed.

It is interesting to note that one very controversial question at the time was the order of planets moving away from the Earth. Which is "above" which? That is to say, which is more distant from the Earth? Pythagoras had adopted (from what data, we do not know) the following order: Moon, Mercury, Venus, Sun, Mars, Jupiter, Saturn. Plato and Aristotle changed that order to: Moon, Sun, Venus, Mercury, Mars, Jupiter, Saturn. This was, seemingly, the tradition of Anaxagoras. Hipparchus came back to Pythagoras. So did, at about the same time, Geminus, Cleomedes, and later Cicero (in *The Dream of Scipio*). Pliny also adhered to this view, which was that of Ptolemy.

The various characteristics of the planetary motions were outlined by Pliny. They differ somewhat from the characteristics enumerated by Ptolemy, as we shall see. It seems that Pliny was in error, for the very reason that he adopted some concepts introduced in astrology, but confused them with the actual characteristics of the planetary motions.

2.6.4
Claudius Ptolemy

Whereas Adrastes, Theon, and Pliny accepted geocentrism without discussion, Ptolemy, the great successor of Hipparchus, advanced the possibility that the Earth was in motion. He suggested that the Earth might be rotating around its axis, and admitted that its immobility was a postulate. He rejected Aristotle's arguments regarding the massive bodies which, in the same manner as the Earth, should move towards the center of the World and remain there. He also discussed the problem of flying objects (clouds, arrows, and stones) which would not follow the Earth's motion if the Earth were mobile. In essence, he partially refuted Aristotle's argument according to which the air would carry the objects in question by a sort of viscosity. This problem has been revisited by Newton, amongst modern physicists.

Claudius Ptolemy was a Roman citizen, not related to the reigning family in Alexandria. By his time, Alexandria was no longer a great imperial capital, but only a provincial town in the Roman realm. Born (100 AD) in southern Egypt, Ptolemy died around 178. He did little observation, quite probably less than he claimed. Nonetheless, he managed to leave us the most remarkable documentation, including a catalogue of about 2000 stars, as well as the characteristics and mechanisms of the motions of all the planets and luminaries, in his monumental *Almagest*. He is rightly regarded as one of the greatest astronomers and geographers of all time, and his influence has been solidly established for almost fourteen centuries.

2.7
Eccentric and Epicycle Circles: The Ptolemaic Mechanisms

2.7.1
Solar Motion: The Eccentric System

We have spoken about eccentric and epicycle circles without precisely defining them. Let us devote this section to understanding these very ingenious and precise mechanisms elaborated by the great astronomers from Apollonius and Hipparchus to Ptolemy, mechanisms which were capable of saving the phenomena or accounting for all the observations as they were then known. The reader's patience, I hope, will be substantially rewarded by the aesthetic pleasure inherent in these geometric constructions.

First of all, Hipparchus's discovery of the inequalities of the seasons led to the idea of eccentric motions, as we have seen. The Sun turns on a circle with a uniform motion, i.e. at a constant velocity. The Earth, however, is *not* at the center of this circle. This circle, therefore, is called an *eccentric*. Its actual center is empty, immaterial. We shall call it the "center of the World." The point of the circle closest to the Earth is called the *perigee*, the point farthest away is called the *apogee* (Fig. 2.14).

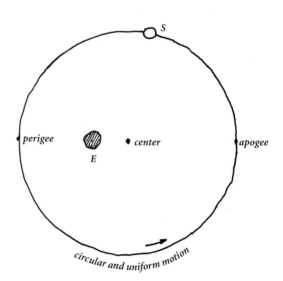

Fig. 2.14. *The eccentric solution.*

These definitions are valid, whether it is the motion of the Sun or any other body, Moon or planet that is under study.

Let us note the significance of having a circle with absolutely nothing at the center. We are now far from Plato's idealistic vision of the planetary motions. We are also far from Aristotle's idea that, at the center of any circle described by some object, there must be a material mass. Saving the phenomena has become more important than any preconception. At all costs! This is a typically Aristotelian idea, even though, in this particular case, some of Aristotle's hypotheses are contradicted. Modern physics is based on this Aristotelian precept.

2.7.2
Planetary Motion: The Epicycle-Deferent System

For any given planet, the other possibility for saving the phenomena is very similar to the Heraclides model (Fig. 2.15). The planet P is presumed to move with a uniform motion, i.e. with a constant velocity, on a circle, the *epicycle*. The center itself describes, also with a uniform motion, another circle, the *deferent circle*, around its center, the Earth. The Sun is located at some point on the line from the Earth to the center of the epicycle and then to infinity. The period of the planet's revolution on the epicycle is called the *synodic revolution*. The period, as seen from the Earth, falls between, say, two successive *inferior conjunctions* with the Sun, the inferior conjunction being the time when the planet is the closest to the Earth, on the line Earth-planet-Sun. The period of the rotation of the center of the epicycle around the Earth is that of the Sun, namely the *zodiacal*

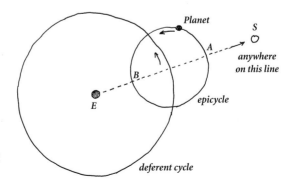

Fig. 2.15. *The epicycle and deferent circles.*

revolution (one year). Points A and B correspond, respectively, to the *superior conjunction* and the *inferior conjunction*. If the bodies are in the order: Sun, Earth, planet (not as in Fig. 2.15), the planet is in *opposition*.

2.7.3
The Equivalence Between the Two Systems for the Sun

Apollonius of Perga, the great master of geometry, was the one who demonstrated the strict equivalence for the Sun between these two descriptions. Let us consider the *eccentric* circle (Fig. 2.16), on which is located the Sun S, of which the center is C, an immaterial point. Now let us consider a circle of the same radius, the *deferent* circle, centered around the Earth E, and trace through the Sun a circle the center of which is located on the deferent. This circle is the *epicycle*. The radius of this circle is constructed in such a way as to be parallel to the line EC. The center of the epicycle is an immaterial point γ. The radius of the eccentric, CS, is equal to Eγ, the radius of the deferent.

During a rotation of the Sun around E at a given time, call it time "zero" or time "origin", at its apogee, S coincides with A. Then S describes the

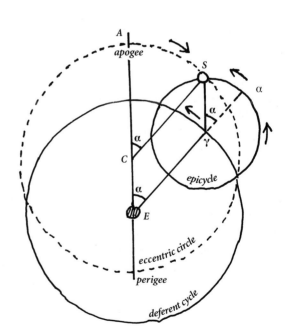

Fig. 2.16. *Equivalence (in the solar case) of the two systems (eccentric circle and epicycle plus deferent circles).*

eccentric circle with a uniform motion in a period of one year, or one zodiacal revolution. During that time, the Sun S describes the epicycle as
well. In the figure, it stays on its top, of course, because of the construction
of the epicycle in which the location of point γ is almost arbitrary. We
must try to visualize that when the Sun was in A, it was also in α on the
epicycle. In the time t, S has moved on the eccentric from A to S, and on
the epicycle from α to S, in such a way that EγSC remains a parallelogram.
Given that the motion of the Sun on the eccentric is indeed uniform, this
demonstrates the uniformity of the motion of S on its epicycle and the
uniformity of the motion of point γ around the Earth on the deferent.

Clearly, angle α, swept in time t, is always the same, be it measured
from C (the angle described by the Sun on the eccentric), from γ (the angle described by the Sun on the epicycle), or from E (the angle described
by the epicycle center γ around the deferent circle). The variation of α
with time measures the motion of the Sun. If one of the three motions is
uniform, all the others are uniform. What we wanted to demonstrate is,
therefore, demonstrated (Q.E.D., Latin for *quid erat demonstrandum*).

In the particular case of the Sun, the zodiacal revolution (ζ) around E
and the synodic revolution around γ are both equal to one year.

2.7.4
⭐ Equivalence of the Two Systems for Mars, Jupiter, and Saturn

For planets, of which the zodiacal revolution and the synodic revolution
may be quite different, the picture is more complicated.

With the epicycle model, it is easy to see that the motion of the planet
is much more complex than the circle that described the Sun's motion
(Fig. 2.17, p. 98) even if we allow that circle to be an eccentric circle ...

Planetary motion may present loops, points where the planet does not
seem to move with respect to the stars as seen from the Earth (stations
Σ). There may also be periods during which the planet seems to move in
the opposite direction from the center of the epicycle (retrogradations
Ψ).The loopy curve is the trajectory of the planet as seen from the Earth.
If we were to adhere to the geocentric system of Hipparchus-Ptolemy,
this would indeed be the real trajectory of the planet. It is easy to see
why this strange notion appeared shocking, more than twelve centuries
later, to Copernicus and his followers.

The zodiacal revolution is always equal to one year, but the synodic
revolution is more than one year for Mars, Jupiter, and Saturn. We shall

Fig. 2.17. *Apparent motion*
of a planet, as seen from
Earth.

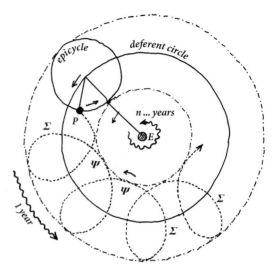

come back (p. 106, 107) to the case of Venus and Mercury, as well as the
case of the Moon which presents interesting peculiarities.

In the case of the three planets under consideration, as in the case of
the Sun, it can be shown that there is a strict equivalence between the
two descriptions, the epicycle-deferent model and the one showing an
eccentric motion.

Obviously (Fig. 2.18), the centers E (of the deferent D) and K (of the
eccentric K), and the planet P cannot stay aligned. If they were so
aligned, the zodiacal and synodic revolutions would have the same period. Therefore, it is necessary to posit an eccentric circle (K), whose center K describes in itself a circle δ around the Earth E. The location of K is
selected in such a way that the figure EKPγ remains a parallelogram, EK
being constructed as parallel to γP. Let us choose as the time zero, the
time at which K, on its circle around E, is in Γ, aligned with C, the center
of the epicycle, and at which the location of the planet on the epicycle is
A, its apogee. In this figure, the location of the Sun is totally irrelevant.
Let us assume (this does not restrict anything) that the motion of the eccentric around the deferent is always uniform, whatever the eccentric-
deferent combination.

The circle (δ) of Fig. 2.18a, centered around the point E (Earth), bears
the point K, the center of the larger circle (K). The circle (δ) is thus the

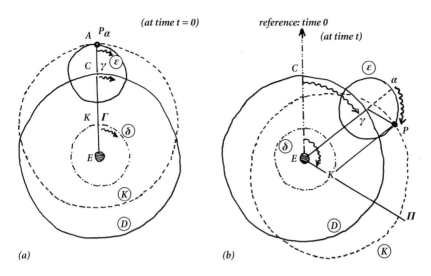

Fig. 2.18a,b. *Equivalence of the two systems.* (Circled letters designate circles, non-circled letters points.)

deferent for (K), which is then seen as an epicycle, by construction, and has the same radius as the deferent (D). It is indeed clear, from looking at Figs. 2.18a and b, that the circle (K) is at the same time an eccentric (with respect to the point-Earth E) and an epicycle circle (centered around the point K, which describes the deferent (δ)). It is also mobile around E. Therefore, it is called a *mobile eccentric circle*. The motions of point γ (around E) and of K (around E) are both the motions of the center of an eccentric ((ε) and (K) respectively) around the center of a deferent circle ((D) and (δ) respectively) both centered on E. These two motions are assumed to be uniform. Let us call Π the point of circle (K) aligned with points E and K.

We can now proceed to the demonstration of the equivalence of the two descriptions of the motion of a planet P around the Earth E, in terms of either (a) the combination of the uniform motion on an epicycle circle and the motion of the center of this epicycle on a deferent circle, *or* (b) the uniform motion on an eccentric circle.

At the origin of time (t = 0, Fig. 2.18a), P and A coincide, as does K with Γ, and γ with C. After an interval of time *t* (Fig. 2.18b), C has moved to γ, A has moved to α. The planet P has moved (on the epicycle (ε)) from α to P located on the mobile eccentric circle (K). Note that angles αγP and γEΠ are equal. From the figure, one thus sees easily that the angle CEΠ (ω_τ) is equal to the sum of the angles CEγ (or ω_ζ) and γEΠ (or ω_σ, and equal to αγP). In other words, one can write the equation:

(1) $\omega_\tau = \omega_\zeta + \omega_\sigma$.

Hence, we must conclude here that if two of the motions (K around E and γ around E) are uniform, the third one (Π around E) is also uniform (see Fig. 2.18b).

We have thus demonstrated the equivalence geometrically (by the very construction itself, implying the parallelogram EKPγ, as explained above), and in terms of motion (with reference to the angular equality written above) of: the motion of P on the "epicycle plus deferent" system (circles (ε) and (D)) and the motion of P on the eccentric circle (K), the so-called mobile eccentric, centered on the "deferent" circle (δ).

Let us consider now the zodiacal period ζ of the planet (revolution of γ around E), its synodic period σ (revolution of P around γ), and finally a third period, τ, that of the revolution of K, or Π, around E. Each of the angles ω_ζ, ω_τ and ω_σ, can be written in terms of the angular velocities Ω_ζ, Ω_τ, Ω_σ, on each of the three circles. The angle ω_τ is equal to $\Omega_\tau.t$, by definition of the angular velocity; or (as the angle 360° corresponds here to the period τ) 360° is also equal to $\Omega_\tau.\tau$ (and the same for ω_σ and ω_ζ). Therefore, in the equation written above, one can replace ω_τ by $\Omega_\tau.t$ or by 360t /τ. The same replacement can be made for ω_σ and ω_ζ. The equation written above (1) can, therefore, be written:

(2) $\dfrac{1}{\tau} = \dfrac{1}{\zeta} + \dfrac{1}{\sigma}$.

As one can see in Fig. 2.19, the motion on the mobile eccentric must also be taken into account. It runs from west to east on the epicycle ε, and from east to west on the deferent (D). The equation for the synodic period may then be changed to measure precisely the motion of point P over the mobile eccentric circle (K). From east to west, it has a period τ defined by

$$\frac{1}{\tau} = \frac{1}{\sigma} - \frac{1}{\zeta}$$

or, from west to east, by

$$\frac{1}{\tau} = \frac{1}{\zeta} - \frac{1}{\sigma}$$

Fig. 2.19. *Properties of the equivalence:* τ, ζ *and* σ.

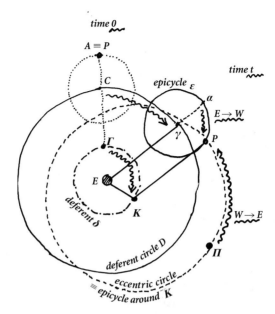

This construction using the deferent and the epicycle thus has the great advantage of being able to reconcile the model with the Platonic principle of locating the Earth at some center and, at the same time, being able to account for observed phenomena, i.e. the inequalities of the seasons for the Sun and the retrogradations and positions of the planets.

It had the disadvantage, for many philosophers of the time who were part of the tradition of Plato and Aristotle, of keeping the Earth, strictly speaking, out of the "center" of the trajectory of either the Sun or the planets. This was difficult to accept, in spite of the fact that the estimates of Hipparchus and Aristarchus had already revealed that the Sun was much larger than the Earth. Bear in mind, however, that from a purely geometrical point of view, there is no need for the Earth to be located at any particular "center."   ✶

2.7.5
Was Ptolemy a Precursor of Copernicus?

Ptolemy arrived at the remarkable conclusion (using the equations given above, the actual values of the zodiacal periods of the planets, and the synodic periods of their epicyclic centers) that the three planets Mars, Jupiter, and Saturn each turn around the center of their epicycles with the same period of exactly one year.

It was a very new, original, and far-reaching idea, which may be, in part, the origin of the Copernican reflections. However, it was more or less, at least in part, an intuition. If we use the Eudoxus determinations for ζ and σ, we do not arrive precisely at the same conclusion as Ptolemy. Rather, we get the following:

	ζ	σ	τ
Mars	2 years	260 days	192 days
Jupiter	12	390	335
Saturn	29	390	376 (instead of 365 days)

This idea is indeed linked to the heliocentric notion that all epicycles are images of one single orbit, that of the Earth around the Sun. Did Ptolemy, then, perpetuate Aristarchus's ideas? Was he a precursor of Copernicus? The historians of science Tannery and Schiaparelli thought so, but Pierre Duhem conjectured otherwise. He claimed instead that Ptolemy's primary focus was his effort to explain the seasonal inequalities by the epicyclic motion of the Sun (Fig. 2.20) and, for Mars, Jupiter and Saturn, by the maintenance of mobile eccentrics. These mobile ec-

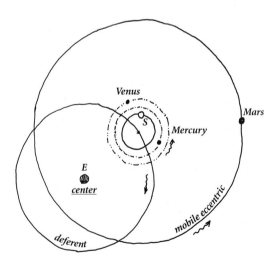

Fig. 2.20. *All deferent circles are the same!*

centrics were assumed to be concentric and centered around the same point. As we know, from the description given for the Sun by Heraclides (and in the Apollonius-Ptolemy conception as well), they would describe the deferent circle in one year. Having the Sun itself on one of these concentric mobile eccentric circles allowed for a representation of the seasonal inequalities.

It is likely, according to Duhem, that this system may have been introduced by Adrastes of Aphrodisias and Theon of Smyrna even before Hipparchus. Note that this system is capable of explaining retrogradations, due to the combination of the motions of γ and of the planet (Mars in Fig. 2.20). This could be the origin of the Ptolemaic idea that τ is equal to 1 year.

The theory of epicycles and deferents was able to satisfy the Platonic double requirement of being capable of explaining the seasonal inequalities and the retrogradations. As we have said, it was conceived by Hipparchus, strongly advanced by the geometrical demonstrations of Apollonius, and clearly formalized by Ptolemy.

2.7.6
* Elaboration of the Epicycle-Deferent System

Using the locations of stars that he had catalogued and the many measurements of the positions of planets that he had made, Hipparchus discovered some "anomalies" in the system described above. One anomaly, the *solar anomaly,* had already been described. It is the consequence of the motion of the planet around the epicycle in approximately one year. Hipparchus went on to identify a *zodiacal anomaly.* In the theoretical system described by Figs. 2.18 and 2.19 above, the retrogradations of the planets continue to have the same angular amplitude, whatever the location of γ on the deferent circle (D), that is, irrespective of the zodiacal constellation they are located in. Hipparchus's zodiacal anomaly is a refinement which describes, from observations, the variation of that angular amplitude.

The evolution of the Hipparchus system, as described above, was, therefore, natural. A fixed eccentric accounts for the apparent motion of the Sun and the seasonal inequalities. An epicycle, the center of which describes a circle concentric to the center of the eccentric (or center of the World), accounts for the zodiacal inequalities. Hipparchus envisioned an epicycle the center of which described an eccentric

Fig. 2.21. *The seven para-*
meters defining the motion
of any planet.

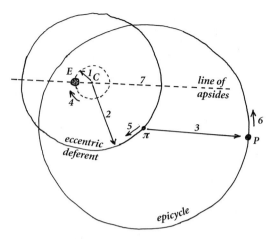

Fig. 2.21. *The seven para-meters defining the motion of any planet.*

(Fig. 2.21)! One can see on this figure that, for each planet, we need to know *seven* parameters, i.e. the three radii of the circles, the three periods of the motion along these circles, and finally the direction of the *line of apsides,* which defines the position in the zodiacal zone, at any given time, of the direction which is defined by the Earth and the center of the World C. Hipparchus was successful in describing the Sun's motion. He did not succeed for the Moon, because he lacked sufficient data on lunar eclipses.

The astronomers who continued in the tradition of Hipparchus used their skills to determine the apogee of the planets, or, more accurately, the apogee of their eccentrics. Many aspects of the planetary orbits were named during this period. The resulting nomenclature shows the influence of Eastern civilizations. For example, the word "aux" came to Arabic from Sanskrit and was latinized during the Middle Ages by the translators. In French, we now use the word "auge." In English, the best translation would be "culmination." It is amusing to compare the location of the "auge" of the planets, and of the Sun's apogee, according some authors. Obviously, it was not such an easy determination:

	Hipparchus	Pliny	Ptolemy
Sun	5°5 (from center of Gemini)	Gemini	Gemini
Saturn		mid-Scorpio	23° (from Scorpio's center)
Jupiter		mid-Virgo	11° (from Virgo's origin)
Mars		mid-Leo	25°30' (from Cancer's origin)
			4°30' (from Leo's origin)
Venus		Sagittarius	25° (from the center of Taurus)
Mercury		Capricorn	10°15' (from center of Aries)

It is clear that it was only at the time of Ptolemy, when the parameters were more fully known, that a determination of all systems was possible. Ptolemy had indeed gathered and listed all the parameters (including the direction of the apsidal line towards some point of the zodiac – one single angle, given in the table above) necessary to predict the positions of the Sun, Moon, and planets. His treatise, *Almagest*, was a monument, very complete, perfectly ordered, and based on the best data. The true title of this treatise was "Μεγάλη μαθηματική σύνταξις τῆς ἀστρονομίας," or, literally "The large mathematical composition (syntax) of astronomy." The first two words were symbolically sufficient to designate the book: "The large," or "Al Majesti," in Arabic, known in modern writings as *Almagest*. The *Almagest* (published in 142 or 146 AD) became the gospel for generations of astronomers over a period of fourteen centuries of observations.

Having a knowledge of the parameters was necessary, but not sufficient. Ptolemy had to eliminate another Platonic principle, namely the belief in the uniform motion of the epicycle around the center of the eccentric, in order to save the phenomena, his chief preoccupation, and one which distinguished his methods and his viewpoint from both numerological Pythagorianism and Platonic idealism.

He did this first by introducing to the construction another immaterial point, the *equans*. The equans is a point C symmetrically opposed to the Earth, along the line of apsides, with respect to the center of the World Δ (Fig. 2.22). The diameter αγβ of the epicycle, invariably linked to the epicycle, turns in such a way as to be always oriented towards the equans. This diameter is moving uniformly in this rotation. The planet P moves with respect to this diameter. Therefore, it is the angle βγP, and not BγP, which is proportional to time. The circle centered around the

Fig. 2.22. *The Ptolemaic equans system.*

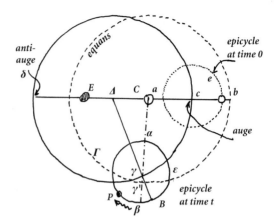

equans is also called the equans (or to be more precise the *circulus equans*).

This construction was sufficient, without the introduction of any new parameter, to determine the motions of the planet Venus, as well as Mars, Jupiter and Saturn. This was good fortune. The construction did not prove successful in the case of the planet Mercury. Nor did it work for the Moon.

In the case of Mercury, the point around which the motion of the epicycle's diameter is uniform did not turn out to be the equans, but a point K, with respect to which the symmetric Z of the Earth E is the center of a circle on which the center Δ of the mobile eccentric is moving. This point Z becomes the real "center of the World." Two more parameters now become necessary: the radius of the circle described by Δ around K and its characteristic period (Fig. 2.23).

In the case of the Moon, the equans is still valid, but the center Δ of the mobile eccentric turns around the Earth (or vice-versa?), which indicates that the line of apsides (nodal line) is not fixed. The period of this motion is the only one additional parameter needed to describe the motion of the Moon (Fig. 2.24).

Fig. 2.23. *The Ptolemaic system for Mercury.*

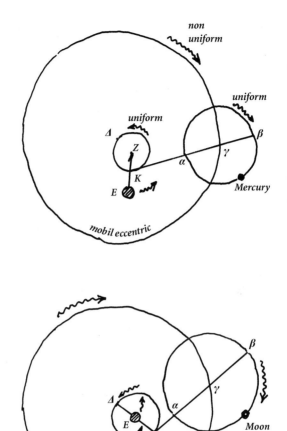

Fig. 2.24. *The Ptolemaic system for the Moon.*

To recapitulate, one needs the following parameters:

	Sun	Moon	Mercury	Venus	Mars	Jupiter	Saturn
no. of circles	2	3	4	3	3	3	3
no. of velocities	1	4	4	3	3	3	3
line of apsides	1	0	1	1	1	1	1

Some common parameters among the seven groups permit a reduction of this description to thirty parameters. This large number, however, should not be considered an argument for or against either Ptolemy or Copernicus. Modern astronomers define the orbits of each of the planets with many necessary parameters, i.e. sidereal period, eccentricity of orbit, semimajor axis of the Keplerian ellipse, and longitude of perihelion, not to mention characteristics such as the longitude of the ascending node, or the inclination of orbit on the ecliptic, the last parameters not being considered in the Ptolemaic construction, which is essentially planar. The "simplicity of the World" is just a myth, much inspired by Pythagorian speculations. Actually, Ptolemy knew, as did Eudoxus before him, that the epicycles have to be inclined with respect to the deferent, i.e. the ecliptic. This inclination may vary along the path of the epicycle's center. This was very disturbing to Ptolemy. He proposed some kinematic hypotheses, adapted to actual computation, and simple enough. Unfortunately, they did not really fit the observations and certainly not the philosophies of Plato and Aristotle.

Having described the Ptolemaic system, one must pause to admire the genius of Ptolemy. He was a positivist, almost in the sense given to that word by the 19th century French philosopher Auguste Comte who coined it. Clearly, just as there was exaggeration in the praise showered on him as the sole keeper of astronomical truth up to the 16th century, he has also suffered unfair treatment at the hands of posterity and modern commentators.   ✶

2.8
The Earth, Sun, Moon, and Planets: Distances and Sizes

2.8.1
The Phenomena

The system Earth-Sun-Moon (our Earth and its two great luminaries) has always had great importance. The phenomena which affect it have been known for ages, even before written history. Cave drawings are evidence of the astronomical observations of prehistoric man. The phases of the Moon were known, although the explanations given for the changes in the appearance of the Moon were often pure fantasy (Fig. 2.25). Eclipses were also observed. In a way, the inhabited world remained more mysterious to primitive man than the heavens.

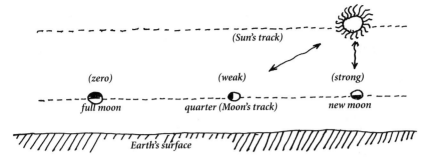

Fig. 2.25. *An ancient view of the lunar phases.* Belossos, a Babylonian priest who emigrated to Cos (ca 280 BC), taught that the globe of the Moon is intrinsically bright on one of its hemispheres. When the Sun is close, its bright light attracts the bright side of the Moon. Then, we see a *new moon*. When the impact of the Sun is weakened, the Moon turns half of its shining face to the Earth and later the totality of it. In intermediate phases, we see arcs of the bright side of the Moon, from the side, so to speak.

From the time of Plato, the need to explain or "save" the phenomena predominated. Saving the Moon's phases (a great interest of Aristotle's) and saving the eclipses had the further value of being useful for measuring the distances and sizes, within the Earth-Sun-Moon system by means of geometry after that perfect branch of classical mathematics was developed by Euclid and others.

2.8.2
The Earth's Radius: a Basic Unit of Length

We have already mentioned the great progress that was made in advancing the knowledge of the Earth. Its sphericity was known for a long time, certainly since Plato. This was not disputed by any philosophers (except perhaps the school of Epicurus, around 300 BC). The next step was to estimate the radius of the sphere. Aristotle gave a value of 400 000 stadia without in any way explaining how he arrived at this figure. Archimedes, following in the tradition of Dichearchus of Messina (?–285 BC), the only known geographer of that time, gave an estimate 300 000 stadia. A real method of accurate measurement, however, was not demonstrated until Eratosthenes (276–194 BC) introduced his method around 235–200 BC. A native of Cyrene, he was the librarian of the Museum at Alexandria,

called there by the reigning kings because of his already substantial reputation.

   The method is indeed quite simple (Fig. 2.26). Erathosthenes lived in
Alexandria. The Egyptian empire extended much further. He was able to
travel due south to Syene (modern Aswan) and beyond. He noted (perhaps by pure chance?) that at Syene at noon on the day of the summer
solstice, no shadow was visible at the bottom of a deep well. You can still
visit this spot on the island of Elephantine. In other words, the Sun was
then exactly at the zenith. The gnomon, to express it in another way,
threw no shadow whatever on the ground. On the other hand, in Alexandria on the same day of the summer solstice, at noon, the shadow had a
measurable length. The angle between the vertical and the direction of
the Sun was not zero but about 1/50 of the whole circumference (or, in
our units, about 7°12'). To transform that measurement into a determination of the radius of the Earth, one needs to know the distance between Alexandria and Syene, which were thought to be on about the
same meridian circle. This distance was determined by trained walkers,
whose step was, so to speak, calibrated (the so-called βηματισταί, or *itin-
erum mensores*). It was found to be equal to 5000 stadia, hence the
Earth's radius was 50 times as large, or 250000 stadia. Eratosthenes later
adjusted this value to 252000 stadia. Of course, a basic question is: what
is, expressed in meters, the length of the stadium used by the Greeks?

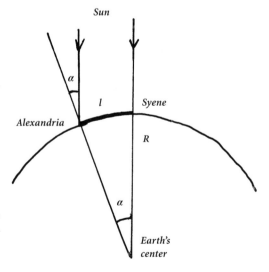

Fig. 2.26. *The measurement of the Earth's radius
by Erathosthenes (ca 241
BC). The angle* α *is measured at Alexandria, from
the ratio of the shadow to
the length of the stylus of
a vertical gnomon. The
distance l is measured by
trained walkers. The ra-
dius of the Earth, R, is
then equal to* αl/2π, *where*
α *is expressed as a fraction
of the circumference, and
R and l are in the same
units (stadia for example).*

Using Pliny's inferences and some other sources about the various units of length used at that time, we arrive at a value for one stadium of 157.50 m. The Earth's diameter is, therefore, found to be inferior by only 80 km to its polar diameter as currently determined (12713.55 km). This agreement is remarkable, but we should not consider that as a proof of precision. There is little doubt that the angle of 1/50 of a circumference is exaggerated, and that Syene and Alexandria are indeed not on the same meridian. It is likely, however, that errors have more or less compensated for each other. Moreover, the value of the stadium, in meters, differs depending on whether you are using the Olympic stadium (185 m), or the Royal Ptolemaic stadium (210 m), still used (by reference to other units used at that time) by Ptolemy. Eratosthenes' value of 252000 stadia was adopted by Hipparchus.

The diameter of the Earth was thus known. From this basic "first astronomical unit of distance," it became logical to try to infer the Moon's distance and the Sun's distance from the Earth, using geometrical constructions. This was the natural next step. Needless to say, speculative thinking preceded real geometrical determinations. For example, (see p. 46) Anaximander put the Sun at a distance of 27 times the Earth's radius and the Moon at a distance of 19 times the Earth's radius. All these ideas more or less emerged from rather vague considerations about the harmony of the spheres. They were speculations, not real measurements, and the Pythagorians were, of course, not the last ones to speculate and numerologize.

It is important to note, incidentally, that these determinations, when expressed in "Earth's radii," were only relative. The measurement of the Earth's radius in some more tangible units (stadia or km) is indeed not necessary to determine the relative distances of the Sun or Moon to the Earth expressed in our basic first astronomical unit.

2.8.3
The Moon

Aristarchus, using scientific methods and possibly influenced by Eudoxus, made some progress with respect to the Moon, as he did in planetary theory. Hipparchus also used the same methodology.

This method basically used observations made during lunar eclipses (Fig. 2.27). During a total lunar eclipse, the Moon enters the Earth's shadow for a long time, about three hours. The total duration of the rota-

Fig. 2.27. *The measurement of the Moon's diameter.* The Moon crosses the Earth's shadow along AB (in the case of a maximum duration total eclipse). If one knows the diameter (lower left) of the shadow on the sky, one knows the ratio of the Moon's diameter to it. The diameter of the shadow is slightly smaller than the Earth's. If we assume it to be equal to Earth's, we find a slight overestimation of the ratio of the Moon's diameter to the Earth's diameter. This overestimated value is 1/3. The true value is 0.27.

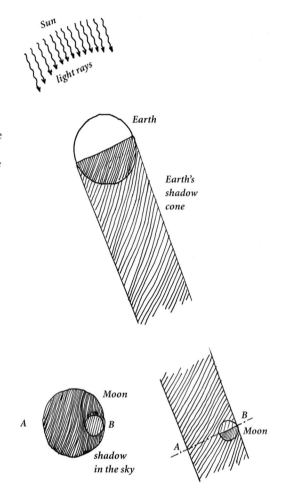

tion of the Moon around the Earth is 28 days, approximatively. Hence, the angle on the sky covered by the Moon's path is about (3 h/28 d) 360°= 1°35'. The apparent diameter of the Moon is an angle easy to determine (during the Moon's set, for example, by measuring its duration), and equal (as that of the Sun) to about 1/2°, 1/720 of the circumference, i.e. $2\pi R$, where R is the Earth's radius). Hence, the diameter of the Moon is about 1/3 of the Moon's path in the shadow cone of the Earth. If the Sun is very far, at an "infinite" distance, this shadow cone is exactly a cylinder and its real size on the sky is equal to the size of the Earth itself. The true diameter of the Moon then would be 1/3 that of the Earth, i.e. 2/3 R. Actually, the Sun is not at infinity. The shadow cone is a real cone, not a cylinder, and the size of the Moon's path in the shadow is smaller than the size of the Earth, thus demonstrating that the proportion 1/3 adopted by Hipparchus is an overestimation. Actually, the ratio is closer to 0.27252, with the modern determinations and using the equatorial radius.

Knowing the size of the Moon, its distance from the Earth naturally follows. It is (with the value 1/3), equal to $(2R/3).720/2\pi = 76\,R$ (an easy geometrical construction). With $R/4$, it is equal to $57\,R$. Hipparchus adhered to a figure of 59.1 times the radius of the Earth, in the extreme case where the Sun is assumed to be at infinity. Now we know it is on the order of 60.5 Earth radii, on the average. It is, however, known to vary with time, as the Moon's orbit is not a circle centered at the center of the Earth (as we have seen earlier in this book, p. 30). Ptolemy, to a great extent used these determinations and regularly adopted the value of 59 Earth radii for the Moon distance, the radius of the Moon being 1/(3.4) that of the Earth. These values were accepted broadly for about 1300 years after Ptolemy, but we should remain aware that several authors discussed these determinations and reached somewhat different conclusions.

2.8.4
The Sun: The Aristarchus Determinations

The Sun is far away. There is no doubt about that! But how far? The apparent size of the Sun and the Moon is the same. Therefore, the ratio of their respective distances to the Earth, D and d, is equal to the ratio of their respective diameters. Is it a small or a large number? Eudoxus considered the diameter of the Sun to be 9 times that of the Moon. Phidias (Archimedes' father) is said, by his son, to have estimated this ratio as 12:1. Archimedes himself adopted the value of 30:1.

Aristarchus made a very astute determination. At the time of the quarter Moon, the Earth, Moon and Sun form a right triangle, the angle of 90° being located at the Moon's center (Fig. 2.27). The apparent angle between the direction of the Moon and that of the Sun, angle $\theta = 90° - \alpha$, measures indeed the angle α, hence the ratio d/D. As Aristarchus found this angle α to be close to 3°, this ratio is (in our modern notation), close to $d/D = \text{tg}\,3° = 1/20$ (Aristarchus suggested 1/20 to 1/18 and used 1/19).

Of course, Aristarchus's value is very much in error, as we now know with the benefit of our current state of knowledge. First, this angle is measured from the surface of the Earth, not from its center. As we can see in Fig. 2.28, notwithstanding the exaggerated sizes of the bodies in the figure, the determination of α is an underestimation. A second problem is that of refraction by the air. In Fig. 2.29, we have introduced the refraction effect of the atmosphere of the Earth, unknown at the time of Aristarchus (although already observed, as a flattening of the setting Sun, but not really measured). It is clear that the refraction incurves the light rays and that celestial bodies are seen at an angle above the horizon larger than they are in reality. The effect is more significant near the horizon where it is known to reach about 30'.

These two effects are present. Many other effects alter the quality of the determination of d/D. Altogether, the determined value of d/D is an overestimation. The correct value for the angle must be definitely smaller than 3° and the ratio d/D greatly reduced. For α = 2°, we found about 1/29; for 1°, it becomes 1/58. The actual value of α, measured in modern

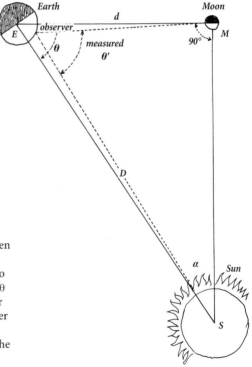

Fig. 2.28. *The Aristarchus measurement of the solar distance.* Note that neither the angles nor the sizes of the astronomical bodies drawn in this figure are realistic. The angle α is very small, wherever the angle θ is slightly smaller than 90°. The distance d is indeed much smaller than D and the triangle ESM has an angle of less than 1°, at the corner S. The measured angle θ' is very near 90°. The sizes of the three bodies are largely overestimated in comparison to the distances between them. The measure of θ is in principle a measure of the ratio d/D, but it is clear also that, if θ were measured from the center of the Earth, it would be smaller than when actually measured from the observer located on the Earth's surface.

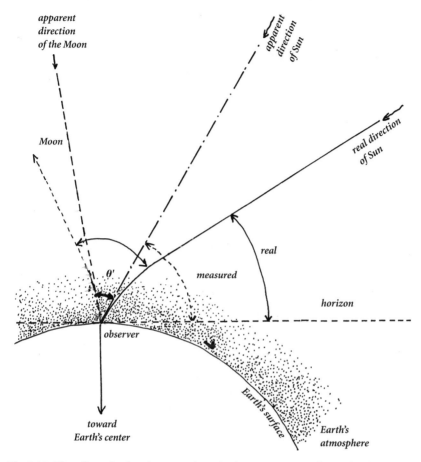

Fig. 2.29. *The effect of refraction on Aristarchus' measurement.* The angle θ' is identical to θ' in Fig. 2.28. However, as light rays are actually curved by the refraction of atmospheric air, the measured θ' is definitely smaller than the real θ thus leading to an overestimation of d/D.

times, is still smaller: 8.9", yielding a ratio $d/D = 1/23\,000$. As measurements of angles close to 90° are very difficult to achieve, we must consider Aristarchus's method to be imprecise, although geometrically correct.

Note that a value D/d equal to 20 already generated a value for the Sun's diameter about seven times larger than the diameter of the Earth, if one takes into account the known distance to the Moon (see Sect. 2.8.3). This point alone should have led astronomers of the time to the understanding (Aristarchus himself was of that opinion) that the Sun, much larger than the Earth, was much more likely to be a center of the World than the Earth. Based on this established fact, the heliocentric system was more natural in a way than the geocentric system.

2.8.5
The Methodology of Hipparchus

Another method, introduced probably by Eudoxus, and adopted by Hipparchus, had a greater capacity, in theory if not in practice, to determine at least some relation between the distances to the Earth of the Sun and the Moon.

Here, we will introduce the word *parallax*, which we will use frequently throughout this book. This is a difficult word. *Parallactic effects* are those which modify the aspects and the apparent positions of anything – a tree, a man, a star, the Moon, the Sun – when viewed against a more distant background from two different points. The two points may be two eyes or two different locations on Earth. The parallax (more precisely, the *diurnal parallax*) of the Sun is the angle ϖ at which, from the Sun, one sees a well-defined distance, in practice the Earth's radius perpendicular to the direction Earth-Sun. The parallax of the Moon is usually defined in a similar way, as the angle at which one sees, from the Moon, the Earth's radius perpendicular to the direction Moon-Earth. The *annual parallax* of a star is the angle at which, from the star, one sees the semi-axis of the Earth's orbit, assumed to be perpendicular to the direction of the star. In actuality, parallaxes are measured using parallactic effects, since the parallactic angles are equal to the difference between the angles of two directions as measured by observers at two different points on Earth. Figure 2.30 describes this methodology.

It is applicable to any parallax, from that of a distant object on Earth (using *binocular* microscopes or binocular marine field-glasses) to that of the Moon, the Sun, or a star.

Hipparchus's method assumed both the Sun and the Moon to be at a finite distance. In Fig. 2.31, the solar parallax is designated by $\odot$, the lunar parallax by $\mathbb{C}$. The apparent radius of the Sun, as seen from the Earth's center, is r; that of the Earth's shadow is ϱ. The very simple geometry of Fig. 2.31 shows that indeed:

$$\odot + \mathbb{C} = r + \varrho.$$

Fig. 2.30. *The parallactic effect.* Seen from two different points, A and B, the point S (Sun, star, Moon, satellite.. anything) is seen at different angles α and β from the local vertical. This is due in part to the curvature of the ground between A and B, separated by a certain length, linked to the Earth's radius by the angle θ. Once θ is taken into account, the difference β – α reflects the angle γ from which, from S, one sees the *base* AB. Simple geometry can link the angle γ and the angle ω under which, from S, the Earth's radius perpendicular to the direction of S is seen.

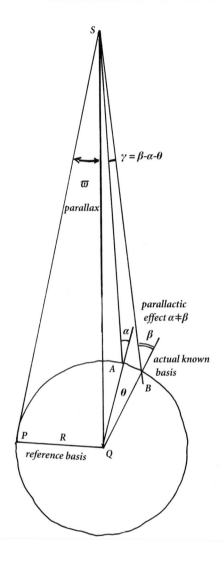

With Aristarchus's assumption that $\odot = 19\ \left(\!\!\!\!\textrm{C}\right.$, $r = 16'36''55'''$ (as determined by Hipparchus, and expressed in modern units), and $\varrho = 2.5\ r$, a value for $\odot$ of $2'\ 54''$ can be found. This value is, of course, affected by exactly the same errors we have earlier mentioned with regard to Aristarchus's determination. It was only useful in that it provided an interesting and simple relation that could be used in the future. Table 2.3 gives

Fig. 2.31. *The Hipparchus method.* Again, as in all these figures, the sizes of astronomical bodies are greatly exaggerated with respect to their mutual distances. The solar and lunar parallaxes are, respectively, $\odot$, and $\mathbb{C}$; r and ϱ are the apparent radii of the Sun and of the Earth's shadow on the sky. It is very easy to derive the relation $\odot + \mathbb{C} = r + \varrho$, established by Hipparchus. Just continue the shadow cone till its summit M, and consider the triangles of which M is one corner. As $\odot$, in practice, is *much* smaller than $\mathbb{C}$, this offers a reliable determination of $\mathbb{C}$, through the measurements of r and ϱ.

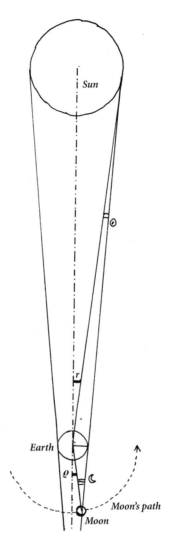

a chronology of the progress made in improving the accuracy of the various measurements and distances relevant to a study of the Earth, Sun, and Moon, from the earliest estimates to the present. We shall come back to this table later in this book when we deal with the precise determinations made in modern times.

Table 2.3. *Distances and sizes of Earth, Sun, Moon* (some selected determinations)

Author	Date (approx)	Earth's radius	Moon's parallax	Moon's distance	Sun's parallax	Sun's distance
Anaximander late Pythagorians	ca 590 BC	1 (unit)		19		27
Pythagorians (as reported by Plinius, etc., see table 3,2)		1 (unit)		1.5		3.5
Plato	ca 400 BC	1 (unit)				9
Eudoxus	ca 370 BC	1 (unit)				9
Aristotle	ca 350 BC	400 000 stades				
Dichearcus	ca 300 BC	300 000 stades				
Phidias	ca 280 BC	–				12 $R_\oplus$
Archimedes	ca 260 BC	300 000 stades				30 $R_\oplus$
Eratosthenes	ca 250 BC	252 000 stades				
Aristarchus	ca 280 BC					18–20 $R_\oplus$
Hipparchus	ca 160 BC	1 (unit)		59.1	2'51"	68 000 $R_\oplus$
Posidonius	ca 100 BC	240 000 stades				
Tycho Brahe	ca 1580				3'	
Kepler	ca 1600				1'	
Vendolinus	1630				15"	
Richer (et alii)	ca 1650				9.5"	
Halley	1676				45"	
Lacaille	1751		57'5"		10.2"; 10.6"	
(various authors)	1761–1769				8.55" to 8.88"	
Lalande	1770				9.18"	
–d–	1774				8.75"	
Laplace	ca 1800				8.6"	
Encke	1822–1824				8.57" ± 0.04	
(various authors)	XIX-th Century				8.8" to 8.9"	
Spencer Jones	1929				8.796" ± 0.002"	
Spencer Jones	1942				8.790" ± 0.001"	
modern data are numerous						
Allen (compilation)	1976				8.79418" ± 0.00003"	
IAU	1976				8.794148" (standard value)	
USNO, RGO	1992				8.794 148" (144)	

We should note at this stage, that the methods used by the Greeks did not really employ the parallactic effect. All of their observations were made from the same place on Earth, i.e. the area surrounding the Aegean sea. They used it rather indirectly, in that when celestial bodies move on the sky, the "center" of their apparent motion would seem to be the observer. A more logical understanding, however, would place the center of their apparent motion at the center of the Earth, as inferred by the astronomers as soon as they realized that the Earth was a sphere. Note also that measurements of angles in the Greek-Alexandrian period were difficult. One generally reduced them to the comparison of durations – duration of a day, of a lunar month, of the year, duration of the total lunar eclipse, of the sunset, of the moonset, etc. Parallaxes, nonetheless, were actually "measured," to the extent that measurements could be made at that time (Fig. 2.32). They were not merely guessed at in some effort to "save the Scriptures" or "save the harmonies". This scientific approach was a true inheritance from Aristotelian methodology. Still, one has to wait until the 17th century to find true determinations of parallaxes (of

Fig. 2.32. *A determination of the Moon's parallax. At the horizon, the Moon is closer to the observer than at some height above the horizon. At the horizon, the distance of the observer to the Moon's center is smaller than when the Moon is high above the horizon. The comparison of apparent diameters at the horizon and at the zenith is indeed a measure of the Moon's parallax. Unfortunately, this method does not give accurate data, as the measurement of the apparent diameter of the Moon was not very accurate in Greek-Alexandrian times.*

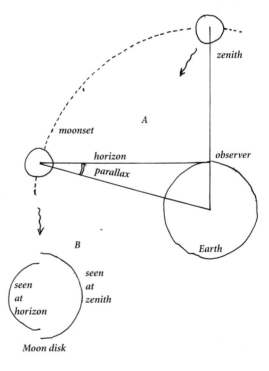

the Moon, planets, and Sun) from a *base* implying the use of two different sites of observation. It was not until the 19th century that astronomers were able to measure stellar annual parallaxes.

2.8.6
Planetary Distances

The beginnings of a consideration of planetary distances were marked, of course, by the Pythagorian fixation on the harmony of the spheres. Plato himself was not insensitive to these considerations.

The observations relevant to an evaluation of the respective distances of the planets to the Earth were the duration of their revolutions with respect to the fixed stars, and the apparent size of their retrogradation loops. Figure 2.33 shows the apparent orbit of three planets (which we call now, from our Copernican viewpoint, *exterior*), namely Mars, Jupiter and Saturn.

As we have noted, measurements were difficult to make and planetary motions were too slow to be properly taken into account. Therefore, the notion of the harmony of the spheres led Plato (see Sect. 2.1.5) to suggest that the planetary orbits were proportional to the numbers 1, 2, 3, 4, 8, 9, 27, – of which at least the first five corresponded to notations in the Greek musical tradition. The musical scales were attributed to Pythagoras, but various later authors arranged them in a different order. According to Pliny, for example, their order was as indicated in Table 2.4, p. 123. If we leave aside the Earth, Pliny's intervals constitute an octave of the Dorian mode. The other authors follow more or less in the tradition of Alexander, the poet of Ephesus (1st century BC).

These were attempts by late authors to adapt their knowledge to preconceived Pythagorian concepts. The true Pythagorians, such as Philolaus, did not follow such a pattern; nor did Eudoxus.

It was not easy to conceive of alternatives to these unproven speculations. The Ptolemaic system could not shed any light on the distances of the planets from the Earth, since the actual sizes of the epicycles and deferent circles were of no importance in the Ptolemaic system, only their ratios. This was perhaps the principal achievement of Copernicus. He related all the epicycles to the Earth's orbit around the Sun. He suppressed one parameter for all but one of the planets, thus giving to all the deferent circles a value expressed in term of the Earth's orbit, enabling the determination of their relative distances (see p. 202, Table 4.2).

Fig. 2.33. *An idea about
planetary distances.* Whatever system we may think
of, be it geocentric or heliocentric, the apparent
sizes of retrogradation
loops, the apparent velo-
city of planetary paths on
the starry vault are estima-
tions of their distance.
One can at least infer that
Saturn is further away
than Jupiter, and Jupiter
than Mars. One could
even, in the Copernican
system, use the apparent
sizes of loops, and the angular distance from one
loop to the following to
determine the planetary
distances.

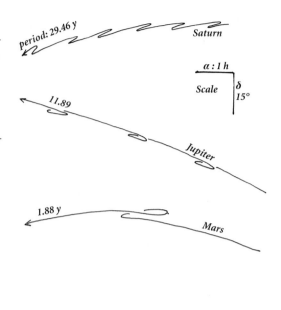

Attempts were made to improve on the Ptolemaic situation with regard to the determination of planetary distances. The distances of the
Sun and the Moon were (supposedly) known. Between them were Mercury and Venus. They were thought to occupy two "volumes." The first,
of internal radius 64 + 1/6 (maximum distance of the Moon expressed in
Earth's radii, as estimated by Ptolemy) and of external radius 177 + 33/60
was that of Mercury. The other, between 177 + 33/60 Earth's radii and
1150 (nearly equal to 1160, the closest distance of the Sun, as estimated
by Ptolemy) was the realm of Venus. Such an idea was defended by Proclus and later by Simplicius. It remained in the literature up to the time
of Copernicus.

The idea of space filled by contiguous volumes, in which the planet's
epicycle could be contained, was very much an idea which emerged after
Ptolemy. The acceptance of this idea was not primarily a function of the
need to save the phenomena, but rather emerged from the Aristotelian
exclusion (see Sect. 2.3.1) of the possibility of the existence of a vacuum.
One can see, for example, in Bacon's construction (see p. 170), the remnants of this attitude, which dates from the post-Ptolemaic Alexandrian
period.

Table 2.4. *The harmony of celestial spheres*

	Pliny Theon	Censorinus	Achilles
Earth	C		C
Earth–Moon: a tone		1	
Moon	D		D
Moon–Mercury: a semi-tone		1	
Mercury	E flat		E
Mercury–Venus: a semi-tone		1/2	
Venus	E		F
Venus–Sun: a minor third		1	
Sun	G		G
Sun–Mars: a tone		1	
Mars	A		A
Mars–Jupiter: a semi-tone		1	
Jupiter	B flat		B flat
Jupiter–Saturn: a semi-tone		1/2	
Saturn	B		B
Saturn–fixed stars: a minor third		semitone 1/2	
fixed stars	D		C

Obviously, the actual size of the planets, which have no apparent measurable diameter (although one finds, in prehistoric times, drawings of "Venus' horns") is not even discussed, nor is their nature. This will not come until the 17th century with the development of optical instruments which could detect planetary disks. The first to undertake this study was probably Galileo, in 1610.

2.9
The Precession of Equinoxes

2.9.1
Hipparchus's Discovery

The precession of equinoxes is a phenomenon of rather long time scale. That is why it was discovered only by Hipparchus, after generations of Chaldeans and Greeks had recorded their observations. It is also the reason why it was not initially regarded as essential in constructing a system of the World. We must, however, take it into account when discussing the false assumptions of astrology. We shall come back to that point at the end of this chapter.

The equinoxes (Fig. 2.34), or more properly the *equinoctial points*, are the points of the apparent trajectory of the Sun around the Earth where the Sun crosses the equatorial plane of the Earth. One is the *vernal* point γ, when the Sun enters the region above the equator. The other is the *autumnal* point γ'.

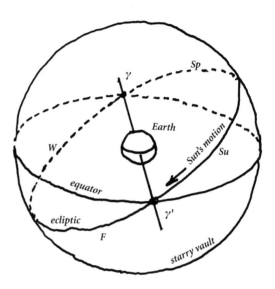

Fig. 2.34. *The equinoxes.* The equator of the sky and the ecliptic plane (or the Sun's apparent orbit around the Earth) intersect along the *lines of nodes*. γ γ', γ being the *vernal equinoctial point*, and γ' the *autumnal equinoctial point*. When the Sun is either in γ or in γ', it is the time of the equinoxes, when the length of night is equal to the length of day.

After the Sun has passed through point γ, at the time of the spring equinox, daylight is longer than night. We enter spring (Sp in the figure) and then summer (Su). After it has crossed the point γ', at the time of the fall equinox and is back in the southern hemisphere, daylight becomes shorter than night. We enter fall (F) and then winter (W). The equinoctial points have an additional importance. On the Earth and sky, they can serve as the origin for measuring longitude.

Calendars are based on this definition of seasons. The spring equinox always marks the beginning of the true year, the seasonal year. It falls, on our legal calendar, within two days of the 21st of March.

In 129 BC, it seems that Hipparchus, possibly after a careful examination of old Chaldean and Greek records, but also using his own observations, discovered that the longitude of some stars (Spica, for example) had changed over a period of some years. The longitude of Spica was then 174°, while when it was measured at the time of Timocharis, about 150 years previously, it was 172°. This implied a rotation from east to west of the equinoctial points (i.e. of the whole ecliptic plane around its axis). This is the phenomenon of precession of the equinoxes, described in Fig. 2.35.

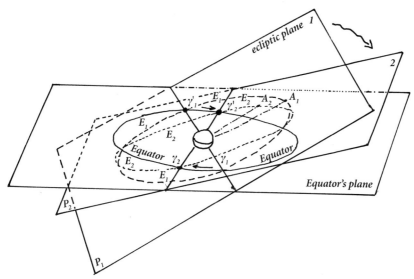

Fig. 2.35. *The precession of the equinoxes.* The plane of the ecliptic is turning from position 1 to position 2, the points γ_1 and γ_1' being replaced by point γ_2 and γ_2', respectively. The ellipses E_1 and E_2 represent, the apparent solar orbit around Earth. The direction of the apogee of the Sun (A) is moving from A_1 to A_2, the velocity of rotation of the *apsidal line* joining apogee and perigee being different from the velocity of the rotation of the nodal line $\gamma\gamma'$.

This discovery had an immediate effect on the governing concept of the duration of the year. The passage of the *sidereal year* was defined as the time during which the Sun, after having described the ecliptic, encountered the same star. It differs from the *tropical year*, which was conceived as the time elapsed between two successive passages through the vernal equinoctial point. Obviously, for men on Earth, bound to the seasons, the tropical year was more important. It became the basis of the legal year. Hipparchus made a very accurate determination of it, as equal to: "365 d + 1/4 d – 1/300 d" (or 365.243 compared to our modern value of the tropical year: 365.24219 d).

The rate of the precession itself was estimated by Hipparchus to be "at least" 1° per century. We now know it to be 1°23'30" per century. If Hipparchus had been confident enough in his own comparison with Timocharis's observation, he would have found a value very close to that one, but with a very weak level of accuracy. It corresponds to a complete rotation of the ecliptic in a period of 26000 years (now known to be 25725 years). In modern times, precession is understood to be a function of the turning of the Earth's axis, describing a cone of angle 23°27' in 25725 years, as shown in Fig. 2.36.

Fig. 2.36. *The modern understanding of the precession of the equinoxes. The axis of the Earth turns, defining a cone of aperture 23°27', the axis of which is perpendicular to the ecliptic plane. The rotation period along that cone is about 26000 years. This description is strictly equivalent to that represented in Fig. 2.35.*

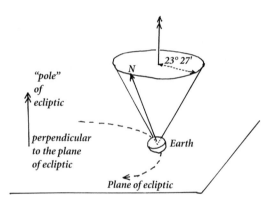

2.9.2
Ptolemy's Precession of Equinoxes

Ptolemy did not adopt blindly the Hipparchus value, in spite of his high respect for Hipparchus's work. On the basis of previous and contemporary observations, he accepted the minimum value given by Hipparchus instead of that deduced from the Timocharis measurements. He then estimated that "stars are moving towards the east by one degree in about a century." This corresponds to a period of precession equal to 36000 years. It seems that Ptolemy himself did not observe as much as Hipparchus. He has even been attacked for merely copying the observations of other astronomers (notably Hipparchus). We shall leave the attack unanswered, as we do not really have the necessary evidence.

Ptolemy naturally had to introduce into his system of the World some explanation for this stellar drift. He assumed that beyond the sphere of fixed stars was a ninth sphere devoid of stars, but responsible for the slow motion (in addition to its daily rotation) of the sphere of fixed stars.

Some disputes among the followers of Hipparchus dealt with the question of whether or not this new motion was transmitted to the fixed inner eight planetary spheres. Most of the predecessors of Ptolemy maintained the position that the planetary orbs were affected by this precession. Adrastes of Aphodisia was convinced that the Sun was indeed affected, but that the five planets were not. Theon of Smyrna, possibly influenced by both Posidonius and Adrastes, advanced another version. He maintained that it was the line of apsides that was affected. His term *anomalistic year* refers to the interval between two successive passages of the Sun at its apogee, but that is another effect indeed.

Ptolemy himself held that the apsidal lines of the five planets were linked to the fixed stars, hence not affected by precession. For the Sun, on the other hand, he believed that the perigee was moving with respect to the fixed stars. Although this is in fact true, Ptolemy's estimates were flawed. Actually, the Sun's apogee moves in the same direction as the fixed stars (Fig. 2.35), along the ecliptic, with respect to the equinoctial points, but more rapidly. The solar apogee actually describes the ecliptic in about 21 000 years (more precisely 20 984 years).

These Ptolemaic statements, well known in the Christian Middle Ages, gave rise to interesting speculations about the need for a ninth sphere and the degree of influence of each successive sphere on those interior to it. This discussion was part of the larger discussion regarding the World system and we shall not say more about it here. We must only note the difficulties that these observations generated for the various clerics of the period (Christian, Arabic, Jewish) who were attempting to arrive at a definition of the year. Something of their ambivalence is still present in the calendars used today in these different religious traditions.

One of the more important effects (although not always accepted as such) of the precession of the equinoxes is that each "astrological" sign, named after a given constellation, is fixed with respect to the ecliptic and linked to the equinoctial points. Meanings are attributed to the constellations and their signs. For example, the sign γ represents the constellation Aries, the Ram, and all things symbolizing spring and the vernal equinox, since the ram is a spring animal. In our time, the difference between the sign and the constellation associated with it can be larger than one sign, larger than one month. This suggests an impossible dilemma. The Ram is no longer in the position in the sky that it was in when the constellations were named by the Babylonians. What, then, does it mean to say that someone was born under the sign of the Ram? The constellation Aries is no longer coexistent with the vernal equinox. What would have happened had the constellations been named not by the Babylonians a few centuries before our era, but much earlier, by the first Egyptians for example? The signs would then have different names and different meanings, associated with different constellations. This would seem to refute any argument in favor of astrological influence and its symbolism.

CHAPTER 3

Ptolemy's Astronomy Questioned

3.1
The Scientific Genealogy of Ptolemy

3.1.1
Ptolemy's Physics

It was well known by the Greeks who were skilled at geometry that one can save the phenomena or the appearances, that is, formulate a theory that would preserve empirical observation, by different combinations of circular and uniform motions. Aristotle had added that geometry must have some basis in physics. As we have seen, Aristarchus (310–230 BC), a geometer, formulated his heliocentric theory in purely geometric terms. His follower, Seleucus (250–180 BC) searched for a theory that would be "physically true" long before Ptolemy. Hipparchus later conceived of the heliocentric and geocentric theories as strictly equivalent.

This equivalence was natural enough. Whether the Sun is moving with respect to the Earth, or the Earth is moving with respect to the Sun is not, physically speaking, relevant, providing that one does not have any fixed point of reference. The stars do not seem to change their respective positions whether we consider them from the point of view of the Earth or the point of view of the Sun. Indeed, unless we can define an "absolute" system of reference without ambiguity, our understanding is limited to the relative motions of objects to one another.

Obviously, the Peripatetic scholars could not accept Ptolemy's theory of non-homocentric motions. Even followers of Ptolemy, such as Theon and Adrastes, had difficulty accepting it, as they did not think it was possible to build a wooden solid model of articulated spheres, embedded in each other, which would describe the whole system of epicycles and eccentric circles.

Fig. 3.1 *The post-ptolemaic representation of the sphere of any one planet.* Adrastes and Theon, followers of Ptolemy, wanted to be able to build a solid (wooden) model for the planets. This model would be based on epicyclic motions. The whole sphere, of which the center is C, is the planetary sphere. It is hollow and contains, solidly associated to it, another sphere, of which the center is C'. The inner boundary of the planetary

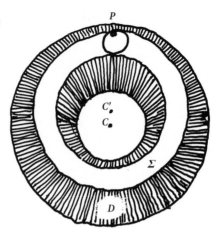

sphere is the deferent circle, concentric to the inner sphere. Between the two spheres, the epicyclic sphere, which carries the planet P, is moving. The inner volume Σ, moving around C', is communicating its motion around C to the external volume D, which in turn, communicates its motion to the orbs of the planets outside this planetary orb. This mechanical system, which imposed limits on the mutual arrangements of the planets, was in use until quite late in history, and was the basis of Ptolemaic constructions and computations still conducted in the 16th century. Note that the initial idea of Adrastes and Theon of not using the part of the Ptolemaic description involving eccentricity is somewhat violated by this system, which contains some amount of eccentricity in the relationship of the two volumes to one another.

Ptolemy attributed to each of the planets an orb of "finite thickness." The planetary orb would touch the orbs of the preceding and following bodies. Between the two spherical surfaces of the orb, the planet would move according to Ptolemaic laws. Ptolemy also claimed to look for the highest degree of simplicity, thereby anticipating Occam's razor (see p. 175). One should make only the necessary hypotheses, no more than that, in order to account for the observed phenomena.

Ptolemy added an interesting statement. He suggested that epicycles were artifacts, necessary to save phenomena with the simplest possible hypotheses, not "real" structures in the sky. The orb of each planet, according to this model, is filled by a fluid matter which does not oppose the motions of bodies within, and in which the planet moves without being guided by any solid sphere.

Ptolemy, therefore, prefered the eccentric models, finding them simpler than those that made use of epicycles, and not so much in conflict with Aristotelian physics. In Ptolemy's system, each planet was thus in a fluid embedded in a system of spheres. A "solid" sphere between these spheres communicated the motion from the inside to the outside sphere (the "orb"). Each planet needed at least two solid volumes inside its orb. Mercury needed three of them because the center of its eccentric is not fixed. It describes a circle the center of which differs from the center of the World (see p. 94).

Adrastes and Theon held to the opposite viewpoint based on epicyclic motions, because of the possibility that this theory would promote construction of a spherical model made of embedded spheres (Fig. 3.1).

3.1.2
The Pagan Followers of Ptolemy

The neo-platonists had some sympathy for Ptolemaic physics, but were less amenable to the geometrical aspects of Ptolemaic theory and the mechanisms of epicycles and eccentric circles. They were also interested in conceiving of "models." Primarily, they concentrated on the "essence" of the astronomical bodies, a very Platonic preoccupation, and as astronomers, claimed that one could not apprehend the nature of the celestial objects. The best one could hope to do was, therefore, to construct an image, i.e. a model, some kind of fictional artifact. Syrianus (380–450) and Proclus (ca 400–480) were the last true pagan astronomers.

The Alexandrian or pagan tradition was actually dying, under the military assault of all the neighboring armies. The destruction of the Library of Alexandria began during the civil wars at the borders of the Roman Empire during the third century. Burned by the Christians in 391, it was finally destroyed completely in 641, at the time of the conquest of Alexandria by the Arabs. This was the symbolic end of the oriental school, but the Christian philosophers, theologians, and thinkers emerged in the same part of the world. The disappearance of the classical Alexandrian school impeded the progress of scientific knowledge. The unstable political situation and the constant warfare acted as a strong brake as well. Nonetheless, both the Christian and Arab cultures were ready to start again. It was not a case of beginning at the beginning, but certainly they were hampered by the great loss of books and data and by the lapses in their oral traditions. It is not surprising, therefore, that the works of Aristotle and Plato, still readily available, made a come-back.

3.1.3
John Philopon

Very much in the oriental tradition, John Philopon, a Christian, still in Alexandria, and Simplicius, a pagan, in Athens, remained very close to Aristotelian thinking. Philopon, a very original thinker, rejected anything but uniform circular motions. For him, it was necessary to reject over-complicated mechanisms, even to save the phenomena (see Fig. 3.2). In one chapter of Philopon's work about the "Creation of the World," he noted that:

> "Hipparchus and Ptolemy had recognized the first sky, the one with-out heavenly bodies; as to the second sky, named the firmament by Moses, the Greeks knew it was also unified; but they subdivided it into parts, some in one way, some in another way, so as to give to each of them an adequate causal description of the anomalies of the wander-

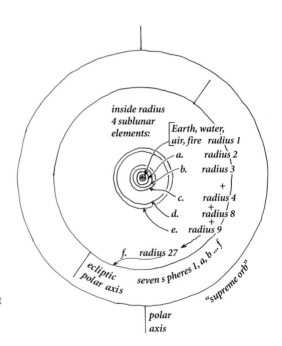

Fig. 3.2 *The World System, as adopted by John Philopon*. It is, in its details, not so far from Platonic and post-platonic concepts.

ing stars; and all of the hypotheses of these astronomers are completely lacking any demonstration."[2]

As a Christian, Philopon's first priority was to identify Hipparchus' ideas
with the biblical Genesis. Secondarily, this point of view allowed him to
remain an Aristotelian.

3.1.4
Save the Bible!

During this post-Ptolemaic period, few astronomical observations were
pursued in the Christian-pagan Mediterranean world. Ammonius (5th
century) invented the astrolabe. Unfortunately, one could not make use
of earlier observations, which went up in the dark clouds of smoke over
the Library of Alexandria. These dark clouds also enshrouded a few
pious minds, who had a very primitive knowledge of the world, in spite
of their often high social or intellectual prestige.

   In the 3rd century, Lactancius was a very popular Christian "scholar,"
who ridiculed the idea of a spherical Earth, just as Cosmas did. Cosmas,
who travelled in India, described a flat and square Earth. A regression of
many centuries! Later, in the 7th century, Isidorus, who became bishop
of Seville, gave the periods of revolution of the Moon, Mercury, Mars,
and Venus, around the Earth as 8, 20, 19, 15, and 15 years, respectively.

   The early Christian thinkers were only interested in commenting on
Genesis. It was not their purpose to save the phenomena by reference to
some model, but rather, to save the Bible by reference to the phenomena.
Ptolemy was more or less forgotten.

3.2
The Church Fathers

3.2.1
The So-Called "Dark Ages"

One of the purposes of this book is to reveal the continuity of science, in
spite of some profound changes in attitudes, in spite of powerful political

---

[2] quoted by Duhem, P., *Le Système du Monde*, Vol. II. (Hermann, Paris 1914) p. 111.
   Translated by the author.

or military events, associated destructions, and other social factors. One could, nonetheless, ask whether the period between, say, the 4th and 15th centuries, did not represent a regression of human lucidity regarding the Universe around us.

It is true that history has often been shaken by the rise and fall of empires. Each time a society was thought to be developing peacefully, some unpredictable forces from within or without proved that this was not the case. As empires crumbled, large bodies of knowledge were buried in the ruins.

Figure 3.3 p. 142 describes the flow of history from Babylonia to Greece (see also p. 17, Fig. 1.1, and p. 41, Fig. 1.31). The golden age of Periclean Greece was followed by a period of military activity, the conquests of Alexander and the formation of his huge empire, and finally by its rapid fragmentation under Alexander's generals and their descendants.

Then came the emergence of eastern Christianity, with the synod of Nicea, on the one hand, and on the other, the birth of Arab culture. These developments correspond more or less to the beginning of the Middle Ages, or the "dark ages," as they were often called (Table 3.1).

With the beginning of the so-called "dark ages," the Arab empire, to the south, embarked on a period of expansion, and bloomed, from the Persian gulf until it covered almost the entire Iberian peninsula. The northern germanic peoples scrambled over the ruins of the Roman Empire, forming kingdoms of short duration and little intellectual life. Between them, Christian philosophy quietly developed. After the conversion of Constantine in the east, Clovis in France converted to Catholicism. The Church, now well established after the synod of Nicea, was installed in Italy, France and Germany. The old civilization of the Greeks, of the Alexandrians, was hardly remembered. Only a few texts remained. The Old Testament and the Gospels were revered. Assorted remnants of the scientific gospels were also revered, primarily Aristotle and some of his commentators, and of course, Ptolemy, if only to be criticized by the Aristotelians.

3.2.2
From Origen to Augustine

Many of the Church Fathers, as the Christian scholars of the first centuries of our era are often called, were primarily engaged in commenting on Genesis. After the reign of Constantine the Great, Emperor from 306

Table 3.1 *The not-so-dark ages*

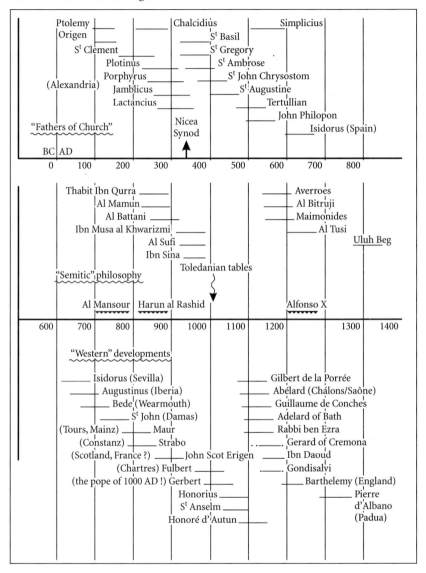

to 337, the Church was almost completely triumphant. We can date the rapid extension of the Christian school back to the synod of Nicea (325).

Origen (ca. 185–254), a contemporary of Ptolemy, was probably the earliest Christian commentator on Genesis. He lived in Alexandria and knew the Ptolemaic constructions much better than many of the Christian commentators who came after him, but we know very little about his opinions on the matter.

Another early Christian, Saint Clement of Alexandria (3rd century) echoed Plato's views about the alternation, ad infinitum, within a Great Year, of storms and floods (the "great winters"), and of fires and heat waves (the "great summers"). It was his contention that this was directly inspired by the story of Noah's ark, as told in the Bible. (Remember that the legendary Deluge is mentioned by Hesiod, together with the Deucalion and Pyrrha story, the Greek or pagan version of the re-population of the Earth.) Origen, himself, it seems, shared this view. He understood, however, that Christian dogma and the concept of periodicity in the life of the Universe were essentially incompatible. The eternal return of the Great Year is in contradiction with the doctrine of human free will, basic to the Biblical tradition. A second Adam would not commit original sin a second time!

In spite of the fact that Origen still read and spoke Greek, the classical Greek language soon became unknown. Chalcidius, a Christian or possibly Jewish thinker, translated Plato's *Timaeus* into Latin at the beginning of the 4th century. He dedicated his translation to Osius, bishop of Cordoba, who had attended the synod of Nicea. In his comments, Chalcidius shows that he knew both the Biblical and Platonic writings. He developed the Platonic vision of the elements, describing the primary nature of matter.

After Nicea, however, the fourth century thinkers Basil (329–379), Gregory (330–400), Ambrose (340–397), John Chrysostome (344–407) and especially Augustine (354–430), all of them later canonized as saints, completely disregarded the old teachings, which they dismissed as pagan superstitions. In particular, they did not retain the idea of saving the phenomena. Let us quote, as an example of this attitude, Saint Basil:

> "... They measure stellar distances; they describe polar stars which do not set; they say which southern stars are visible to us, why some other southern stars are invisible; they analyze the northern sky, and divide the zodiac in a thousand parts; they observe carefully retrogra-

dations and stations of planets, and their motions with respect to stars; they determine the time during which planets are revolving with respect to stars; of all the resources of invention, only one escapes them: the one which reveals God, the creator of the Universe, the fair judge, who applies to all who have lived the right fate."[3]

Obviously, Basil was not ignorant, but there is in this attitude the beginning of an "anti-science" worldview still very much evident in the 20th century. This text can also be held up as evidence that the Church thinkers were basically anti-Aristotelian, but perhaps open to Plato's views. After all, Plato had always emphasized the role of a *demiourgos*, a Creator.

Even as late as the 6th century, as fine a scholar as John Philopon expressed a similar view when he wrote:

"... No man will ever be able to account for the number of heavenly bodies, for their positions, their velocities, their colors ... We know only that what God has created, he has created well; nothing is missing; nothing is superfluous ... There are only a few things of which we fully know the causes. If we are not able to tell the physical cause of sensible facts, we are no more able to name the cause of unobservable things."[4]

Quite clearly, we are treading very close to the Platonic path regarding the observable and the unobservable, though Plato does not really think that we cannot know the causes of observable facts. In John Philopon's text, there is a warning against any possibility that scientific methodology could be employed to reveal the nature of reality.

3.2.3
Augustine

One of the most influential thinkers was Augustine, bishop of Hippo. He wrote, in Latin, three treatises concerning Genesis. In his two most significant books, *The Confessions* and *The City of God*, he also commented on the cosmogonic meaning of the Holy Scriptures. One could argue that indeed he was perhaps more concerned with cosmology than with cos-

---

[3] quoted by Duhem, ibid, II, p. 396. Translated by the author.
[4] quoted by Duhem, ibid, II, p. 499. Translated by the author.

mogony. The suffix *gony*, and its variants, implies the concept of evolution – of the Universe, in the present case. The suffix *logy* implies only a general description of the Universe as it is now. Today, these two problems are intertwined, because 20th century scientists know that one cannot distinguish between time and space. The description of the Universe "as it is" now implies the description of its evolution over time. Augustine was only concerned with description, not with evolution.

In Augustine's thinking, nothing of the astronomical culture of Origen remains. He seems to have acquired almost nothing more than a layman's knowledge, as we see in the following passage:

> "About the motion of the sky, some of our brothers ask: Does it move? or does it stay still? If it moves, how can it be the firmament of the Bible? If it is still, how could the stars, which are, according to the common belief, attached to the sky, be turning from east to west, along circles of different sizes, the more septentrional stars describing the smaller circles, so that the whole sky seems to turn like a sphere or a disk? I reply to these brothers that the answer requires a very elaborate reasoning ... I have no time to enter into such elaboration, and they should not either. It is our wish that they be informed only of what will grant their salvation, and of what is useful to the Church. They should know only that we do not believe that the word firmament forces us to assume the sky to be motionless ... In any case, one can show that all that has been observed can be understood by the motion of the stars, without asking the sky to move."[5]

In other words, the only valid physics is that found in the Bible. It is not desirable to encourage the acquisition of knowledge. It is certain that Augustine knew more than he admits to. It is even possible that Augustine himself knew, and understood, that the concordance between the Ptolemaic astronomical models and the actual celestial phenomena constituted strong evidence for the operationality of scientific methodology. He did not, however, want to give scientific knowledge any weight. Serving God was the only thing that really mattered.

Nonetheless, he used his knowledge, in his own way, to combat certain heresies which were proliferating at that time. In particular, Augustine attacked the Manicheans. These primitive sectarians presented an outlandish description of the motions of celestial bodies. It was actually un-

---

[5] quoted by Duhem, ibid, II, p. 396–397. Translated by the author.

related to their theological doctrines, but Augustine, employing a strategy which we might regard as dishonest, used his knowledge to criticize the Manichean astronomical propositions as a means of undermining their basic doctrine.

3.2.4
Astrology

Under the influence of the Chaldeans, using Ptolemaic tables, astrology flourished everywhere in the Middle East. It was very popular indeed. Augustine himself was violently opposed to horoscopes, which he considered a deliberate swindle for a variety of reasons. Astrologers, he reasoned, may imprison the people who believe in their horoscopes in the net of a pact with the devil. It was one thing to monitor the tides by the Moon, quite another to interfere with the "free will" of man, given to him by God, as described in Genesis.

Some Christian writings on astrology were less clear than Augustine. Theodoret (390–458) mentioned that there are "signs" in the sky to help men to choose the most propitious time for planting, or seeding, for building houses, the best time to undertake a voyage by sea, to know when grasshoppers will invade the area, or when the cattle will suffer from a high rate of mortality. Theodoret is obviously an extreme case. Augustine was quite firm, and most of the Church Fathers were with him.

In their battle against astrology, considered a pagan heresy and condemned as such, the Church Fathers were struggling against the hyperdeterminism which characterizes the logic of astrology. In doing so, one could almost say that they were laying some of the cornerstones of modern science. They also discussed other aspects of astrology, perhaps more metaphysical than physical, but this discussion nevertheless opened the door to critical thinking. Take for instance the case of "primary matter," considered eternal; or the domination of the astral world over the sublunar; or the rhythm of the World marked by the eternal return and the Great Year. In all these instances, the Church Fathers were really taking issue with some ideas that we criticize as well today.

3.2.5

The Debate Over Creation and the Meaning of Time

Can the Church Fathers then be associated with any of the traditional philosophies originally elaborated in Greek times? They seem to be determined to undermine the cosmologies of the neo-platonic philosophers and of the Peripatetic school, not to mention the fantasies of the Pythagorians. They belong only to the Church. So they say. However, one can see the dominant influence, in a new costume, of Platonic doctrine. The commonality of the Latin of neo-platonic and Church writings was one factor. In addition, the neo-platonic philosophers argued that God, a unique God, was the Creator, or the demiourgos of Plato. Of all the neo-platonic thinkers quoted by Augustine (Plotinus, Jamblicus, Porphyry, and even Apuleius), it is likely that Chalcidius was the determining influence on him, since Chalcidius was the translator of the *Timaeus* into Latin.

However, Augustine did not follow all the neo-platonic views of Chalcidius. Three central questions were under discussion. Is there primeval matter in the Universe? Was it created or has it always been in existence? Is Creation the origin of time, or are these two different notions?

Primeval matter was called ὕλη. The word has been used in modern times, with the same meaning, by Gamow and followers, using the spelling "hyla." All schools of philosophy believed in the existence of this primeval matter, different from matter that we can observe now, but no one arrived at a clear definition of it.

For Chalcidius (and Plato) ὕλη was, basically, created by God out of nothing and transformed by God into real matter as we know it. Augustine held the same view of the origin of ὕλη but did not touch on the question of its transformation into ordinary matter or κόσμος (cosmos).

For Aristotle, on the other hand, hyla existed for all eternity and is essentially identical to κόσμος (cosmos).

Clearly, the Platonic view, implying a Creator, had much more appeal in the Christian world than the Aristotelian view where God exists, but not as a Creator, only as a sort of "manager."

The notion of a "beginning" of time, coexistent with the Creation, does not necessarily identify time itself, a physical quantity, with the astronomical motions. Augustine rejected this identification because of his need to formulate an idea which would not contradict the biblical text from the Book of Joshua in which the Sun was said to have stood still during the course of a battle. Astronomical motion came to a halt but human life in time continued. The essential meaning of time remains an open question, even if some people believe in the advent of Creation at the beginning of "time." This is a problem that philosophers and scientists continue to grapple with.

There is little doubt that, up to the 12th century, Augustine was well-known throughout the Christian world as a master, whereas neither Plato nor Aristotle were sufficiently well-known. Much has been said about the destruction of the Aristotelian paradigms during the Renaissance, but this was well under way in the work of Augustine.

After Augustine, the Christian Mediterranean world was violently shaken by invasions from the north, by economic disasters (exhaustion of the Spanish gold mines, in particular), by frequent epidemics (such as the pestilence brought to Rome in 188 by the armies after their Asiatic campaigns), and by despotic and corrupt emperors exerting a mad power ... in essence the death throes of antique culture. During this period, Christian scholars progressively forgot even the teachings of Augustine. They lived secluded in monasteries, in a climate that became colder than it had been previously (the Little Ice Age). The best minds were exiled and isolated. Simplicius, for example, the popularizer of Platonic and Aristotelian ideas, was exiled in 529 (after the dissolution of the Academy by Emperor Justinien) to Persia by the Church. Both the Roman and Eastern Orthodox hierarchies were completely preoccupied with asserting their power and defending against the numerous threatening heresies. They regarded everything that departed even slightly from the Scriptures, taken literally, as heretical. Isidorus, the bishop of Seville, for example, whose astronomical ignorance we have already noted, was a "literal creationist." For him, the world was only a few thousand years old. Biblical chronology was to be taken literally.

3.2.6
The Seeds of a Reborn Science

Nonetheless, any sign of knowledge, of scientific knowledge, that appeared during the early Christian period was indeed a seed for the future. In this respect, the blooming of Arabic culture, which took much from the Christians, the Indians, and the dying Ptolemaic tradition beginning in the 7th century, kept the flame alive and transmitted it again, during the 9th, 10th, and 11th centuries, to western scholars.

Some scholars, apart from Augustine, were determined to keep the ancient knowledge alive. A good example of that continuity is the work of Anicius Boethius (480–524), a Roman Christian philosopher (who did much of his work in prison!). Not only did he write a treatise about the basis of music, but he compiled, in Latin of course, the works of Euclid, Nicomachus (the logician), Ptolemy, and one part of Aristotle's logical treatises.

We shall follow that inheritance later, when modified, transformed, and rejuvenated by the new Christian schools in western and northern Europe, by such scholars as Bede, Alcuin, and Rhaban Maur.

3.3
The Contribution of the Arabic World to Astronomical Knowledge

3.3.1
Indian Astronomy

The conquest of India by Alexander (see Fig. 3.3) spread an understanding of Greek philosophy and astronomy to the subcontinent. Under the Gupta dynasty, in the Industan (ca 400–650), several books were written, notably the *Siddhantas*, describing the Universe. Aryabhata (born 476) and Brahmagupta (born 598) were the best known of these authors. Later, one must note the work, in that tradition, of Bhaskara (born 1114).

The Indians did not know Ptolemy. The link to Greco-Egyptian astronomy was lost after the conquests. By then, the empire of Alexander had been destroyed, so the Indian scholars did not go as far as Ptolemy. Still, they noted the rotation of the Earth around its axis, and they also created some mathematical tools, which were later developed by the Arabs, namely the number zero, and the decimal notation of what are now called Arabic numerals.

Fig. 3.3 *The wanderings of science.* Over time and influenced by political and military events, the scientific center of gravity moved. (1) A first move from Greece to India (2nd century BC) carried science on the backs of the armies of Alexander. From India, it returned to Alexandria, where Ptolemy was active (2nd century),

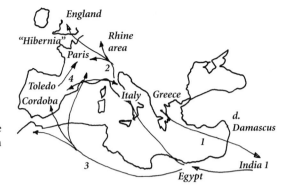

and, later, to the Arab world. From there, after the fall of Alexandria (3rd century), and the synod of Nicea, it entered the newly organized Christian world in the eastern part of the Mediterranean; and from there (2), progressively traveled north to England and Ireland in the 6th and 7th centuries. In the meanwhile, Islam, born in the 7th century, spread rapidly (3). By the 9th-10th centuries, it had reached Spain, where it retained its power until the14th century. Exchanges between the Arab-Jewish world in Spain and the Christian world in France and Italy (4) led to the emergence of western science, which then spread throughout the whole of north-western Europe.

3.3.2
The Development of the Islamic Knowledge of the Sky

The great Moslem Caliphs were interested in the development of science, perhaps for economic and political purposes, and because of the prestige of astrology. The first to be well-known for this interest was influenced by the Indian writings. In 773, al-Mansour, the founder of Baghdad, then the intellectual capital of Islam, is said to have learned astronomy from an Indian astronomer visiting him. He then apparently ordered the translation of Indian books into Arabic. In the 9th century, al-Khwarizmi (a corruption of his name led later to the word "algorithm") gave a general account of Indian astronomy. He published accurate astronomical tables, obviously inspired from Indian models, as they base their system of longitude on the meridian of Udshain (a location in Central India), which was the seat of a real observatory, probably since the 5th century.

Later, Caliph Haroun-al-Rashid (766–809), and his son, Caliph al-Ma'-mun (end of 8th century-833?), became interested in Greek science. The

Fig. 3.4 *Trepidation as understood by ibn Qurra.* The axis of the sky rotation (N) (perpendicular to the ecliptic) describes a small circle around its average direction, deviating from the axis perpendicular to the equator by an angle of about 23°. One axis is revolving around the other with a long period on the order of 26 000 years, the period of the precession of the equinoxes.

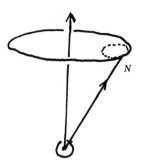

latter got hold of Greek manuscripts on the occasion of the signing of a peace treaty with the Byzantine Emperor. Among them, he found Ptolemy's writings and immediately translated into Arabic (827) the work we know as the *Almagest* (see p. 105), from the Arabic *al* for "the" and the Greek *megiste* for "greatest."

Several Arabic and Moslem scholars then studied and commented on Ptolemy, forming the missing link between the end of the Alexandrian school, and the emergence of western science. One was al-Farghani (Alfraganus, 9th century) who summarized Ptolemy. Abû Ma'shar (Albumazar, 9th century) even developed the astrological implications of it, having a knowledge of the Babylonian corpus.

Thabit ibn Qurra (829?–901) went further. To the knowledge of the precession of the equinoxes, he added the idea of *trepidation* (Fig. 3.4), a variation in the velocity of the motion of the equinoxes, and of the angle between the axes of the ecliptic and the equator (23°51'20" was Ptolemy's value; the Arabs found 23°34' or 35'). This difference can be accounted for by an oscillation of a period of 4000 years and an amplitude of 4°, according Thabit ibn Qurra.

Can trepidation be the same thing that we know now as *nutation*? Probably not. Nutation is the phenomenon which is described by a periodic change of the angle of the Earth's axis with respect to the perpendicular to the plane of the ecliptic (period 19 years, amplitude 9".210). This is now quite an accurate measurement. Trepidation is probably only a

Fig. 3.5 *Nutation: a modern concept.* The plane of the ecliptic contains the orbit of the Earth around the Sun, described in 1 year. The axis of the Earth is directed toward the North Pole of the sky, at an angle of 23°51 20" (Ptolemy's value) from the axis of the ecliptic. It rotates around it with a peri-

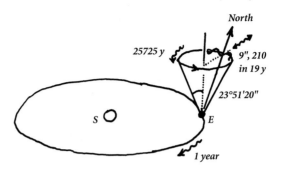

od of 25725 years (*precession of the equinoxes*). In addition, it oscillates around its mean path during the precession with a period of 19 years and an amplitude of 9".210 (20th century values). This last phenomenon is the "nutation".

device to rationalize the inaccuracy of the determinations. The characteristics of nutation (Fig. 3.5) are actually quite different from those of the alleged trepidation.

Nutation seems, in fact, to have been known to Hipparchus and forgotten by Ptolemy. The invention of trepidation shows that Moslem scholars were independent thinkers indeed! Al-Ma'mun ordered the measurement of a degree of latitude in the plain of Palmyra, in order to measure the sphericity of the Earth. The accuracy achieved by this measurement was quite comparable to that of Erathosthenes. Quadrants, armillae, and astrolabes, often beautiful pieces of copper art, enabled accurate determination of stellar positions (Fig. 3.6a, p. 146).

Among Arab astronomers, al-Battani (Albategnius, 850?-929) was one of the most active. In the town of Rakka, he made observations which often deviated somewhat from those of Ptolemy, but he did not question the Ptolemaic system. Ibn al-Haitam (Alhazen, 965–1039) was an important follower of al-Khwarizmi. He developed optical astronomy, in a way, as did al-Kindi (?–873) before him. He even described the principle of an optical telescope, made of spherical mirrors and lenses. He wrote a great deal and developed ideas that were very close to the Eudoxian theses. He held that the World must be a solid sphere, because there is no such thing as a vacuum. His description, however, was quite simplified (Fig. 3.7, p. 147).

He described a true supreme orb that was without stars and created diurnal motion. Still, al-Haitam maintained the Ptolemaic mechanism.

(a)

(b)

Fig. 3.6a,b *Some instruments used by the Arab and western observers:* (**a**) an Arab astrolabe, (**b**) a sundial (without its style directed towards the celestial hole).

Fig. 3.7 *The system of al-Haitam.*

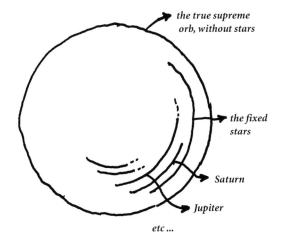

the true supreme orb, without stars

the fixed stars

Saturn

Jupiter

etc ...

For each planet, there would be a typical deferent-epicycle system. The deferent was the "bearing sphere" (falak kharidj el markaz), the epicycle a "sphere in rotation" (falak el tadwir).

During this period, the Arabs made much progress in many fields of science: in optics, in mathematics, even in the analysis of series, and elementary algebra. They did not progress to any great extent in theoretical astronomy, until they started actively to question Ptolemy (at a time when the Christian commentators had already completely forgotten about him).

In observational astronomy, however, they progressed rapidly. In Spain, Maslama ibn Ahmed (?–1008) transferred all astronomical tables available to him, referring them to the meridian of Cordoba (i.e. Ptolemy's longitude + 12°42'). Al-Sufi (the Wise), in the 10th century, discovered the Andromeda Nebula, as we can see on the beautiful sky maps he drew (Fig. 3.8).

Al-Sufi determined stellar magnitudes on his own. Even Ptolemy did not do such a thing, relying as he did, entirely, upon Hipparchus' determinations. At the same time, in Egypt, Sultan al-Hakim (Cairo, 977–1007) ordered new tables in support of his own work. In Turkey, Prince Sharaf al-Dawla ordered the construction of an observatory, with several new instruments. New observations of obliquity of the ecliptic were made by Abu'l Wefa, who also measured the locations of the equinoxes and solstices on the starry vault.

The intellectual center of Arabic culture became Spain, primarily in Seville, Cordoba and Toledo. But remember Bishop Isidorus, the "literal creationist" was also in Spain! Ibn Sina (Avicenna, 980–1037), who was famous as a physician, wrote books on Aristotle's ideas, as well as a compendium of the Ptolemaic doctrine. Ibn al-Zarqali (Arzachel, 1029–1087), a very prominent Toledan, constructed the famous *Tables of Toledo* and built an astrolabe. Jabir Ibn Aflah, at the same time, made similar observations. Modern historians know of about 400 Arab astronomers and mathematicians. So we have necessarily limited ourselves to a few examples of this tide of revived science, washing up on the shores of the occidental world. We should note that part of this progress was due to many Jewish scholars who lived in Spain in good harmony with their Arab colleagues. Even after the Catholic Kings of Spain cleared the Arabs and Jews out of Cordoba, astronomy continued to develop in both western Africa and the French southern provinces where many exiled Jews found a new home.

All these astronomers, following their Indian teachers, contributed much to improve the Ptolemaic determination of the precession of the equinoxes. Al-Battani, Thabit ibn Qurra, al-Sufi of course, and Rabbi Abraham ben Hiyyia (12th century) contributed to establishing good values for the parameters of the solar system and setting the pace for a re-evaluation of Ptolemaic ideas, based upon a comparison of theory and data. We shall return to this questioning.

In Persia, the grandson of Gengis Khan, Hulago il Khan, founded the observatory of Maragha, where al-Tusi organized an extensive library and built a large quadrant with a radius of 10 feet. He built the *Ilkhanic Tables* around 1274.

In Samarkand, the grandson of Tamerlane, Ulugh Beg, was an astronomer himself and a great one indeed. He built (ca 1420–1437) a large observatory, still standing in Samarkand with a large quadrant with a radius of 60 feet.

These early constructions, not far from the achievements of Tycho, paved the way for the 17th and 18th century larger observatories in India, such as those in Jaipur and in Delhi. The fact, however, that they did not use optical devices became a serious limitation in the 17th century when astronomy was flourishing in western Europe.

We should also mention, in this context, that in the valley of the Rhone in the 14th century a contemporary of these far-eastern observers, Rabbi Levi ben Gerson, developed the cross-staff as a measuring device which was a step towards more accurate visual determinations (Fig. 3.9).

Fig. 3.8 *A fraction of al-Sufi's map.* One sees the Andromeda nebula in front of the mouth of the big fish.

Fig. 3.9 *The Jacob's staff, or cross-staff of Levi ben Gerson.* The mobile part, along a gradation, permits the instrument to point in two directions. One side is pointed towards the horizon, the other to the star. The angle of the star above the horizon is measured directly on the gradation.

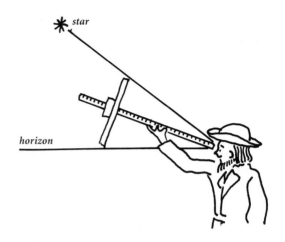

3.3.3
Alfonso the Tenth and His Time

So much for observations. Coming back to the Spanish school, before the expulsion of Jewish and Arab thinkers and the exclusive domination of the Roman Church, we should first state that, in spite of the realism of Islamic astronomers, as noted by P. Duhem[6], the Ptolemaic combination of solid spheres had never been considered very seriously. In the 11th and 12th centuries, Ibn Badja (Avempace), ibn Tofail (Aboubekr), ibn Rushd (Averroes), Moses ben Maimon (Maimonides), and, pre-eminently, al-Bitruji (Alpetragius) seriously questioned Ptolemaic thought, using Aristotle as the basis for their attemped refutation.

An important turning point was the reign of King Alfonso the Tenth "the Wise" (?–1284). It was the apex of the Arabic-Jewish school in Spain, and its last flowering. Alfonso gathered together several astronomers under the leadership of Isaac ben Said to construct new astronomical tables, the best possible ones. These were the so-called *Alfonsine Tables* (1282), and they were indeed better than the *Tables of Toledo* of al-Zarqali, already two centuries old.

These tables were, so to speak, the epilogue in the assault of the post-Aristotelians of the Arabic world against Ptolemaic doctrine in the name of Aristotle. In a way, this paralleled a similar trend of thought pursued

---

[6] quoted by Duhem, ibid, III, p. 126. Translated by the author.

by the Church Fathers, notably Augustine. Ptolemy was actually much ahead of his time, as were, for different reasons, Heraclides, Aristarchus, and Seleucus.

Strangely enough, the attraction to the idea of the "solid spheres," and the associated rejection of the concept of the epicycle-deferent mechanisms, were directed against Ptolemy. No one (neither Averroes, nor Maimonides, nor al-Bitruji) seemed to be aware that the system of al-Haitam was still Ptolemaic, albeit using Ptolemaic spheres in the Eudoxian way, as did Ptolemy himself. The synthesis was at hand, but some new observations again produced a new wave of skepticism about the best way to save the phenomena, if not on the very use of circular or spherical motions or of solid spheres.

The so-called "fifth" motion of the Moon, although known by Ptolemy, had been forgotten by the 12th century. Al-Haitam doubted it, but Averroes insisted upon it. He wanted to reveal the imperfections in al-Haitam's system. Nasir Eddin, a contemporary of al-Haitam noted that: "Al-Haitam said, moreover, that by using disks [circles] instead of spheres, he could demonstrate [the possibility of accounting for the variation of the inclination of the epicycle plane on the plane of the eccentric-deferent system.] But [says Nasir Eddin] a non-spherical system does not conform to the principles of astronomical science."[7]

Ibn Badja was born in Saragoza, functioned as a doctor in Seville until 1118, a vizir in Morocco until 1138, and was then assassinated. He wrote that "the existence of epicycles is inadmissible." The world must have a physical center. The motion of an epicycle is neither centrifugal, nor centripetal, nor is it around any material center, hence it is impossible! Al-Tafail also rejected the eccentric, for similar "Aristotelian" reasons.

Ibn Rushd (Averroes, 1126–1198) was also definitely Aristotelian. He wrote: "If the center of the Universe is not the center of the revolution, then another Earth must exist, and this contradicts the basic principles of Physics."[8] Averroes was looking for a system both consistent with the "principles of physics" (i.e. Aristotelian physics, as described above, p. 81) and the observed motions of the stars and planets (his effort to save the phenomena). This eliminated the Ptolemaic descriptions and the systems derived from them.

---

[7] quoted by Duhem, ibid, III, p. 129. Translated by the author.
[8] quoted by Duhem, ibid, III, p. 134. Translated by the author.

The considerable influence of Averroes was equalled by that of Maimonides (1135–1208). It is not certain that he even knew Averroes.
Expelled from Spain by the persecutions, he found refuge in Cairo
(ca 1160), but this was 30 years before Averroes left Spain for Morocco.

As the French philosopher Ernest Renan has noted, the thought of
Maimonides is quite different from that of Averroes. He is closer to the
idealism of Plato than to the positivism or realism of Aristotle. Moreover,
he was strongly influenced by the Talmudic tradition of giving weight to
opposing arguments. He thus "concludes" a debate less easily, and less
abruptly. For him, at first, the concept of a vacuum could not be ruled
out. Therefore, a system of solid or full spheres, centered around the
Earth (earth, water, air, fire, quintessence) where many embedded
spheres were uniformly moving in circular motions, was a conceivable
model. Nevertheless, he suggested, and this is a Ptolemaic idea, that
some spheres could have a center which differs from that of the World.
These centers would have to be inside the Moon's orbit, i.e. sublunary.
Maimonides considered the pros and cons of that notion and concluded
that it did not really save the phenomena (for instance the center of the
eccentric for the Sun is clearly outside the Moon's orbit).

The real attitude of Maimonides is that of a skeptic. His position was
that the sky belongs to God; the Earth belongs to the sons of Adam. We
can only understand sublunary physics; hence, his involvement with
the study of the properties of the four basic elements and their mix-
tures and his lack of interest in the sky, which man could not hope to
understand.

Averroes was more optimistic than Maimonides about the possibility
of describing the sky, in spite of his criticisms of the Ptolemaic systems.
The hopes he expressed were more or less fulfilled by a real astronomer,
al-Bitruji (Alpetragius) who was probably of Christian origin, a convert
to Islam. Along with Averroes, al-Bitruji criticized Ptolemaic ideas, finding them contrary to strict Aristotelian principles. He built a system of
his own, diagrammed in Fig. 3.10.

As in Maimonides, al-Bitruji initially described the spheres of earth,
water, fire and air. In his thought, there are then eight spherical shells,
contiguous to one another, and centered at the center of the Universe
(not necessarily the Earth!). The eighth sphere contains the Milky Way
and the fixed stars. The seven internal spheres are, from bottom to top
the Moon, Mercury, Sun, Venus, Mars, Jupiter, and Saturn. Finally, there
is a ninth sphere, without any celestial body. It moves, but although it

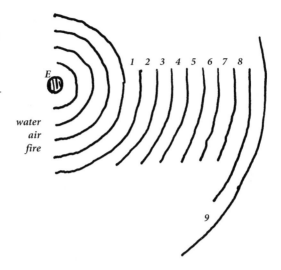

Fig. 3.10 *Al-Bitruji's geo-centric system.* This shows only the relative position of each sphere. Al-Bitruji, however, (see text) proposed a much more elaborate mechanism, in order to save the phenomena.

communicates its motion to the other spheres, its motion is independent of them. This is diurnal motion; its period is the sidereal day.

This model is not complex enough to save the phenomena, but it contains some rather astute ideas, in the direction of complexity, and in order, of course, to save the phenomena. For instance, al-Bitruji noted that each of the eight spheres "desires" to follow the perfect and simple motion of the ninth one, but cannot do it perfectly. In the same way, a stone is willing to follow the impetus which initiated its motion, but this tendency gradually fades until finally the projectile falls, when it is sufficiently far away from the motor which gave it its impetus! Each sphere, therefore, has a particular polar axis. These poles are mobile and describe small circles around the ecliptic pole. Some planets (Mars, Mercury) are located on their sphere, but not on the equator of these spheres. Unfortunately, al-Bitruji completely failed to explain why the distance Moon-Earth, or Venus-Earth changes during one revolution of either Moon or Venus. Let us note also that, along with Aristotle, al-Bitruji linked the motions on Earth (motions of seas, of continents) to the motion of the eighth sphere.

Up to the time of Copernicus, al-Bitruji's book was the new gospel and it was an anti-Ptolemaic one. It supported the arguments of all of the adversaries of Ptolemy and this continuing dispute made room without a doubt for the new and original thinking of Copernicus two centuries later.

There were many other important authors, Jewish and Arab, who contributed to the debate, but it was within the Christian world in the West that their tradition continued. This was probably for several reasons. The Arabic world was divided. The Crusades really brought the two cultures together. The political weakness of the Arabic world left the door wide open to the Ottoman Empire, whose main aim was to conquer eastern Europe. (Constantinople fell in 1453, a date which has often been considered, rather arbitrarily, as marking the end of the Middle Ages.) On the other side, some Catholic princes (such as Alfonso) had learned a great deal from the Arabs and Jews before expelling them; and some Jews went north, while most of the Arabs went south. Finally, the end of the period of the Crusades ushered in a time of peace in the 11–12th centuries in western Europe where monasteries and religious orders developed. The monasteries were the home of many distinguished and original minds. The times were hard for everyone. Only in a monastery could thought freely develop and discussions take place which mixed natural philosophy or scientific speculation with metaphysics.

Arabic civilization then plunged into some long period of silence and science developed henceforth only in the western Christian world. We are now, at the end of the 20th century, perhaps, and perhaps not, at the end of this tendency. Science seems to have now become the property of the whole world community.

In the 12–13th centuries, we see the center of gravity of intellectual life moving from the now sterile south-east of the Mediterranean to the north-western part of Europe, to Italy, France, England, and Germany.

So science travelled from Alexandria to Toledo or Cordoba; from the eastern Christian world to Padua, Marseilles, Paris, Oxford, and Heidelberg. The merging, and the mutual impregnation of the two traditions, which had evolved separately for a few centuries, was complete by the 12th and 13th centuries, through translations, and new migrations associated with the victory of the Spanish over the Arabs.

Each move, from Babylonia to the Aegean sea, from Greece to Alexandria, from Alexandria to Toledo, or to Oxford, from Oxford and Paris to Germany and Italy corresponded to a loss of information, and to the return of some strange older ideas. Still, sometimes painfully, sometimes more easily, the past was resuscitated. If the Middle Ages were dark, so far as the accessibility of human memory was concerned, they were also illuminated by the pace of intellectual activity which continued in spite of wars, in spite of starvation, in spite of epidemics, in spite of poverty.

Mistakes were made but memory was, in essence, rebuilt. The progress of ideas has been continuous. It is indeed remarkable that, in spite of the terrible difficulties of the times, so many scholars, clerical and secular, monks and teachers (Table 3.1, p. 135), in the protective walls of their monasteries, or in the newly built universities, were reading, writing, working, publishing, never ceasing to work harder and harder, never losing their eagerness to understand. They had students. They were often very influential and their ideas were passed on.

3.4
The Western World Up to Copernicus

3.4.1
The Continuing Heliocentric Temptation

As we have seen, Augustine, and to an extreme degree, Isidorus, had established the fixed dogma of the western Christian world. Augustine was one of the few open channels through which the old Greek tradition could enter that Christianizing world. He was deeply influenced by Plato, more certainly than by Aristotle, and by Plato's successors, the neo-platonists, such as Plotinus. There is little doubt that the *Timaeus* was one of the key sources of the Augustinian concepts. He held to the notion that ideas, exist quite outside any material objects, as eternal essences in the world. Augustine was deeply impressed by the truth implied by mathematics, a sort of truth that no one could dispute. In that sense, he was also impregnated by Pythagorian ideas. Augustine, however, did not deal with astronomy. He devoted most of his thoughts to Scripture and his cosmological ideas are primarily based on the Book of Genesis, not on any consideration of the observable sky. In that sense, he was certainly not an "Aristotelian," as we have defined the word in the introduction to this book, p. 10 ff.

Little by little, however, ideas began to permeate the obscure tissues of metaphysical discussion. Ptolemy's observations became known, either through the work of Hipparchus and others, or by a reading of Ptolemy's treatise, and with them a huge body of well confirmed facts. Some Latin authors (Pliny the Elder for instance) were very widely read. Typical of this tendency were the achievements of the Venerable Bede (672–735) in

Table 3.2 *The principal sources of ancient science in western christendom between 500 and 1300 AD (after Crombie). The spelling of names is that of Crombie and Crombie. It may differ from that given in the text of the book*

Author	Work	Latin translator and language of original of translation	Place and date of Latin translation
(1) Early Greek and Latin Sources			
Plato (428–347 BC)	*Timœus* (first 53 chapters)	Chalcidius from Greek	4th century
Aristotle (384–22 BC)	Some logical works *(logica vetus)*	Boethius from Greek	Italy 6th century
Dioscorides (1st century AD)	*Materia Medica*	from Greek	by 6th century
Anon.	*Physiologus* (2nd century AD Alexandria)	from Greek	5th century
Anon.	Various technical *Compositiones*	from Greek sources	earliest MSS 8th century
Lucretius (*c*, 95–55 BC)	*De Rerum Natura* (known in excerpts from 9th century; full text recovered 1417)		
Vitruvius (1st century BC)	*De Architectura* (known in 12th century)		
Seneca (4 BC–65 AD)	*Quœstiones Naturales*		
Pliny (23–79 AD)	*Historia Naturalis*		
Macrobius (*fl.* 395–423)	*In Somnium Scipionis*		
Martianus Capella (5th century)	*Satyricon, sive De Nuptiis Philologiœ et Mercurii et de Septem Artibus Liberalibus*		

Author	Work	Latin translator and language of original of translation	Place and date of Latin translation
Boethius (480–524)	Works on the liberal arts, particularly mathematics and astronomy, and commentaries on the logic of Aristotle and Porphyry		
Cassiodorus (*c.* 490–580)	Works on the liberal arts		
Isidore of Seville (560–636)	*Etymologiarum sive Originum* *De Natura Rerum*		
Bede (673–735)	*De Natura Rerum* *De Temporum Ratione*		
(2) Arabic Sources from *c.* 1000			
Jabir ibn Hayyan corpus (written 9th–10th centuries)	Various chemical works	from Arabic	12th and 13th centuries
Al-Khwarizmi (9th century)	*Liber Ysagogarum Alchorismi* (arithmetic)	Adelard of Bath from Arabic	early 12th century
	Astronomical tables (trigonometry)	Adelard of Bath from Arabic	1126
	Algebra	Robert of Chester from Arabic	Segovia 1145
Alkindi (d. *c.* 873)	*De Aspectibus; De Umbris et de Diversitate Aspectuum*	Gerard of Cremona from Arabic	Toledo 12th century
Thabit ibn Qurra (d. 901)	*Liber Charastonis* (on the Roman balance)	Gerard of Cremona from Arabic	Toledo 12th century
Rhazes /d. *c.* 924)	*De Aluminibus et Salibus* (chemical work)	Gerard of Cremona from Arabic	Toledo 12th century
	Liber Continens (medical encyclopædia)	Moses Farachi from Arabic	Sicily 1279

Author	Work	Latin translator and language of original of translation	Place and date of Latin translation
	Liber Almansoris (medical compilation based on Greek sources)	Gerard of Cremona from Arabic	Toledo 12th century
Alfarabi (d. 950)	*Distinctio super Librum Aristotelis de Naturali Auditu*	Gerard of Cremona from Arabic	Toledo 12th century
Haly Abbas (d. 994)	Part of *Liber Regalis* (medical encyclopædia)	Constantine the African (d. 1087) and John the Saracen from Arabic	South Italy 11th century
	Liber Regalis	Stephen or Antioch from Arabic	*c.* 1127
pseudo Aristotle	*De Proprietatibus Elementorum* (Arabic work on geology)	Gerard of Cremona from Arabic	Toledo 12th century
Alhazen (*c.* 965–1039)	*Opticæ Thesaurus*	from Arabic	end of 12th century
Avicenna (980–1037)	Physical and philosophical part of *Kitab al-Shifa* (commentary on Aristotle)	Dominicus Gundissalinus and John of Seville, abbreviated from Arabic	Toledo 12th century
	De Mineralibus (geological and alchemical part of *Kitab al-Shifa*)	Alfred of Sareshel from Arabic	Spain *c.* 1200
	Canon (medical encyclopædia)	Gerard of Cremona from Arabic	Toledo 12th century
Alpetragius (12th century)	*Liber Astronomiæ* (Aristotelian concentric system)	Michael Scot from Arabic	Toledo 1217
Averroes (1126–98)	Commentaries on *Physica, De Cælo et Mundo, De Anima* and other works of Aristotle	Michael Scot from Arabic	early 13th century
Leonardo Fibonacci of Pisa	*Liber Abaci* (first complete account of Hindu numerals)	using Arabic knowledge	1202

Author	Work	Latin translator and language of original of translation	Place and date of Latin translation
(3) Greek Sources from _c._ 1100			
Hippocrates and school (5th, 4th centuries BC)	_Aphorisms_	Burgundio of Pisa from Greek	12th century
	Various treatises	Gerard of Cremona and others from Arabic	Toledo 12th century
		William of Moer-beke from Greek	after 1260
Aristotle (384–22 BC)	_Posterior Analytics_ (part of logica nova)	Two versions from Greek	12th century
		from Arabic	Toledo 12th century
	Meteorologica (Book 4)	Henricus Aristippus from Greek	Sicily _c._ 1156
	Pysicia, De Genera-tione et Corrup-tione, Parva Nat-uralia, Metaphysi-ca (1st 4 books) _De Anima_	from Greek	12th century
	Meteorologica (Books 1–3), _Physica. De Cælo et Mundo, De Generatione et Corruptione_	Gerard of Cremona from Arabic	Toledo 12th century
Aristotle (384–22 BC)	_De Animalibus_ (Historia _animali-um, De partibus animalium, De generatione ani-malium_ trans. into Arabic in 19 books by el-Batric, 9th century)	Michael Scot from Arabic	Spain _c._ 1217–20
	Almost complete works	William of Moer-beke, new or revised translation from Greek	_c._ 1260–71
Euclid (_c._ 330–260 BC)	_Elements_ (15 books, 13 genuine)	Adelard of Bath from Arabic	early 12th century

Author	Work	Latin translator and language of original of translation	Place and date of Latin translation
		Hermann of Carinthia from Arabic	12th century
		Gerard of Cremona from Arabic	Toledo 12th century
		serveral revisions; revision of Adelard's version by John Campanus of Novara	c. 1254
	Optica and Catoptrica	from Greek	probably Sicily
	Optica	from Arabic	
	Data	from Greek	
Apollonius (3rd century BC)	Conica	perhaps Gerard of Cremona from Arabic (of this translation only a short fragment of Book 1 is now extant, as the introduction to Alhazen's De Speculis Comburentibus; but Book 2 was known to Witelo in the 13th century)	12th century
Archimedes (287–12 BC)	De Mensura Circuli	Gerard of Cremona from Arabic	Toledo 12th century
	Complete works (except for the Sandreckoner, the Lemmata, and the Method)	William of Moerbeke from Greek	1269
Diocles (2nd century BC)	De Speculis Comburentibus	Gerard of Cremona from Arabic	Toledo 12th century
Hero of Alexandria (1st century BC)	Pneumatica	from Greek	Sicily 12th century
	Catoptrica (attributed to Ptolemy in Middle Ages)	William of Moerbeke from Greek	after 1260
pseudo-Aristotle	Mechanica (Mechanical Problems)	from Greek	early 13th century

Author	Work	Latin translator and language of original of translation	Place and date of Latin translation
	Problemata	Bartholomew of Messina from Greek	Sicily *c.* 1260
	De Plantis or De Vegetabilibus (now attributed to Nicholas of Damascus 1st century BC)	Alfred of Sareshel from Arabic	Spain, probably before 1200
pseudo-Euclid	*Liber Euclidis des Ponderoso et Levi* (statics)	from Arabic	12th century
Galen (129–200 AD)	Various treatises	Burgundio of Pisa from Greek	*c.* 1185
	Various treatises	Gerard of Cremona and other from Arabic	Toledo 12th century
	Various treatises	William of Moerbeke from Greek	1277
	Anatomical treatises	from Greek	14th century
Ptolemy (2nd century AD)	*Almagest*	from Greek Gerard of Cremona from Arabic	Sicily *c.* 1160 Toledo 1175
	Optica	Eugenius of Palermo from Arabic	*c.* 1154
Alexander of Aphrodisias (*fl.* 193–217 AD)	Commentary on the *Meteorologica*	William of Moerbeke from Greek	13th century
	De Motu et Tempore	Gerard of Cremona from Arabic	Toledo 12th century
Simplicius (6th century AD)	Part of commentary on *De Cælo et Mundo*	Robert Grosseteste from Greek	13th century
	Commentary on *Physica*	from Greek	13th century
	Commentary on *De Cælo et Mundo*	William of Moerbeke from Greek	1271
Proclus (410–485 AD)	*Physica Elementa (De motu)*	from Greek	Sicily 12th century

England, who, finding his source in Pliny the Elder, described an Aristotelian Universe, modified in the light of the Ptolemaic system, i.e. having perigees and apogees, non-uniform apparent motions, and the like. In other words, Bede knew Ptolemy's astronomy, and the astronomical data known to Ptolemy, at least qualitatively. After Isidorus, this must be regarded as enormous progress!

Both traditions continued in the western world simultaneously. Some were attracted to the Aristotelian views of Bede, who was not only a transmitter of the old culture. He made considerable progress with respect to Pliny. Rhaban Maur (Mainz, 776–856), for example, was a follower of Bede. Others were followers of the Platonic heritage of Augustine and Isidorus. Strabo (born in Swabia), was faithful to the primitive views expressed by Isidorus.

This binary tradition continued, almost in a vacuum, till the great era of the translators. One speaks of the "Copernican revolution," but the revolution of the translators, and soon after, the revolution of the printing press, had a much broader impact, although my inclination would be not to call them "revolutions," but "necessary steps in an evolution."

The need for translation became essential at the time of an event which was of paramount importance in the history of the period. In 827, Michel, the Christian emperor of Constantinople, sent Louis, king of the Franks (one of the sons of Charlemagne) several books dealing with the Universe, as a gift. These books were certainly in the neo-platonic tradition. A little later, Charles the Bald, Louis's son, who reigned over the western part of the Empire (France), asked John Scot Erigen (801–886) to translate them. John Scot was possibly Scottish, as his name would indicate, but he may have been Irish. He was a very remarkable man, discerning and devoted. He used, when translating these Greek texts into Latin, a system very similar to that of Heraclides, which he found in the commentary on Plato's *Timaeus* written by Chalcidius. His followers worked along similar lines, adding the idea of a Creator, which gave to universality the faculty of "being." This universality is the set of "principal causes" of phenomena. To this Platonic-Christian foundation, these thinkers added the Aristotelian theory of the four elements. There is no doubt that these scholars had considerable knowledge. It is clear also that John Scot knew about the observations of Erathosthenes, and that the latter had measured the size of the Earth. It is strange, therefore, that John Scot equated the Earth's diameter with the distance from the Earth to the Moon. On what basis? We do not know really why Ptolemy's Universe had shrunk so much, but there is a certain sense that God, the Christian God, who is conceptualized on the scale of man, is only at ease with mankind in a small Universe, of human proportions. In any case, this was the first time that Christian scholars touched upon these questions. In Augustine, they were not mentioned; in Isidorus, they were considered absurd.

Scot, like Bede, mentioned the colors of planets, which he related to the nature of the fluid in which they are embedded. Cold makes them pale, while heat makes them bright and red. Scot did not seek systematically, as did Isidorus, to "save the Bible" at any cost. In essence, he argued. For instance, he did not accept the mosaic idea of supra-celestial waters (see above p. 43). He went so far as to say, "Jupiter, Mars, Venus and Mercury will ceaselessly turn around the Sun, as Plato teaches us in the Timaeus. They look bright when above the Sun, red when under ...",[9] And this long before the end of the 19th century!

Ideas were flowing. Their dissemination was rapid. Eminent people contributed to this growth, if not of knowledge, than at least of debate. One thinker of the period was Gerbert (930–1003). Under the name Sylvester II, he was the French pope in the year 1000 and was an influential presence in the millennial year when there was widespread fear regarding the fate of mankind. Among other contributions, he made some advances in the development of the astrolabe. His ideas were widely known, if not extremely original, as they were very much in tune with the scholarship of the Byzantine Macrobe (400–450) who proposed two different categories of men, the *antipodes* and the *antichnes* (Fig. 3.11) Another scientist whose work flourished during what is erroneously regarded as a "dark age" was Manegold (1060–1103) who worked in Alsace. He opposed the idea of different kinds of men.

Helperic (1050?–1080) was, on the contrary, a defender of classical Macrobian ideas, inherited from even pre-platonic thinking. Helperic was indeed a true astronomer. He engaged in a lucid discussion about the motion of the Sun, Moon, and planets, and he fought energetically against astrologers. Guillaume de Conches (11th century–1091) tried to reconcile Aristotelian physics with the thinking of Democritus. He believed, contrary to the opinion of John Scot Erigene, that there are "elements" associated with qualities that are only attributes, not essences. Guillaume de Conches knew of the eccentricity of the solar orbit, but he

---

[9] quoted by Duhem, ibid, III, p. 61 Translated by the author.

Fig. 3.11 *Antipodes and Antichnes.* In position 1, Europeans. In position 3, the antipodes (their feet are on top, their heads on bottom). In position 2, the antichnes (their front is in the back, their back in the front).

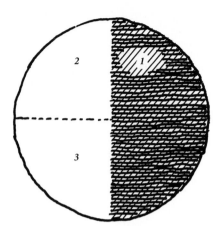

ignored the seasonal difference between north and south, which implies that he did not know about the inclination of the ecliptic. He believed that the solar perigee is in summer. In all other respects, he made a sound series of arguments for a heliocentric theory, as had Heraclides and John Scot Erigene. He even insisted upon the "attractive nature" of the Sun, which forces the planets to their retrograde motions. Many other scholars of the period thought along the same lines.

It even became fashionable in some circles. The Heraclidian theory (rarely referred to in elementary books about the history of science!) was indeed flourishing. Rabbi Abraham ben Ezra (1119–1175, in Toledo), a well-known kabbalist, discussed this matter at length, and concluded, in a manner consistent with talmudic thinking, that several descriptions may be equally "true." In this, he differed from other scientists of the Arabic-Jewish tradition, as he probably knew the commentary to the *Timaeus* written by Chalcidius and considered Heraclides' model quite plausible.

Barthelemy the English (1200–1275?) compiled an encyclopaedia which was largely reproduced and read, in Latin and in French, up to the 16th century. It was even reprinted in Frankfurt in 1601. This book, *De proprietarus rerum*, or, *About the owner of the things*, is a good example of the new fashion. Barthelemy had not read the *Almagest*, but he had certainly read Macrobe, and he adopted the Heraclidian system. Had Copernicus, or Osiander, read Barthelemy? ... Possibly !

In other words, a fluctuation between the heliocentric and the geocentric systems was ongoing during this period, in spite of the strong

Aristotelian flavor of many books and teachings of the time, in spite of Christian-Platonism, and for reasons linked more with fashion than with mere logic. The Arabs had the *Almagest;* the Christian world had Pliny and Macrobe, but they also had Augustine.

3.4.2
The Era of the Translators

Through the translations from Arabic, the ideas of Ptolemy entered into western philosophy more deeply. This was later accentuated by the easy dispersion of the translated works through the technology of printing.

We have already mentioned the tremendous importance of the translators. Many migrations, some of them more or less under constraint, led people to speak several languages. Translation became a necessary component of monastic activity. It also promoted the unification of Christianity by making Latin the universal language of Roman Catholicism. A college of interpreters was even created around 1140, in Toledo, by Archbishop Don Raimond. The structure of the work was quite interesting. It is of note that a similar organization is still used in some groups specializing in the history of science. A team was composed consisting of an Arab scholar with a wide scientific knowledge who translated Arabic works into Spanish (many of them being earlier translations of Greek treatises, lost by then) and a latinist who retranslated the Spanish into Latin. This was the common method of working in Toledo. Many translators, who learned their technique in Toledo, started to spread their knowledge throughout western Europe.

Table 3.2, reproduced after Crombie is an account of how, in the 11th and 12th centuries, this enormous task developed. Several great scholars brought the Arabic inheritance of Ptolemy to the western world. Gerard of Cremona, Adelard of Bath, Guillaume de Morbaecke, and several others were engaged in this work. Not only did they translate Ptolemaic works but also Aristotle. Previously, only his *Logica Vetus* had been translated – by Boethius in the 6th century. Now, also the *Physica,* the *De Coelo et Mundo,* etc. became accessible. Euclid, Apollonius and Simplicius' comments on Aristotle provided new generations with the mathematical, geometrical, and astronomical tools they were badly lacking.

Another dramatic change in the scholar's life of that period was the emergence and development of printing techniques. Whereas, the Chinese had used engraved wooden plates since about the 6th century, similar techniques first appeared in western Europe only during the 12th or 13th centuries. In 1147, the first "lettrines" (initials of texts, often beautifully drawn, and enriched by colored illuminations) appeared, engraved on wood. In 1289, one notes the appearance of the first wooden engraved plates. If we return now to the beginning of the12th century, we can see how the influence of the translators, stimulated by the distribution of written literature, led to the development of the influence of Aristotle.

Thierry de Chartres, in Paris (1100?–1155) returned to the Augustinian methodology of combining Aristotelian ideas with Biblical teaching regarding Genesis. Actually, he was quite daring, and attempted a purely physical description of the Universe and its evolution. According to his work, the initial fire (i.e. the external sphere of elements) must have, at some time, begun to rotate. Each rotation was a "day," producing various effects linked with the "two virtues of fire: splendour, heat." This was a statement, however incorrect, within the realm of physics, not metaphysics.

Gilbert de la Porrée (1070–1154, in Poitiers, central France), like Guillaume de Conches earlier, came back to the Aristotelian categories. In particular, he discussed the difference between location and the "located," or, to put it another way, he asked the question: ubi? (where?), which related the located to the location. The "supreme orb," ultimate surface of the world, is not a location, as something else must necessarily occupy the place beyond. This concept has no meaning. Even to ask the location of the supreme orb is meaningless. Therefore, what is the definition of the supreme orb? Such were the questions debated.

A very important work, although we know little about its author, was the *Tables of Marseilles*, published in 1140, in which the astronomical debate held in the Phocean town of Marseilles is described at length. This author was indeed well informed about astronomy and medicine. His tables make corrections in preceding tables, on the basis of errors discovered concerning the conjunctions of Mars and the Sun (and also the other planets). His work was strongly influenced by Ptolemaic thinking and he often quotes Ptolemy.

Gerard of Cremona translated the *Almagest*, in Toledo, and commented on it. He added observations of his own, tables and even a theory of planets. He was a very distinguished scholar whose translation was very much in use after his lifetime.

Alair de Lille (?–1203) wrote a long and broad Latin poem, something like a precursor to Dante's *Paradiso*, in which he indicated rather strange properties for the planets. It is very similar to the myth of Er, as described at length by Plato in the *Republic*. All the planets are made out of precious stones. Saturn is diamond, Jupiter agate, etc. They are ordered in the Platonic way, Mercury and Venus being "above" the Sun, contrary to the Ptolemaic description.

One century after the *Tables of Marseilles,* the *Tables of London* (1232) were published. In that particular year, we know that William, an English scholar, was teaching in Marseilles, but it is certain he was not the author of the tables. The new tables described the system of planets with the same desire for improved accuracy as the *Tables of Marseilles.* In particular, they insisted upon the importance of specifying the longitude of the town where the tables were made, and for which they are valid. The author gave the latitude and longitude of London (resp.: 51° North, and 57° West of Arim, the point of reference employed in the *Almagest*, and located in India, according to tradition). This was remarkable since the author did not know the *Almagest*. He also gave a great deal of attention to the calendar. The great disturbances introduced into the computation of the liturgical calendar by the use of the Julian year of 365.25 days needed correction. The *London Tables* were instrumental in the Gregorian reform, among many other contributions.

Joannes de Sacrobosco (John of Holywood, now Halifax) was an English scholar and a contemporary of the author of the *Tables of London*. A skilled and knowledgable mathematician and the author of some textbooks much in use, he was a reader of Ptolemy. He noted the precession of the equinoxes with the Ptolemaic value of 36 000 years for its period, instead of 26 000 years, closer to the true value, thus ignoring the work of Thabit ibn Qurra.

Again in England, Michael Scot, a translator of al-Bitruji, popularized the Eudoxian homocentric spheres and the work by Arabic scholars which followed from it.

The translation by Scot and others of Aristotle and Averroes reopened the debate widely. Two dilemmas appeared clearly, then, in the eyes of scholars in Latin Christendom : one in metaphysics, the other in astronomy.

In the realm of metaphysics, there was Aristotle on one side (and with him the Greeks, the Arabs, the Jews) with his Universe essentially without Creation. On the other side, there was Catholic doctrine, based on the Book of Genesis, which condemned some of the essential aspects of Aristotelianism such as the eternity of the astral world.

In astronomy, the same situation obtained. Should celestial bodies, following Aristotle, be permitted to be characterized by theoretical circular, uniform, and geocentric motions? Or, should observations be made paramount so that phenomena could be saved in the manner of Ptolemy? In other words, would it not be wise to be more Aristotelian than Aristotle?

Aristotle became, for entirely different reasons, the common enemy of both experimental science and the Catholic Church. However, the many scholars who had read Aristotle and adopted his views did not give up so quickly. Moreover, the borderlines of the doctrines were certainly not as well delineated for the people of the 12–13th centuries as they may appear to us. Their expression was often fuzzy, imprecise, ambiguous. Indeed, we may be oversimplifying the old systems to fit our Cartesian minds! In any case, the 13th and 14th centuries resounded with this triangular battle. One side of the triangle is represented by Aristotle and Ptolemy, saving the phenomena first. A second side stands for a Christian-Platonic conception, saving the Scriptures more than the phenomena. The final side represents the occasional Pythagorian temptation to save nothing but numbers and polyhedrons, based on an aesthetic view of the world.

3.4.3
The Priests and Monks of the Church. The Triangular Battle

At first, the priests of the Catholic Church were on the side of Platonic-Christian orthodoxy against Aristotle. This idea is sometimes difficult to see clearly, in the light of the Catholic Church's later embrace of Aristotle.

We can cite the ideas of a few outstanding priests of that time, which will show that, at the very least, ideas of the various philosophical currents permeated each other, through a process of intellectual percolation.

Guillaume d'Auvergne, for example, who was Bishop of Paris (1180–1248) wrote *De Universo*, a book along neo-platonic lines which was re-edited several times up to the end of the 17th century. Guillaume was a partisan of al-Bitruji's system. He ordered the planetary orbs more or less in the Aristotelian way. The eighth sphere is that of the fixed stars. The ninth one, mobile, stays outside the stars. It is the *Aplanon* (the crystal sphere of Aristotle). Above the nine mobile spheres, one actually finds the Aristotelian aether, between the nine mobile skies and the Empyreum, which is at perpetual rest. Was this very primitive description of the sky really consistent with the thought of Aristotle? Not really! Plato, and now Guillaume d'Auvergne with him, put a single "soul" in the world, that of God. Aristotle did not do anything of the sort. His followers put a soul in each of the celestial bodies rather than in the world as a whole. Guillaume was faithful to both Plato and the Bible in this respect.

The most interesting contribution of the period was that of Roger Bacon (1214–1294) – not to be confused with his namesake of the 16th century, the famous Francis Bacon. Bacon taught at Oxford and perhaps in Paris. It is seemingly there that he started to be involved in astronomical thinking. One of his earliest works was the *Questions*, a very dense book. Puzzled by the very existence of celestial motion, and motions, Bacon asked the questions: What is the cause of that? How is it powered? Who, or what is doing the powering? Is the soul of a metaphysical nature? of a mathematical nature? of a physical nature? In other words, should we follow Plato, Pythagoras, or Aristotle?

Bacon counted eight spheres, similar to the structure described in Fig. 3.1, above (Fig. 3.12). The eight spheres contain the stars and communicate their motion, diurnal, to the whole sky. Bacon, whose knowledge was still pre-Hipparchian, then disregarded the precession of the equinoxes, which was the reason why others had introduced a ninth sphere. To his other question, he replied, in a strict Aristotelian way: "As there are eight orbs, there are eight motions." In his first work, Bacon thus gave little recognition to Ptolemy, but he learned a great deal after this book, as we shall see.

At about the same time, Robert Grossesteste (Robert GreatHead?, 1175–1253) had the incredible courage to disregard the usual translations of Aristotle. Robert, instead, seems to have been a very original thinker, first preoccupied by practical questions. He was interested in the calendar, of course, but also in astrology. He recomputed the motion of the ninth sphere, in order to account for the precession of the equinoxes and was able to discuss the calendar problems with precision.

It is impossible to quote in detail all of the thinkers of the period. The schools of Marseilles and Montpellier were flourishing. Several English teachers and also Jewish scholars exiled from Spain taught at a high level. The schools of Austria and Italy were quite good. Campanus, a follower of Thabit ibn Qurra, engaged in scholarship there and was an excellent astronomer. These schools achieved an extensive knowledge of all as-

Fig. 3.12 *Bacon's model of a planetary orb.* Compare this figure with the post-ptolemaic model in Fig. 3.1, p. 130. Point C is the center of the world. The space D, centered in C', is the *deferent orb.* It has a width equal to either the diameter (in the case of the Sun), or the epicyclic sphere E, a full sphere, which bears the planet P. The volumes A and α are "solid." But at N, contact is possible between the deferent of planet P and the orb of an external planet.

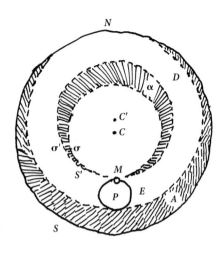

Similarly at M, contact is possible with the deferent orb of an internal planet. The existence of points M and N permits a continuity of "free space" which was not present in the old model of the Ptolemaic period.

pects of astronomy, theoretical and observational. Still, the debates on the theory of the sky developed primarily in the quiet peace of the cloisters, notably in the Dominican and Franciscan monasteries where various philosophical systems were under consideration. These debates did not involve the scholars who must already be considered professional astronomers and who were primarily preoccupied with the calendar and with observing the planets and stars.

Among the Dominicans, the most famous was perhaps Albert the Great (1206–1280). He was the Bishop of Ratisbonne. He may be noted for his detailed and learned discussion of al-Bitruji and the rejection of homocentric spheres, in spite of his admiration for the Arab scholars. Again, we see the influence of neo-platonism. He postulated one single motor for all the spheres. Albert was a defender of eccentric motions, but he did not attempt to explain them. He accepted Averroes' idea, according to which "epicycles or eccentric circles are working hypotheses that cannot be demonstrated." Nonetheless, he also wrote that "there is no reason to reject the system of epicycles and eccentrics, even with an

Aristotelian point of view,"[10] thereby departing from Averroes. As one can see, Aristotle slowly became the frame of reference, even if only for the purpose of criticizing his views.

The brightest pupil of Albert was St. Thomas Aquinus (1225–1274), who worked in Paris. He was a very original thinker, under whose banner many philosophers were and still are willing to fight, "Thomists" at that time, "neo-Thomists" now. I will paraphrase him for one prophetic sentence, which previews some of the modern cosmological dispute: "Do not confuse the problem of the origins" – i.e. a problem in physics – "with that of the Creation" – a problem in metaphysics.

St. Thomas, like Robert GreatHead before him, stressed here and elsewhere, the difficulty of reconciling the two methods of astronomy, Aristotelian physics based on principles that everyone could accept but which were insufficient to save the phenomena, and Ptolemaic geometry which was based on principles that were unacceptable or even unlikely from the point of view of Aristotelian physics.

Both the Dominicans and the Franciscans wavered. Among the latter, we should note Bonaventure (1221–1274), a lifelong opponent of St. Thomas. At about the same time, Roger Bacon, whose early work the *Questions* we have already considered, went to Oxford to study astronomy, primarily with Robert GreatHead, but also with Adam of Marsh, another English scholar. Bacon became, under the influence of Robert, very involved in speculations about the calendar. He had also learned by that time about the precession of the equinoxes, and this problem, closely linked with that of the calendar, puzzled him very much. The question of precession represented one of the key problems for determining the calendar, and, more seriously still, understanding the mechanism of the planetary system, not to mention its implications for the standing of Babylonian zodiacal astrology. Bacon, as an alternative to precession, which he interpreted after the manner of Hipparchus, introduced the idea of trepidation (Fig. 3.4, p. 144, above).

Knowing the complicated motions involved, he adopted Ptolemaic constructions. He did not, however, see the contradictions between Ptolemy and his fierce adversaries, Averroes and al-Bitruji. He, at least, made "saving the phenomena" his highest priority.

---

[10] quoted by Duhem, ibid, III, p. 334, translated by the author.

In still further publications (some of them unfortunately lost), Bacon showed himself to be an exacting physicist. In his *Opus Tertium*, probably his most important contribution to astronomy, he outlined again, in a very clear way, the debate between the Ptolemaic eccentric-epicycle system and the al-Bitruji system. He seemed more reluctant than previously to accept the Ptolemaic construction and his arguments tended to follow the Aristotelian course of Averroes. Still, instead of reproducing the work of earlier scholars, he introduced his own system of solid spheres as if Ptolemy's followers had not already done it (Fig. 3.12).

There was no doubt that Bacon's spheres saved the appearances. Still, he was not fully satisfied. He argued that the fact that the distances to the Earth of the Moon, the Sun, and the planets are not constant must be taken into account. He made no final conclusion, but a very good presentation of the argument. The main contribution of Bacon, although suspended between Ptolemy and al-Bitruji, was the primacy he gave to "experimental science." This is an Aristotelian idea, but also a very modern one, close to our positivism. It demanded that the astronomer carefully select the constellations used as references for the study of planetary motions. This is very good advice indeed. Bacon had a good understanding of the importance of experimental science, but he never really understood how to arrive at an experimental methodology, which left him in a permanent state of doubt.

Others hesitated to a somewhat lesser extent. Bernard de Verdun (end of the 13th century) offered a quantitative discussion, favorable to Ptolemy, and strongly argued against Bacon's hesitations, as well as Averroes' arguments. He represented the Ptolemaic eccentric-epicycle system, as did Ptolemy himself, as materially realized with solid and contiguous orbs.

In Paris, a very scholarly astronomy developed in the 13th and 14th centuries. One can really speak of a Paris school, the *nominalist* school of Albert of Saxony and Buridan. Mostly observers and calculators, the members of the school were hard workers, wise in many ways, and deeply impregnated with Bacon's doubts, on the one hand, and with the thinking of the Toledanian school which used the *Alfonsine Tables* on the other. Guillaume de Saint-Cloud (end of 13th century) seems to have perceived the main arguments of the sky observers of the past, whether they were Greek, Alexandrian or Arab. He did a great deal of observing himself. Leon the Jew (also quoted under the name of Rabbi Levi ben Gerson, 1288–1344) did not fall under the expulsion decree, by which Philip the

Good in 1306, forbade the Jews, previously expelled from Spain, to stay in France. Probably, this was because he lived in the papal state in Avignon. He had a strong influence on the Paris school, notably through his 1341 study of the Saturn-Mars-Jupiter conjunction. He was indeed a very astute observer, the inventor of the Jacob's staff, or *arbalestrillo* (Fig. 3.9, above). He also tabulated, with good accuracy, the apparent orbits of the Sun and the Moon. Typical of this cosmopolitan school was Themon (?–1371) who came from Münster in Westphalia and was a staunch defender of the Aristotelian idea of an incorruptible sky, going so far as to dispute the ideas of spots on the Moon and of its rotation. His books on Aristotle's *Meteorologica* and on the Sacrobosco sphere were well known.

During this period, it was under the influence of the Paris school that the need for a reform of the calendar became not only an obvious necessity, but a realistic possibility. Jean de Murs and Firmin de Belleval determined the exact length of the (tropical) year, i.e. 365.2422 days. They suggested the reform to Pope Clement VI in 1345. Had the papal authorities been clever, they would have accepted the reform which would have been indeed a rather simple one. The calendar suggested to them was a modification of the Julian calendar. One leap-year (normally one every 4 years) would have been omitted every 134 years, making the calendar wrong by less than a day over a period of 3028 years! It was not until the 16th century, however, that Pope Gregory XIII finally put a reform into operation, at a time when Europe was divided by religious wars and Protestantism was expanding. If the reform of the calendar had been accepted by Clement VI, not only would it have been a very good one, but it would have been accepted easily by a united religious world. We see the vestiges of the use of the Julian calendar even now with the observance of Easter in the Orthodox rite one week later than its observance in the Roman church.

The Paris school also contained physicists and philosophers. Let us mention only two representatives of this school. Jean Buridan (1295–1370?) was perhaps, but not certainly, the same who was later mentioned by François Villon (1431–1465) in his *Testament* ("... Semblablement, où est la Royne / qui commanda que Buridan / fût jeté en un sac en Seine? ...").[11] He was more likely also the one who allegedly intro-

---

[11]   Villon F., in: *Le Testament, Ballade des danses du temps jadis*, according modern sources, for example, Georges Pompidou, *Anthologie de la Poésie Francaise*, 1961, Hachette, p. 59.

duced the famous philosophical argument of Buridan's donkey and he was certainly also the Rector of the University of Paris. Buridan noted that some apparent motions of the sky are so slow (such as the precession of the equinoxes) that one cannot apprehend them unless one knows all the history of astronomy, all the records of previous observations. Again, as with Bacon, (but Buridan was not as involved in physical science as Bacon) he vacillated between Ptolemy and al-Bitruji. Clearly, he was against the use of an epicycle system for the Moon. If this system were true, the "man in the Moon" would appear to turn, when that is not the case. We now know that this is a specious argument. There is only a "libration" of the Moon. We see more than 50% of its real surface, actually 59%, because, as seen from the Earth, the Moon oscillates slightly due to the non-centered orbit of the Moon around the Earth. Moreover, there is no relation, a priori, between the duration of a revolution and that of the rotation. Buridan, not a physicist, was more of a mathematician. A devoted scholar and teacher, he was quite familiar with Apollonius' treatise on conics and he redemonstrated, for the benefit of his students, the equivalence of the geometrical descriptions, as Hipparchus had done earlier

Another remarkable scholar was Nicolas Oresme (1320?–1382), a student in Paris and finally bishop in Lisieux (Normandy). He was a subtle theologian, but also an excellent mathematician, protected by the King of France, Charles the Fifth. He translated Aristotle into French and wrote commentaries on him. He wrote, notably, "The Earth is moving; hence I do not agree with Aristotle."[12] He probably knew Heraclides' theory. He even stated, explicitly, regarding the rotation of the Earth, "I therefore conclude that one could not by any conceivable experiment show that the sky is animated by a diurnal motion, and that the Earth is not so moved,"[13] thus setting the stage for Galilean relativity. In essence, however, Oresme remained Ptolemaic, believing in the representation of "solid spheres." As with Buridan earlier, he defended the idea of the ninth sphere as a way of explaining the precession of equinoxes.

There was progress in England and Paris but also in Italy, and Germany.

---

[12] Nicolas Oresme, *Le Livre du Ciel et du Monde*, Bibl. Nation. Fonds Francaise, MS No. 1082, translated by Albert D. Menut, Madison, U. of Wisconsin Press, 1968.
[13] ibid.

Dante Alighieri, the great poet (end of 13th–14th century) discussed Ptolemy in *Il Convivio* (1297–1314). In *Paradiso,* however, later in life, he shifted to the system of Thabit ibn Qurra. Strangely enough, many astronomers in Italy, notably in the great center of Padua, did not really enter the battle. It is only beginning in the 15th century that we find a trace of the debate. Only Pietro d'Albano seemed to be interested, but by the time he was a student in Paris, the struggle was ended and Ptolemy the provisional winner, due primarily to the strong influence of Oresme.

The astronomical discussion thus went back and forth among the various systems we have discussed. At the same time, the metaphysical debate was ongoing everywhere, storming around Aristotle and, more or less, obscuring the astronomical issues for many commentators. Although like Thomas Aquinus I am determined to avoid intermingling the two fields, one must be aware that, in fact, the interaction between them was strong. Mention must at least be made of William of Occam, an English Franciscan who taught in Paris (1300?–1349?). Occam was an important theologian whose philosophical influence is still noticeable and explicit at the end of the 20th century.

Occam was a firm defender of the principle that one should always stick to a minimum number of hypotheses. We must shave all models of the superfluous. *Occam's razor* is therefore identified with the principle of simplicity, that can be expressed now as follows: "It is unnecessary to accomplish by many ways what a lesser number of ways is sufficient to produce. If, without a given hypothesis, we can save all that one pretends to save with it, then this hypothesis must be rejected." This is a good physical principle, but it sometimes leads to excesses. What is more simple indeed than the one simple hypothesis: "God made it, as it is, – and that is enough!"? Still, in the realm of physics, as opposed to metaphysics, Occam's principle defended a methodology which spoke to Bacon's preoccupations. It is what we would now call a positivist attitude, similar in some ways to Buridan's.

We must at this point underscore the remarkable work of the scholars of this period. We have mentioned, so far, primarily the astronomical aspects of their work, their references to the universal organization of astronomical bodies. Their discussions (in particular in the Paris school of the14th century) also dealt with several other aspects of Aristotelian physics, then strongly shaken by astronomical observations on the one hand and by Christian faith on the other.

The notion of infinity was subject to a lengthy discussion. Is the "infinitely large" a conceivable idea? Again Buridan, Oresme and others were involved. Can we define the "location," separately from the "located?" Ernst Mach and Einstein echo this question in modern times. Duns Scot, William of Occam, Buridan, and Oresme took up this issue. Is motion a succession of realities, or the evolution of a permanent reality? Again Duns Scot, Occam, and Buridan were also taking this subject into consideration. We may even note in the work of F. de la Marche and Nicolas Bonet the appearance of the analogy of time and location (relativity!). Another interesting problem was the question of the "latitude of the forms," i.e. the meaning of width vs length of an object. Nicolas Oresme made an essential contribution to the mechanical debate. The inventor of analytical geometry, he constructed graphs, studied motion, and established the analytical representation of motions in which the velocity of moving bodies varies in a uniform way. There was indeed a total change in the points of view. Before Oresme, one could conceive of motion as a succession of abrupt small steps accomplished at a constant velocity, which had the property of being subject to change abruptly between each step. Instead, Oresme introduced, largely with graphs, the notion of continuous motion, and, implicitly, of acceleration (Fig. 3.13).

Oresme's views set the stage for Galilean and Newtonian mechanics.

Another much debated question, was that of the vacuum. Was it possible? Was it impossible? In 1277, the Aristotelian position on the impos-

Fig. 3.13 *The meaning of Oresme's geometrical constructions.* In this figure, a motion represented by successive steps, as small as conceivable of a given velocity can be, according to Oresme, replaced by a continuous curve, on the dotted line. Now, the movement has a constant acceleration, the slope of the dotted line. The words longitude and latitude might be considered equivalent to ordinate and abscissa.

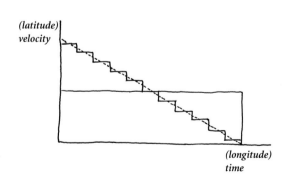

sibility of infinity and of a vacuum was solemnly condemned by Etienne Tempier, Bishop of Paris. This stern condemnation had an enormous impact. This position was defended by Duns Scot (arguing about bodies falling in a vacuum), Buridan, Oresme, Occam, and several others. It was opposed, on the contrary, by Bacon, following the Arab tradition. We are on the way to the experiments (17th century) of Torricelli and Pascal, which will still not really solve the problem. We are also on the way to Dirac and Einstein's modern speculations about the ether.

The motion of projectiles had been studied by al-Bitruji, but was also taken up in the dynamic theories of Occam, Buridan and Oresme. The theory of *impetus* was widely held to be true. To move a body, some initial effort, the impetus, is necessary. John Philopon had already given this consideration but he was ignored by the Aristotelians. We would probably now call it *momentum*. It is more characteristic of the initial velocity than the force at the origin of the motion. Aristotle (and later Bacon) maintained that the mobility is included, so to speak, in the moving body, as substantially intrinsic to it. Referring now to the classical equation of dynamics, as it is known today by schoolchildren: $F = m\gamma$. Here, m typically represents the "intrinsic" property of the body, its inertia, while F represents the origin of the motion. The m is an Aristotelian notion, the F perhaps an Occamian one. The fall (accelerated) of heavy bodies was at the heart of this discussion and was studied by St. Thomas, Bacon, Robert, Buridan, Bonet, and Oresme. The impetus could explain the fall. What could be said about the role of the air put in motion by the very motion of the falling bodies? What about the "station" between ascent and descent of a projectile? What about inertia, complementary, as implied by Buridan, to impetus? One can see in this discussion the painful birth of modern mechanics. It will reach maturity only at the time of Galileo, Descartes, and Newton.

This discussion about impetus had repercussions, of course, in the field of heavenly bodies. Buridan attributed the motion of celestial orbs to a combination of impetus and inertia. A very modern idea indeed!

Other problems such as the theory of tides were dealt with in the discussion (see Bacon, Robert GreatHead, and Duns Scot).

The theory of the plurality of inhabited worlds was considered. Condemned by the Catholic tradition (St. Thomas), it was nevertheless strongly debated during the 13th century.

Other subjects under discussion were the equilibrium between waters and emerged lands, geology and the motions of the Earth's crust (Buridan) and again the idea of relative motion and the question of the rotation of the Earth around its axis, as opposed to the rotation of the sky around the Earth. We should note that this problem was approached differently from the manner in which it was studied by Heraclides, Aristarchus *et al.* Their system implied the revolution of the Earth, but not necessarily its rotation, although the latter was more or less explicitly thought to exist. Actually, this problem was described much more clearly by Oresme than it was later by Copernicus. According to Oresme, the air and sea move together. This was an assault on the classical argument of the arrow which asserted that the Earth did not move at all.

At the margins of astronomy, the fight over astrology was active. Although Augustine and the earlier Christians rejected astrology as a violation of the free will granted to human beings by God, western Christianity later developed a more accepting attitude toward the possibility that there might be some truth in it. In 1277, astrology was indeed condemned, along with Aristotelianism. Albert the Great, Bacon, and especially medical scholars such as Mondeville, on the other hand, had their doubts. The correspondence between the macrocosm (the sky) and the microcosm (the human body) was a very popular idea, often included in the beautiful illustrations of the religious and medical books of the period. An excellent example of this is the beautiful microcosm-macrocosm painting in the *Très Riches Heures du Duc de Berry*, painted by the brothers of Limbourg (1411–1416). Alchemy was based on the four elements theory of Plato and Aristotle. All these conceptualizations were based on the perception of the deep unity of the Universe, of God's world. In spite of this "unscientific" approach, these thinkers contributed extensively to the progress of knowledge. The condemnations of 1277 attacked only "illicit" astrology, which implied an action of the celestial bodies upon the human soul, a sort of negation of free will. The prognostication of the "natural future," as it was called by Pierre Abélard was quite acceptable. Tempier also condemned the eternal return, again on the basis of the theological status of the free will of human beings.

Nicolas Oresme was firmly entrenched in the anti-astrology camp. Noting the incommensurability of celestial circulations (remember the difficulty of computing the length of the Great Year), Oresme concluded that, as one can only judge the future by the past, and as nothing has ever happened before that is identical to anything that will happen tomorrow, there can be no possibility of eternal return. This was more or less intuitive. Oresme did not demonstrate incommensurability by that reasoning, as any data can be approximated by a number with a defined number N of figures. Yet, it seemed to him a probable truth. This left the door open for the astrologers to object, on the basis of the inaccuracy of the measurements.

The development of the Oxford and Paris schools should quite naturally have led to the Copernican system and it should have done so much more quickly than it actually did. We must remember, however, that the Hundred Years War was violently shaking and weakening England and France, precisely at the end of the 14th century. The Black Death (1347), the wars (ca 1340–1450), the War of the Roses (1450–1500) in England, and the religious wars in France (1550–1590), all served as distractions to the progress of philosophical thinking. Monks and scholars became soldiers. The center of gravity moved again to the Germano-Italic world in the 15th century. New universities developed in the Austrian Empire, in Prague, in Heidelberg, in Vienna, and in Leipzig and with them a renewal of the philosophical battles.

On one side, some philosophers and theologians, like Nicola da Cusa (Cusanus, 1401–1464) went very far. God has the attribute of infinity and the Universe is "unbound," unlimited, but these are quite different notions. For da Cusa, as the Universe is unbound, it could have no center whatever. Hence, the Earth could not be at the center of the Universe. Later, in the 17th century, Blaise Pascal arrived at a similar idea in his reflections concerning the situation of Man between two infinities, and deduced from his thinking that the Universe is like a sphere of which the circumference is nowhere and the center is everywhere. What then is the Earth the center of? The question had already been asked a millennium before by Plutarch! Again echoing Oresme, da Cusa showed that no absolute motion could be detected. The question then arose that if the Earth was not at the center, why not accept the plurality of inhabited worlds? of "lunatics" or some solar beings? Nicola, a Cardinal, continued to speculate and play games with the sacred books.

3.4.4
The Progress of Observations

While thought and reflection were battered by wars and catastrophes, there was continuous progress in the area of accumulated observations by skilled astronomers, continuing in the Arabic-Jewish tradition, transmitted to the northern part of Europe through Levi ben Gerson and

Fig. 3.14a–c *A few typical instruments used in the western world:*
(a) a vertical quatrant. For (b) and (c) see next pages.

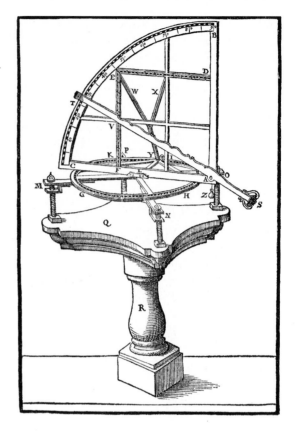

others. A few typical names out of this long list of able observers should be mentioned. Guillaume de Saint-Cloud, for instance, used astrolabes and evolved staffs (Fig. 3.14, 3.15).

Interestingly, Geoffrey Chaucer (1328–1400), better known for the *Canterbury Tales*, published a very carefully documented treatise on the astrolabe (Fig. 3.14c).

Paolo Toscanelli (1397–1482), an adviser to Columbus, noted many cometary observations. Georg Peurbach (1423–1461) defended the Ptolemaic theory, complete with solid spheres. Johann Müller (Regiomontanus, 1436–1476) worked with Peurbach initially. He matured into a very productive observer of eclipses, comets, the positions of planets, the

Fig. 3.14b An armillary sphere. (Bibliothèque nationale de France, Paris)

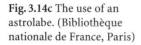

Fig. 3.14c The use of an astrolabe. (Bibliothèque nationale de France, Paris)

Moon, and the Sun. Müller corrected the *Alfonsine Tables*. He developed new instruments, such as the three-staff instrument (Fig. 3.16).

The first movable type, in melted metal, appeared in Limoges in 1381, then in Antwerp in 1417 and Avignon in 1444. Gutenberg, in Mainz, in the period 1447–1455, is world-reknowned for having developed lead type and the method of perfect alignment of letters, notably in his printed Bible. Copernicus' *De Revolutionibus* (1543) was not one of the first books well printed and widely distributed, but there is little doubt that its fame and reputation benefited enormously from the newly developed technology.

At the same time, the known world was becoming much larger. Between the end of the 15th century and the beginning of the 16th century, 100 years at the most, the concept of unlimited space invaded everyday life. Diaz discovered the Cape of Good Hope in 1486. In 1492, Columbus

Fig. 3.15 Various uses of the Jacob's staff. (Observatoire de Paris)

discovered the New World. No human frontier was out of reach any longer. As soon as the Atlantic was crossed, the newly discovered American continent was explored, exploited, sacked, robbed. The knot was tied and all the world's cultures, material and intellectual, were enmeshed in one another. The spice trade created affluence; gold circulated. Native American peoples were soon conquered, enslaved and converted. Their Very Catholic Majesties reigned now, forever it seemed, in a world which was limitless. Everything revolved around the bifurcated empire, from Vienna to Madrid, dividing the Protestant Reformation, in the north, from the anxious Holy See, which, recovering after the Council of Trent, launched the Counter-Reformation. The evolution of developments on the world stage corresponds to the evolution of points of view in the philosophy of science. In astronomy, we progress from Copernicus to Newton.

Columbus himself made the hypothesis that the Earth is divided into two equal parts: Ocean and Earth. This was, of course, not quite in agreement with the estimated proportions and Erathosthenes' value for the Earth's radius. But never mind! It was a new beginning. Apianus, in 1520 and Mercator, in 1569, made the first reliable maps of large areas,

Fig. 3.16 *The three-staff in-strument of Regiomonta-nus.* The observer points at the star through the alignment of two holes on the first staff (A). The top end of this staff moves around its axis (on its top), while the bottom end is kept on the second staff (B). The motion is the re-sult of the action of the second staff (B), manipu-lated by the observer. The whole system turns around the third staff (C) which is vertical. Proper gradation on a fixed base provides the azimuth while proper gradation on the second staff (B) pro-vides the angle of the star's direction with re-spect to the horizontal plane.

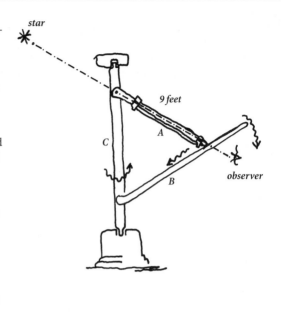

including the Atlantic Ocean and its shores, and they greatly improved the methodology of cartography. The need, for purposes of safe naviga-tion, of a precise determination of longitudes, soon became a strong sti-mulus for the development of astronomy.

Regiomontanus had a printing press of his own and sent a regular cir-cular letter to all scholars explaining the progress of his knowledge. He was very keen on re-publishing old classical books such as the treatises of Ptolemy (the *Almagest*), of Theon, of Euclid, Archimedes, and Apollo-nius, and made them widely available (including to Copernicus!). He also published the writings of Peurbach and his own works, of course. In 1475, Johannes Müller was consulted by the pope about the reform of the calendar. Unfortunately, Müller died too soon to achieve his goals, in particular his own translation of Ptolemy, a *"new translation in Latin."* He had many students. Bernhard Walther also made many observations, later used by Tycho, as did his pupil, Johann Schöner. Their data were of a better quality than those of Hipparchus and the Arabs. It is not long at this point until the appearance of Copernicus, in a widening world, where printing, translation, navigation, and communications had made the debates more public, and had spread knowledge throughout the entire scholarly realm, establishing the fertile ground for a still better astronomy.

CHAPTER 4

The Period of the Renaissance

4.1
From 1450 to 1600: An Overview

There is a tendency now, particularly in the light of Kuhn's ideas about "scientific revolutions," to reduce this period to the so-called *Copernican revolution*, thereby relegating the discoveries of Tycho, Kepler, Galileo and other astronomers to the status of an appendix of those of Copernicus. At the very least, many scholars have limited their consideration of the progress of science between the triumph of Aristotelianism and the Newtonian enlightment to the achievements of these four men. As we have previously noted, however, this scheme is oversimplified. Indeed, a great movement of ideas was taking place, just as during the centuries prior and the centuries yet to come (Table 4.1). Our "four musketeers" were at one time young students. Later, they reached maturity and became teachers of a new generation of students. This era was characterized by a great continuity and a remarkable spirit of intellectual debate, quite comparable to the debates of our times, taking into account the slowness of travel and the primitive level communication systems. Patient and determined men exchanged much information all across Europe.

The death, in 1464, of Nicola da Cusa whose ideas we have previously considered, provides a convenient transition from medieval to Renaissance science (Table 4.1).

Let us begin with Copernicus himself, born in 1473, and destined to become a very learned scholar in many fields including Greek and economics. Copernicus was a student of Albertus Brudzewo (or Brudzewski, 1445–1495), and of Domenico Maria Novarra (1454–1504). Through them, he knew of the old masters, German and Greek, including his near contemporaries Georg Peurbach (1423–1461) and Johann Müller (Regiomontanus, 1436–1476), both excellent observers, and in many ways in-

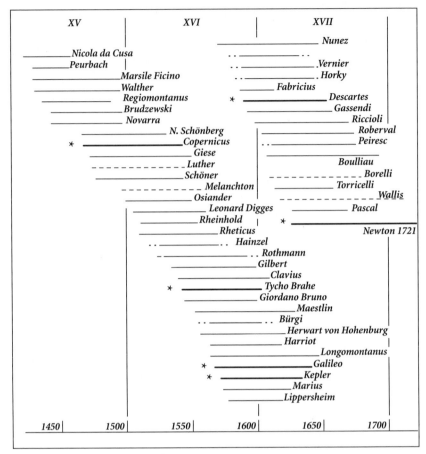

Table 4.1 *The "great" period of the Renaissance*

novators. Copernicus' friends and students, such as Tiedemann Giese (1480–1550) and above all G. Joachim Rheticus (1514–1576), played a large role through their discussions and correspondence with Copernicus. He was also influenced by Johann Schöner, a friend and mentor of Rheticus (1477–1544) and Andreas Osiander (1498–1552) who is often said to have written the preface to *De Revolutionibus Orbium Celestium* (hereafter abbreviated *dR*). Among the ardent critics of Copernicus were, for theological reasons, the reformers Martin Luther (1483–1546) and Philipp Schwartz (better known under the name of Melanchton, from the Greek translation of his German name). At the same time, Bernard

Walther (1430–1504) was a good astronomer and observer and Erasmus Rheinhold computed the (Copernican) *Prutenic Tables,* which superseded the *Alfonsine Tables* (1511–1553). Much later, Clavius (1537–1612) was a firm defender of Copernicus' skills, if not of his ideas.

Later, but not much later, Tycho Brahe (1546–1601), himself strongly influenced by Paul and Johann Hainzel (at their apex in 1569), hardly knew Copernicus' works, and in any case, disagreed with Copernicus' concepts. Others were implementing Tycho's ideas. There were his paid assistants (the rich man had a large staff) such as C. Rothmann, a Copernican in Cassel who was a very skilled observer and Joost Bürgi, a Swiss scientist, who was very good with mechanical devices and clocks. Another good observer, Longomontanus (Christian Zevern, 1562–1647) was also observing with Tycho. Finally, Tycho attracted the great Kepler for a while. Around 1625, the instrumentation was further improved along the lines initiated by Tycho by many other scientists such as Pierre Vernier in France and Pedro Nunez in Portugal, who improved the measuring devices.

During the same period, Michael Maestlin (1550–1631) played an essential role, as did Harwart Von Hohenburg (1553–1622), through their lengthy correspondence with Tycho and Kepler.

Johannes Kepler (1571–1630) himself, although a sort of lone wolf in his social behavior, worked extensively with Tycho, who inspired his most productive research in Prague. That was a most remarkable combination of geniuses. Kepler also had students – Martin Horky for example – and he corresponded with Galileo Galilei (1564–1642).

There were many others who made real and significant contributions of whom we shall hardly speak! Everyone knows the name of Galileo, but he was not the only one who was involved in the scientific debates of the time. We must also note Ismaël Boulliau (1605–1694) and Gian Alfonso Borelli (1608–1679), remarkable in the field of dynamics, and Giordano Bruno (1548–1600), who believed in the multiplicity of inhabited worlds and was burned for his heretical views. Of great importance was René Descartes (1590–1650), of whom we shall speak further. Others who made contributions were William Gilbert (1540–1603), who was one of the first to understand magnetism; Pierre de Gassend (Gassendi, 1592–1655), a philosopher and observer; Riccioli (1598–1671), Roberval (1602–1675), C. Scheiner, s.j. (1575–1650), the active observer of the solar surface; Nicholas Schoenberg (1472–1537), Harriot (1560–1620), Torricelli, (1608–1647), John Wallis (1616–1703), and even Isaac Newton

(1643–1727) who was not born much after Galileo's death (1564–1642). Leonard Digges (1520–1559) invented some optical devices, as did Hans Lippershey (1570–1619) who is also credited with having invented the telescope some time before Galileo. Thomas Digges (1546?–1595), son of Leonard, is said to have used a telescope for terrestrial observation. Simon Marius (1570–1624) also observed sunspots, as did Galileo, Scheiner, and John Fabricius (1587–1615).

For the sake of simplicity in teaching and for the fact that we deliberately limit ourselves to a discussion of ideas leaving out for the most part methodology and the detailed results of observations, we will speak only about the very few people about whom everyone else writes. This will enable us to keep our perspective in view, tracing the development of the main new phenomena discovered, those relevant to man's views about the world.

We must never forget, however, that this was a very active intellectual period, like many others, but perhaps richer. Scientists travelled throughout Europe and corresponded extensively. There was, in effect, a single and coherent European school, which contributed enormously to improved observations of the sky and heavenly bodies, allowing ideas to be modified in depth. In depth, but not by a drastic revolutionary upheaval instigated by the genius of one single man, a sort of lone lighthouse above the sea. We have a distorted picture of this history. Lighthouses and a few rocks emerge, but the bottom of the sea is much deeper, full of noticeable rocks and surrounded by splendid beaches. Whenever the reader reads simplified statements about this period, he is, therefore, advised to be very careful. Remember: "It ain't necessarily so!"

4.2
Copernicus and the Determination of Planetary Distances

Nicholas Copernicus (1473–1543) was a learned scholar, an influential priest, a skillful administrator and manager, an active observer and calculator, and an artist as well. This man deserves our great admiration, certainly not the irony directed at him by Arthur Koestler who called him the "Timid Canon" in his book *The Sleepwalkers*.

The following is a short abstract of his well-known biography, concentrating primarily on Copernicus' astronomical ideas.

He studied at the University of Cracow (at that time the largest in Central Europe). He concentrated on astronomy (from 1491 to 1495), working with professor Albert Brudzewski, author of a commentary on the Peurbach planetary theory, but also studied dialectics and philosophy. Then, he attended the University in Bologna, Italy (from 1497 to 1500) where he was a student of Domenico Maria Novarra. He learned Greek and was in contact with the neo-platonic and Pythagorian schools. From 1501 to 1503, Copernicus studied medicine and canon law in Padua and Ferrara.

Back in Frauenburg, in the capacity of a canon, Copernicus, although engaged in administration, medicine, and monetary reform, found enough time to think, to write, to observe, and little by little, to construct his monumental treatise *De Revolutionibus Orbium Celestium*.

Copernicus observed the sky more than it is usually noted. At least 63 detailed observations are known. The first one, on March 9th, 1497, was an observation of the occultation of Aldebaran by the Moon. In 1535, he composed an *Almanac*, similar to astronomical tables, based on his heliocentric system.

The most noteworthy aspects of the man, as mentioned by his faithful friend and pupil Rheticus in his *Narratio Prima*, were a very systematic mind and the Platonic-Aristotelian desire to use all available observations, from all periods, together with his own, so that phenomena could be saved in a complete manner.

4.2.1
The Heliocentric System as a Working Hypothesis

The central idea in the Copernican system is that putting the Sun at the center of the planetary system is sufficient to explain all the appearances. Saving the appearances was a Platonic-Aristotelian necessity. The heliocentric system had a priori the merit of simplicity, an almost Pythagorian argument, at least an Occamist one.

Strangely enough, however, Copernicus' system was not more accurate than Ptolemy's. From the point of view of mathematics or geometry, there is not much new in Copernicus' system. The only innovation was the use of the sine, an Arab construct, instead of the cord, used by the Greeks, to measure angles (Fig. 4.1).

Nor was the innovation of Copernicus in the realm of physics. Putting the Sun at the center of the system was an idea known since Heraclides and Aristarchus, and, as we have seen, never completely forgotten. The perception that there was a need to locate something at the center, in-

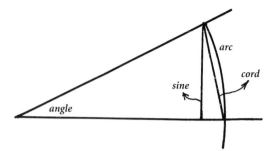

Fig. 4.1 *The sine of an angle.* If the radius of a circle is taken as unity, the length of the arc sustained by an angle α is equal to α. The *cord* is a function of this arc, as is the sine (or sin α), both defined in the figure.

stead of nothing, had driven some thinkers such as Averroes and Maimonides to use Aristotle against Ptolemy. Copernicus' system can be seen as a successful attempt to put some thing at the center (the Sun!), while at the same time saving the phenomena as successfully as Ptolemy. The achievement of Copernicus was, therefore, his successful reconciliation of Aristotle and Ptolemy. The Aristotelian paradigm which posited some thing, a material body, at the center of the orbs, and the Ptolemaic paradigms pertaining to circular motions around points that are not necessarily material bodies and to combinations of several circular motions, are not only preserved in Copernicus, but improved, and somewhat refined.

In other words, Copernicus followed the tracks of Ptolemy, and those of al-Bitruji and Regiomontanus in suggesting an alternative to the system of solid spheres embedding the orbs of the planets with their eccentric and/or epicyclic systems. It is likely that the *dR* was written in 1530–31. It was abstracted in the *Commentariolus,* not dated (before 1514 in any case) and popularized by Rheticus in his *Narratio Prima* (1540). The *dR* was not published until 1543, the year of Copernicus' death.

The heliocentric hypothesis was used only as a guide. Copernicus' actual system was indeed much more complicated, as we shall see.

The *Commentariolus* presents the main points of Copernicus' work. They are essentially physical and they are presented as "hypotheses," – we might say postulates – of the Copernican theory. They can be expressed as follows:

(i)   No one unique center is needed for all planetary orbs or spheres.

(ii)   The center of the Earth is not the center of the World, but only the center of weights and of lunar orbit.

(iii) The center of the World is located near the Sun, as all orbs are surrounding it.

(iv)   The ratio of the distance Earth-Sun to the distance of fixed stars is smaller than the ratio of the Earth's radius to the Sun-Earth distance. (Comment: There are no measurable stellar parallaxes.)

(v)   Every apparent motion seemingly pertaining to the starry sphere comes from the Earth. Hence, we must assume the rotation of the Earth around its axis in one day, whenever stars are "fixed." (Comment: The diurnal motion is a consequence of the rotation of the Earth.)

(vi)   Motions that seem to be solar are really due to the Earth's motion, as the Earth turns around the Sun like any other planet. The Earth, therefore, has several motions.

(vii) Retrograde and direct motions observed in the case of planets are explained by the motions of the Earth, which, alone, explains a considerable number of oddities observable in the sky.

Let us note that the reason to put the Sun at the center is not expressed, and actually not put in concrete form in the system. The Earth turns around a point different from the Sun (because of the eccentricity of the Sun's apparent motion) and no one dared to give the Earth an epicycle (as one did to the Sun!), which was the next logical step. The *Commentariolus* contains no computations. It gives only their results. There are seven circles for Mercury, five for Venus, three for the Earth, four for the Moon, turning around the Earth, and five each for Mars, Jupiter and Saturn, "hence 34 circles are sufficient to describe the total structure of the world, and the dance of the planets." They are not only sufficient, according to this system, they are, in fact, necessary. The model was, therefore, not simpler than the constructions of Eudoxus, Callippus, Aristotle, and Ptolemy. This complexity was a priori so shocking that the *dR* described it differently from the *Commentariolus* and it will be necessary for us to expand our discussion of Copernicus' system.

In 1524, Copernicus wrote an important text, the *Letter against Werner*. Werner, a friend of the painter Albrecht Dürer, published, in 1522, a treatise about the motion of the eighth sphere. This letter is a step towards the refinement of Copernicus' ideas.

Fig. 4.2 *The annual paral-*
lax of a nearby star. The
distance Earth-Sun is, by
definition, the *astronomi-*
cal unit of distance
(1 A.U.). From a star at a
distance *d* from the Earth
or the Sun, the astronomical unit is seen at the angle
ϖ, called the *annual par-*
allax of this star. If the
Earth is assumed to be
fixed, the star describes in
the sky an orbit exactly
identical to that described
in reality by the Earth
around the Sun. The unit
of distance often used in
astronomy is the *parsec*
(the distance of a star of
which the annual parallax
is 1 arcsec). The light-year,
i.e. the distance travelled
by light in 1 year is also
used. See appendix 1 for
correspondences.

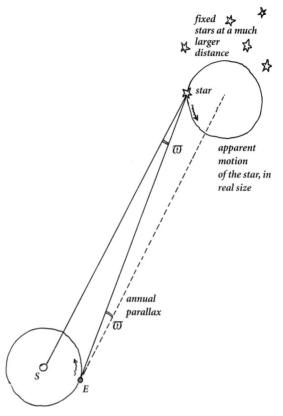

The next important step was Rheticus' *Narratio Prima,* in which Rheticus, a fervent admirer of Copernicus, constructed a very lucid and clear
exposition of Copernicus' ideas. This was actually the first printed version of the Copernican viewpoint. We must also credit Giese with whom
Copernicus and Rheticus had regular and stimulating contact. Rheticus
situated Copernicus in a comfortable place, in the center of the triangular discussion taking place among the Pythagorians, the Platonists, and
the Aristotelians, possibly somewhat closer to the last two. The new astronomy was faithful to the Aristotelian ideas of circular and uniform
motions of celestial bodies. In a way, Copernicus' system avoided the
need to introduce the *equans* (see above, p. 105 and 106). In addition, it
avoided the problem of planetary irregular motions, by assuming, – hypothesis iv – that even the nearby stars are indeed very far (Fig. 4.2).

At the same time, the geocentric-religious Platonic idea is also preserved. "There is something divine in the fact that the safe understanding of celestial affairs must depend upon the regular and uniform motion of only the Earth." Rheticus was also aware of a Pythagorian influence. By putting the Sun at the center of the system, Copernicus returned to a clearly Pythagorian idea, to wit, that the Sun, because of its brightness, because of its "dignity," is at the center ("below" the rest, in the old phraseology), like the fire in the Philolaus system. In the mediaeval world, the Earth (and the sublunary world) is at the center (the lowest position) because the Earth is imperfect. Perfection is above. Hell is inside the Earth; paradise beyond the stars.

It is clear that Copernicus reintroduced heliocentrism in a way that was not contrary to the teachings of Pythagoras, Plato, and Aristotle, the old masters of antiquity.

While we are still considering Rheticus' work, which indeed introduced Copernicus to learned society more than his own book, the *Commentariolus*, there are strange considerations in it, that may seem quite foreign to us. For example, Rheticus extrapolated astrological consequences from the Copernican description, notably by comparing some instances of the solar orbit with historical events such as the rise of the Roman Empire, of Islam, and later the End of the World as predicted by the prophet Elijah. The year was 1540. At about the same time, the *Centuries* of Nostradamus were published; poison was lavishly distributed by magicians and necromancers; popes and kings were guided by astrologers. So perhaps we must forgive Rheticus his delusions!

4.2.2
The Distance Scale in the Planetary System

At this point a pedagogical pause seems necessary. We shall assume here, for the sake of simplicity, that all planetary orbits are circles centered exactly on the Sun, and have uniform motions on these orbits. One can, to begin with, consider the Ptolemaic and Copernican systems in a primitive way, as Kepler did (Figs. 4.5 and 4.6, p. 199 and 200). In Fig. 4.3, we represent the Sun, the Earth, and Mars, together with distant stars. The orbits of both Mars and the Earth around the Sun are circular; and the motion of both Earth and Mars on their orbits are uniform.

Therefore, as shown by a geometrical construction, the path of Mars as seen from the Earth appears irregular and shows retrograde loops, as

Fig. 4.3 *The great success of the Copernican model: an explanation of the retrogradation of planets.* In this highly simplified and highly schematic figure (not drawn to scale), the orbits of the Earth and Mars are represented. Points 1, 2, 3, ... 10, 11, 12 represent successive locations of both planets at regular intervals of time. From the Earth, the apparent trajectory of Mars on the sky has loops, of several degrees extent, creating an appearance similar in a way to that of the orbit of the Earth around the Sun represented in Fig. 4.2.

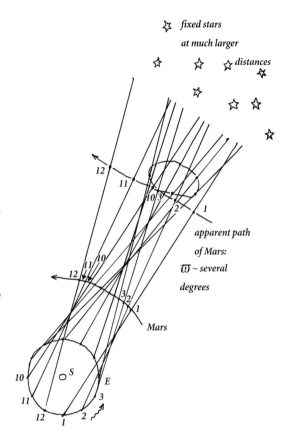

observed, on the background of fixed stars. This is the essential aspect of the Copernicus system, but an oversimplified view of it, as we shall see.

In a way, the apparent sizes of retrograde loops are a way to estimate their distance. In Fig. 4.2, the stellar annual parallax ϖ is defined, but it could just as well apply to any planet. Only for the star, there is no motion around the Sun. The phenomena are more pure. The nearby star (but still much more distant than the planets) appears to describe a circle in the sky identical in real size with the Earth's orbit around the Sun. The figure defines the annual parallax ϖ of the nearby stars as the angle ϖ under which one sees, from the Earth, the semi-diameter of the real trajectory of the star as seen from the Earth, assuming the Earth is fixed, as in Ptolemaic theory. At the time of Copernicus, could we detect parallaxes? Stellar parallaxes? Planetary parallaxes? Planetary parallaxes can be deduced precisely from the loops of the apparent tracks of planets, corrected for the velocity of the planet on its orbit deduced from the interval between two successive loops. These phenomena had been known for a long time. Loops are reproduced in Figs. 1.20, p. 36, 2.33, p. 122, and 4.3. They are clearly much larger than a degree. To the naked-eye, without any sophisticated instrumentation, a parallax of 1° can be seen, rather easily, but no apparent yearly displacement of stars with respect to each other is observable at that level of accuracy. What does that mean? This parallax of 1° corresponds to a certain distance Sun-star d of $1/\varpi$. If one uses as the *astronomical unit* the distance Earth-Sun, abbreviated 1 A.U. (remember: in the old days the Earth's radius was used as the astronomical unit), then $\varpi = 1$ A.U./d , if d is expressed in A.U., and ϖ in arcsecs. In rough values, one estimates now the distance of the Sun to be 150 000 000 km. If a star is located at 1 *parsec* (abbreviation pc), its parallax is, by definition of the parsec, equal to 1 arcsec, or 1". For a parallax of 1° (3600"), the distance is 1/3600 pc (easy computation, knowing that 1 pc = 3.3 light-years or ly and 1 light-year = 30 000 000 × 3600 × 24 × 365 km = 946 000 000 000 km, or about one thousand billion km; 1 pc = 3 090 000 000 000 km), i.e. as large as about 8 700 000 000 km (50 times the solar distance). Still, everyday knowledge in Ptolemaic times should have shown that all visible stars are further away than that distance. Somewhat before Tycho, the possibility for accuracy was greater, so a parallax of perhaps 10' could be detected, but it was not. All fixed stars were located beyond about 14 500 000 000 km, 1000 times the solar distance. Tycho could have detected a parallax of 2', hence reached a distance of more than 72 250 000 000 km, 5000 solar distances, but he still did not detect any parallax, although he seems to have looked for it. The first parallaxes measured in the stellar realm were measured when an accuracy of better than 1" could be achieved, during the 19th century. Modern techniques permit astronomers to observe parallaxes of 0.001", hence to measure the distance of stars located as far as 3300 ly. The closest stars we know are located at more than 1 parsec (about 200 000 solar distances, or A.U.), their parallax being smaller than 1".

Note, first of all, that if stars are that far, it really creates a problem for the geocentric hypothesis. Their diurnal motion would be incredibly fast, as clearly noted by Kepler, and even by the anti-Copernicans, such as Clavius, who attributed this phenomenon to the omnipotence of God.

Note also, that the reality of the motion of the Earth around the Sun, and not the reverse, could be demonstrated through observation of the

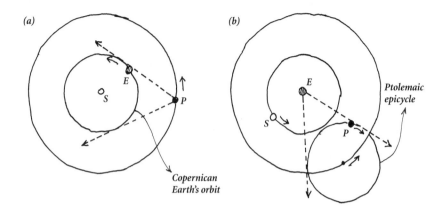

Fig. 4.4a,b *The Earth, the Sun, and a planet in the two systems.* On left, the three objects S (Sun), E (Earth), and P (planet), in the Copernican system. Both the Earth and the planet turn around the Sun. On the right of the figure, the Ptolemaic description of this system, at the same epoch in time. The Sun turns (uniform motion) around the Earth; so does the center of the epicycle around the deferent and the planet on the epicycle. The angle at which one sees, from the planet, the Earth's orbit (left) is equal to the angle at which one sees, from the Earth, the epicycle of the planet (right). This equality is basic to Kepler's construction of Fig. 4.5.

parallactic motion of these nearby stars only in the 19th century. Whether the Sun revolves around the Earth or vice-versa was, in Copernicus' time, quite irrelevant, as it could not be proven. The remark attributed to Osiander in the preface to the *dR* (but possibly in the mind of Copernicus himself) is that the heliocentric system cannot be proven. What can be demonstrated is only that, geometrically and kinematically speaking, and by way of describing the apparent trajectories and motions of planets, it is equivalent to the geocentric system (Fig. 4.4).

Actually, this is not completely true. As we have said, the apparent size of the retrograde loops of planets, the study of the planet's velocity between loops, can be used to reconstruct the parallactic orbit of the planet, equal in real size to the Earth's orbit around the Sun. In essence, this parallactic orbit is the epicycle. We do not, however, know the distance of the epicycle, only the ratio of its radius to its distance. The fact that, in the heliographic construction, each of them, for all planets, is in a way an exact image of the same single circle, the one described by the Earth around the Sun, decreases the number of parameters necessary in Ptolemaic theory. To our way of thinking, this fact, in addition to the exis-

Fig. 4.5a,b *The two systems according to Kepler (in the* Mysterium Cosmographicum*). Right (a) The Ptolemaic system. Epicycles are traced and have (strangely enough, see text) different sizes. The angle at which each epicycle is seen from the Earth is drawn, as on Fig. 4.4. At the bottom (b), is the Copernican system. The same angles as in (a) appear. They are the angles at which one sees the Earth's orbit from Mars, Jupiter, and Saturn. For Mercury and Venus, they have the same significa- tion as in (a). Scales on (a) and (b) are of course dif- ferent.*

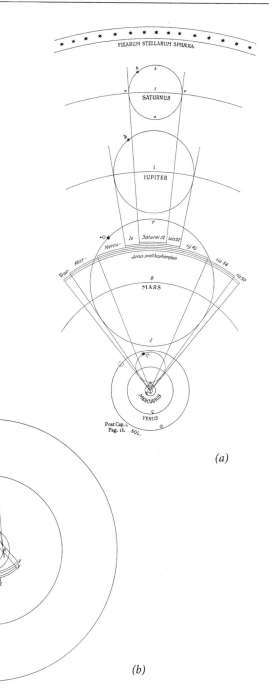

(a)

(b)

Fig. 4.6 *Kepler's demon-stration of the equivalence of the two systems.* At the left, the Copernican construction; at the right, the Ptolemaic representation. This figure is similar to Fig. 4.4, but Kepler chooses three moments, A, B, C, of the Earth's orbit, as seen from the planet P, the Sun being assumed to be fixed (left). At these moments, if the Earth is now assumed to be fixed (right), the Sun is at A, B, C, and the planet is at A, B, C, on its epicycle, the point P (not a planet!) being now, on the deferent, the center of the epicycle. The 3 successive apparent positions of the planet, as seen from the Earth, are unchanged, as can be seen from the obvious geometrical equalities of the quadrilaterals PABC and EABC.

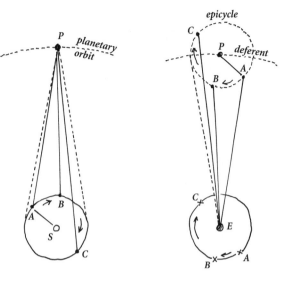

tence of stellar parallactic closed motions, constitutes a strong arguments in favor of the value of the heliocentric system. Every appearance is an image of the Earth's motion around the Sun.

Most inequalities and oddities of the planetary motions are thus only appearances, caused by the motion, or motions, of the Earth. This point of view is also exemplified later by Kepler, in his discussion of the Copernican system (Figs. 4.5 and 4.6).

Let us look at the simplified views of the planetary system (Fig. 4.5), as described by Kepler, in his *Mysterium Cosmographicum* (1596). We see there the simplicity of the Copernican system, compared to the complexity of the Ptolemaic system. Kepler redemonstrated their equivalence (Fig. 4.6).

Strangely enough, however, the young Kepler, probably still impressed by the idea that orbs must pile above each other without free intervals, without vacuum (an old idea similar to Adrastes', or Bacon's, see p. 130, and later, p. 170) represents the Ptolemaic system with epicycles of different sizes, at least for Mars, Jupiter and Saturn, different from the size of the Earth's orbit. Had he constructed the figure in order to achieve an

Fig. 4.7 *Figure 4.5a corrected.* This figure, drawn to the same scale as Fig. 4.5a, makes use of the fact that for Mars, Jupiter, and Saturn, the epicycles, as demonstrated in Fig. 4.6, have the same size as the orbit of the Earth around the Sun. The angles being the same as in Fig. 4.5a and b, the distances of these three planets can be derived in astronomical units. Kepler should have indeed drawn this figure, instead of Fig. 4.5a, but the *Mysterium* was his first book, the book of his youth.

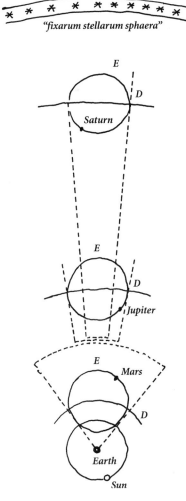

equality among the four orbits (leaving aside for the moment the cases of Venus and Mercury), he would have found Fig. 4.7, and determined the relative distance of planets. Actually, Copernicus did certainly achieve the correct construction and obtain (Table 4.2) distances within the solar system expressed in astronomical units, in the sense we have defined the meaning of this term.

Figure 4.7 demonstrates that, indeed, the relative distances of the planets to the Sun are closely linked to the apparent sizes of the retrograde

Table 4.2 *Planetary distances*

	from Copernicus	same (in A.U.)	modern values (in A.U.)
Mercury	9.24	0.38	0.387
Venus	18	0.72	0.723
Earth	25	1	1
Mars	38	1.52	1.524
Jupiter	130	5.20	5.203
Saturn	230	9.20	9.539

(distance to the "mean Sun")

loops, or of epicycles. Kepler clearly expressed finally this idea by saying, in essence, that epicycles are nothing but the projection in the heavens of the proper motion of the Earth.

The ability to determine these relative planetary distances based on the heliocentric hypothesis is perhaps the major contribution of Copernicus to our description of the world.

4.2.3
The Copernican System Elaborated

We will now discuss the Corpenican system, in all its complexities, as described in the *dR*. Note to begin with that one additional hypothesis, which is not mentioned in the *Commentariolus,* and which we could actually label "hypothesis viii" is explicitly mentioned in the *dR*: "The orbits must be circular; the motions must be uniform."

If we follow the basic premise of all epicycles (at least for the outer three planets) describing exactly one year and being of the same radius, the A.U., we should be able to decrease the number of Ptolemaic circles. We can eliminate from the model the radius of all epicycles save one, as shown in the two figures from Kepler shown above (Fig. 4.5a,b), and in Fig. 4.7. However, the other motions linked to the eccentricity of the deferent (or the mobile eccentric circle) are still there. Physically, this is associated with the *anomalies* of the motions of the planets, with their true elliptical motion, which was discovered by Kepler at the beginning of the 17th century. Since the Ptolemaic system required 28 parameters, we should be surprised to see that, in the *Commentariolus*, 34 parameters are necessary, whereas one would have thought that with the elimination of 5 epicycles, 23 parameters would suffice for Copernicus, not 34!

This disparity arises from the fact that many anomalies were not known to Ptolemy, at least not described in his system. It underscores the role of medieval observers, and of Copernicus himself in observing the sky. These anomalies were accounted for by new epicycles added by Copernicus, without indeed substantially improving the description of the planetary apparent motions. However, there were some definite improvements, which encouraged Copernicus to extend his work. In particular, Ptolemaic theory with regard to the Moon was far from satisfactory, as we have seen. Copernicus' theory is far better, without using the equans. In some cases, that of the Sun for example, Copernicus' description is essentially identical to the Ptolemaic one without any significant improvement.

Previously, we have identified the epicycle with the Earth's orbit around the Sun. This is valid for all *exterior* planets, i.e. Mars, Jupiter and Saturn. In the case of Mercury and Venus, the problem is more complicated. The Earth's orbit is identified with the deferent. Fig. 4.8 gives an example of this identification and of the demonstration, in both cases, of the equivalence between the two systems, according J. de S. Price.

The reason for the differential descriptions of the motions of the *interior* planets, Mercury, Venus and the exterior planets, Mars, Jupiter, Saturn, comes from the observed fact that whereas the interval of one year

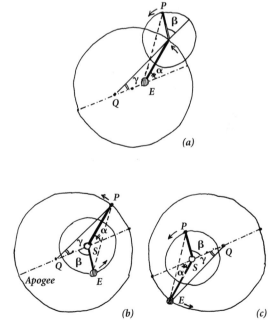

Fig. 4.8a–c *The equivalence of the two systems (according M.J. de S. Price).* On top **(a)**, the strict Ptolemaic construction, using the equans. On bottom left **(b)**, the Copernican construction for Mars, Jupiter or Saturn; on bottom right **(c)**, the Copernican construction for Mercury or Venus. The constructions at the top and bottom are clearly equivalent to each other, the angles α, β, γ having the same value and representing angles varying uniformly with time.

defines more or less the separation of two successive loops in all cases, the orbit of the exterior planets is larger than the orbit of the Earth, whereas the opposite is true for the interior planets. This leads to the observation (already noted by Heraclides) that Mercury and Venus seem to accompany the Sun in its motion. The loops always envelop, if one may use this word, the Sun, which is at their center. The loops largely dominate the apparent motion of these two planets and the deferent is identified with the Earth's orbit itself, described by the Sun in one year, the Sun being the center, moving on the deferent, of the epicycle.

However, at least one complication arises. The center of each deferent circle is now different from the other, or from the exact center of the Earth's orbit. We have something here which demands a device equivalent to the equans. The apparent velocity is not constant, even though motions must be uniform. This statement is, of course, false, but regularly applied by Copernicus, in the following form: "Any observed motion of a planet may be represented by a combination of uniform motions on circles of which the center and the radius can be determined from the motions." (hypothesis ix) In other words (i.e. ours!): "Can an ellipse be described, with any given motion along it, by a combination of such uniform motions?" And indeed, the reply to this question is most generally (see however Fig. 4.29, p. 224): "no!" ... Which explains some of Copernicus' failures.

The detailed description, planet by planet, of the system differs from the *Commentariolus* to the *dR*. In the *Commentariolus*, Copernicus held that the Sun has a fixed apogee in the sky. Later, he believed (incorrectly on a small time scale; the motion of the line of apsides is very slow) that it was definitely moving. For that reason, in the *dR* he used an eccentric linked to another eccentric.

Figure 4.9 demonstrates the case of the Sun, according the *Commentariolus* (a) and the *dR* (b).

In (a), there is a simple eccentric. Its center is the "mean Sun." This is really equivalent to the Ptolemaic description, especially with respect to the fact that, there is no way, by looking at the stars only, to detect whether the geocentric description is better or worse than the other. There is no absolute reference system! In (b) according to the *dR*, the motion is more complicated. Since Copernicus wants the line of apsides (joining the aphelion and perihelion) to turn, he introduces a small eccentric turning around a moving point on the main eccentric.

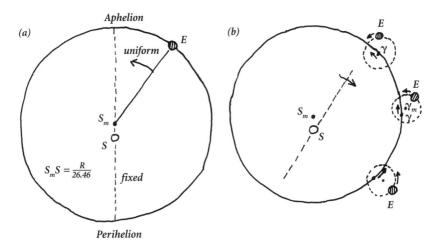

Fig. 4.9a,b
(a) *A first Copernican view of solar motion.* This is a simple eccentric. By trial and error, Copernicus computed the distance SS_m of the Sun to the *mean Sun*, the center of the Earth's orbit, of radius R, the astronomical unit. He found $SS_m = R/26.46$. The motion is uniform around the Sun. This eccentricity explains the seasons.
(b) *The second Copernican view of solar motion.* In order to allow the apsidal line, joining the Sun, the perihelion, and the aphelion to turn around point S, Copernicus invented this complicated system. The circle centered on S_m played the part of a deferent, and the Earth moved on an epicycle, itself eccentric with respect to the point moving on the main circle. A very complicated construction indeed but successful at keeping all motions uniform and circular, and saving the phenomena (see also Fig. 4.32).

In the case of the Moon (Fig. 4.10), Copernicus uses one deferent circle and two epicycles. There is no equans. The model is identical in essence to that of Ibn Ash Shatir. It correctly predicts the different types of eclipses.

In the case of a planet (Fig. 4.11), Copernicus uses as a deferent the eccentric (different from planet to planet) and two epicycles. Note the apsidal line differs from planet to planet.

The case of Mercury (Fig. 4.12) is very complicated. Around the eccentric deferent, there is a revolving epicycle, the center of which is oscillating.

To improve a given solution, the only means that Copernicus had at his disposal was to add circles (eccentric, or deferent and epicycle) to those already there. This accumulation of epicycles, deferents and eccentrics, makes the whole system look really *ad hoc*. As if, while progres-

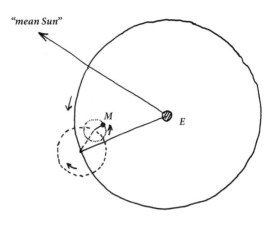

Fig. 4.10 *The Moon-Earth system according to Copernicus.* This complicated construction requires one deferent and two epicycles around the Earth. The deferent is described in exactly one synodic month. The first epicycle, centered on the deferent, is described by the center of the second epicycle in one anomalistic month. The Moon, on the second epicycle, goes through two revolutions in one anomalistic month. There is no equans, but the construction implies an explanation of eclipses of different natures (as described in Chap. 1, Fig. 15, 16 and 17). A similar system was earlier proposed, within the Ptolemaic frame of reference, by the Arab astronomer Ibn Ash Shatir.

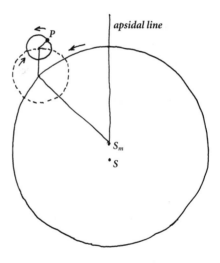

Fig. 4.11 *Planetary motions according to Copernicus.* Two epicycles and one deferent, with the proper revolution times, are necessary to fit the observations. Note that the apsidal line changes in space from planet to planet.

sing in his studies, Copernicus always put (implicitly) the motto "save the phenomena" before any other consideration, notably before the need for simplicity, the Occamist point of view, or any deistic conception, such as Plato's. Indeed he became, with time, more and more "Aristotelian," as we have defined this term.

Fig. 4.12 *The case of Mercury according Copernicus.*
The strong eccentricity of the motion of Mercury led Copernicus to propose this strange combination of circular and uniform motions. The first epicycle is centered around the mean center E_1, which oscillates along a line (dotted) linked to the apsidal line (located by the angle α) as shown in the figure. The point E_1' is actually the real center of the real epicycle (not drawn here), and moves together with the epicycle along the deferent; in the same time they both oscillate around the mean center E_1.

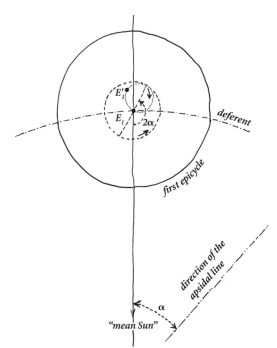

We will not consider too many details of the complicated constructions of the *dR*, intended primarily, as we have just said, to "save the phenomena." We will, rather, abstract the three main objects of the Copernican study, as we can see them now.

(α)   By attributing to the Earth's revolution the appearance of the retrogradations, Copernicus obviously went further than Ptolemy. The latter had the feeling (from Eudoxus' data) that the epicycles of Mars, Jupiter, and Saturn should be described in exactly one year. However, they did not have the same radius (see Fig. 4.5a). Copernicus assigned to them the same radius, of 1 A.U. This allowed him to determine the distance of planets to the mean Sun, as seen in Table 4.2.

(β)   Point (α) is linked closely to the idea that the stars have to be very far away, so that the fact that they do not have visible parallaxes is compatible with the reality of the motion of the Earth around the Sun (see Fig. 4.3).

(γ)   The Aristotelian-Ptolemaic combination, advocated later by Bacon, was still a potent notion for Copernicus. The Earth, revolving around the Sun, is carried by its "orb." Its rotational axis should describe in one year a cone of angle $\varepsilon = 23°$ (Fig. 4.13), but this is not really so. The Earth's axis always points toward the same pole in the sky. To account for that fact, one must add a new motion to the Earth's axis with respect to the "solid Earth-ghost" (as I suggest we call this strange Earth glued to the solid orb), a conical motion described in one year, and corresponding exactly to the conical motion introduced by this strange solidity.

We can see, through all these constructions, an image of Copernicus which makes him a good follower of the traditions: the Aristotelian tradition, the Ptolemaic tradition, even anticipating the Baconian tradition of the next century.

4.3
The Progress of the Observations; Tycho Brahe and the Nature of the Universe

4.3.1
The Supernova of 1572

Tycho was born only three years after Copernicus died. A nobleman, surrounded by a rich family, adventurous and often wild, he was accustomed to a good living, good food, and good drink. He was eager to learn but inclined to be domineering with his friends and parents, and belligerent with his fellow students. With his irrepressible enthusiasm, he quickly became devoted to science, and in particular to astronomy – after seeing a partial solar eclipse, so it seems. The aristocrat became a fanatic observer of the sky and his contribution to astronomy has been of paramount importance. We shall not here give a detailed account of the active life of Tycho, from his birth in Denmark to his death in 1601 in Prague.

The contributions of Tycho are two-fold. On the one hand, he developed a *Tychonian system*, which departed from the heliocentric ideas of Copernicus, while at the same time displaying interesting features. Above all, however, he was a great inventor of instruments and a great developer of accuracy in classical instruments, such as quadrants.

Fig. 4.13 *The Copernican device to insure the fixity of the direction of the Earth's axis.* On top, an attempt to represent the "third motion of the Earth." As the Earth is linked to the spherical body carrying its orb, it should revolve on its axis in one year, always at the angle ε (23°30' off the axis of the orb), thereby describing a cone around point O. In reality, however, this is not what happens … therefore, the necessity for a device or trick (figures on the bottom). Copernicus assumed that the Earth's axis describes a cone in the opposite direction to that of the revolution, with a very slightly smaller velocity, to account for the precession of equinoxes. The axis of this cone, its aperture, is parallel to the cone of summit O in the figure at the top.

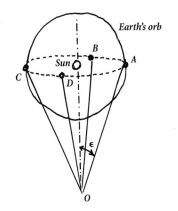

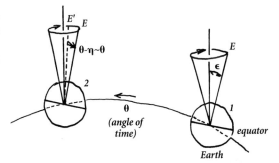

One dramatic observation he made was that of the supernova of 1572. On the evening of November 11, 1572, Tycho saw a star, brighter than Venus, very near the constellation Cassiopeia, which was not present before. It was a "new" star, observed by many other witnesses, at the request of Tycho, in his neighbourhood, but also confirmed by astronomers such as Maestlin and Thomas Digges. The new star was observed night after night for months.

It did not appear to move with respect to the fixed stars. Therefore, it had to be further away than the Moon which moves from night to night. Its parallax was stellar; hence, as we have seen, its distance must be very great. This obvious "explosion" belonged, therefore, to the astral world, not to the sublunary world. With that observation (Fig. 4.14), the principle (an Aristotelian principle) of the incorruptibility of the aether, of the astral world, as opposed to the sublunary world where things were ephemeral, changing, mortal, was finally abandoned.

4.3.2
The Comet of 1577

The explosion in the heavens was not easily accepted, but five years later, Tycho observed a comet, and was able to determine its distance. By Tycho's time, it was possible to make measurements to an accuracy of 1–2'; hence *diurnal parallaxes* for the Moon, or for objects somewhat more distant, could be determined, by the method (Fig. 4.15) used by Tycho. Tycho was then able to show that the distance of the comet was at least six times that of the Moon, another "stone in the marsh" of the astral aether.

This discovery was linked with the development of instruments, which is not the primary subject of this book. We do show, in Fig. 4.16, the construction and the principles behind some of Tycho's instruments. Their size and the accuracy of the adjustments promoted a high degree of efficiency.

Carrying on the efforts of Copernicus to better estimate the relative distances within the solar system and to compare them with the stellar distances, Tycho, having improved the accuracy of angular determina-

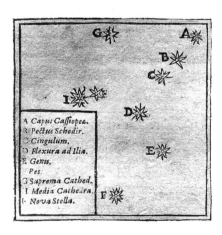

Fig. 4.14 (after Tycho Brahe's *De Stella Nova*). The drawing by Tycho of the constellation Cassiopeia in which he observed the 1572 supernova. It is the brightest star of all in this drawing.

A *Caput Cassiopeæ.*
B *Pectus Schedir.*
C *Cingulum.*
D *Flexura ad Ilia.*
E *Genu.*
   Pes.
G *Suprema Cathed.*
I *Media Cathedra.*
F *Nova Stella.*

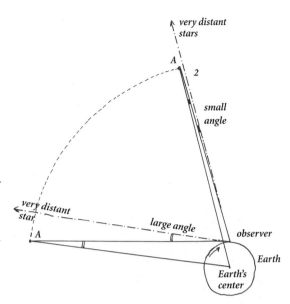

Fig. 4.15 *Tycho Brahe's model of the diurnal parallax effect.* A near-by object A, observed twice in the same night, at its culmination and at its setting, occupies a different place with respect to the very distant sky. The Moon's distance may be estimated this way. Although it moves slightly over a period of a few hours, a comet's distance is also measured this way, even though Tycho's comet was at least six times more distant than the Moon.

tions, was able to prove the "astral" and not "sublunary" character of such strange objects as a "new" star, or a comet. These objects were unpredictable. They were both regarded with terror as evidence for the inadequacy of the essential Aristotelian distinction between the eternal and perfect fate of the astral world as against the unstable, destructible, putrescible world of the sublunary realm. The unity of the Universe, of its very nature, was thus emerging as a key idea of modern astronomy, but was this not, in a way, implicit in Platonic philosophy?

It is curious that in this case we see an Aristotelian idea upset, whereas, in the Copernican system, some Aristotelian ideas were used. Here we see, again, that there is no such thing as the so-called Copernican revolution! Various Aristotelian paradigms and Ptolemaic paradigms were discarded little by little, not all at once. Some of them were even used to disprove others and some are still with us even now. The logic of the path is not always very easy to follow if one obscures it behind overly defined concepts. At each step, there is first progress in observational techniques, accompanied by progress in the accuracy of the observations. Some improvements to the models come next. These are not necessarily completely internally consistent. Nevertheless, they increase our ability to "save the phenomena," both the phenomena observed for many centuries and the newly discovered phenomena.

Again, this important conceptual progress was made possible by Tycho's extraordinary skill in measuring angles (in measuring everything actually, time intervals as well). Before him, the accuracy was to several minutes of arc. Tycho's instruments were accurate to one minute of arc or so (Fig. 4.16b). As always, the accuracy of observations is the determining factor in the progress of knowledge. Tycho could not have determined the distance of the Comet of 1577 without this accuracy and later, Kepler could not have determined the elliptical law without it either (Fig. 4.30).

4.3.3
The Helio-geocentric System of Tycho

Tycho was not preoccupied by the problem of geocentrism, perhaps because he realized that, without being able to measure stellar parallaxes, he could not reach any positive conclusion. He was, nonetheless, certainly sensitive to the arguments of Luther and Melanchton. So, while working to save the phenomena, to account for the several retrogradation loops, Tycho did not wish to adhere to a generalized heliocentrism. Without going into as detailed a description as we did in the case of Copernicus, we should note that Tycho's system is indeed geocentric and at the same time heliocentric: all planets revolve around the Sun, but the Sun, and the Moon revolve around the Earth (Fig. 4.17).

It is a rather good compromise, equal to Copernicus' system in its ability to save the phenomena. This model reinforced the strange status of the preface of the *dR*. It was supposedly written by Copernicus, assumed later to have been actually written, or at least rewritten, by Osiander, but perhaps, after all, representative of the true Copernican ideas. In this preface, Osiander expressed the idea that the heliocentric system is only a mathematical tool, not the expression of a physical reality. Rheticus objected, but since Copernicus had died by that time, we shall never know his real thoughts on this subject.

4.4
Kepler and the Death of Circularity

In a sense, Kepler's ideas followed from Tycho's observations in that they debunked another Aristotelian dogma by ruling out the complicated sys-

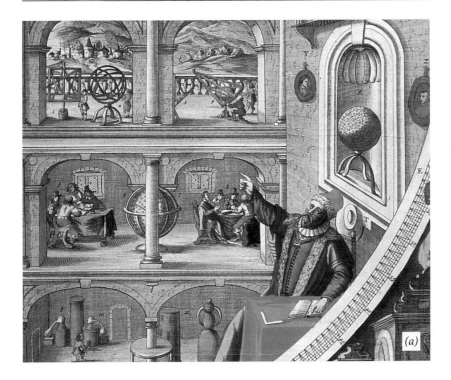

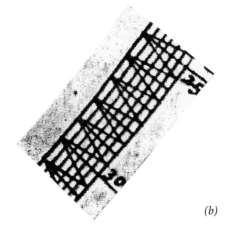

Fig. 4.16a,b *Some of Tycho's instruments.* On the top (**a**), the *large quadrant*, allowing a determination of the height of a star above the horizon. Note the clocks which measure the time of each observation. At the bottom (**b**), an oblique scale, invented by Levi ben Gerson, but used extensively by Tycho. Each degree is divided in sectors of 10'. In each one, an oblique line, divided in ten segments by points, allows a minute of arc to be read easily. Looking at 90° of the quadrant, we can see a complete oblique scale on the figure at the top left.

tems of Ptolemaic-Copernican epicycle-eccentric combinations. The word "revolution" is probably more appropriate here than in the case of Copernicus, but even for Kepler, it is not really correct. After all, Kepler made use of the thinking and observations of Copernicus and Tycho, as well as those of their predecessors. Like every thinker of that period, Kepler exploited the best observations of Tycho, notably the Mars data. In that respect, he was no doubt more lucky than Copernicus, who died before the birth of Tycho.

Kepler was a very prolific writer, interested in everything. He had bad health but tremendous energy and will. He was always fighting against everything and everyone, including himself and his own ideas, turning a somewhat erratic style of thinking into "serendipitous" discoveries. We shall limit our study of this thinker, philosopher, poet, writer, astrologer, musician, and novelist to his astronomical investigations.

4.4.1
Planetary Distances

Strongly influenced by the Pythagorians, the young Kepler tried to find a way to explain the arrangement of the planetary orbits, to find a law which the planetary distances would obey. First, he tried locating the orbits within each other, separated one from the other by an interval containing regular polygons in a progression from the triangle for the largest to the hexagon for the smallest as explained in Fig. 4.18. Note in this figure (drawn from Kepler's original) that the he does not go as far as Mercury (which would have required a heptagon).

But this construction did not answer a question asked by Kepler, in his Pythagorian mood. Why were there six planets and not 20 or 100, just as physicists now ask why there are six quarks and not more or less? Indeed in plane geometry, the same construction yields a much larger number of planets than six. In fact, an infinity of regular polygons can be inscribed in a circle and surround an inscribed circle (Fig. 4.18).

Under the influence of Maestlin, Kepler adopted the essence of the Copernican heliocentric system. Unlike Tycho, Kepler held that the Earth was a planet like all the others, so the obvious question remained: why six planets? Between the successive orbs of these six planets, there are five intervals, five spaces actually, since the circular orbits are borne by spheres. As the model required five of them, Kepler was naturally in-

Fig. 4.17 *The helio-geocentric system of Tycho Brahe.* Every planet turns around the Sun, but the Earth (E) and the Moon do not. The Sun (S) and the Moon (M) turn around the Earth. The Earth is fixed. Completely saving the phenomena would require an elaboration of this system, with epicycle and eccentric circles.

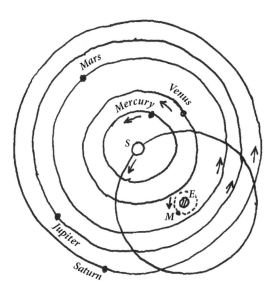

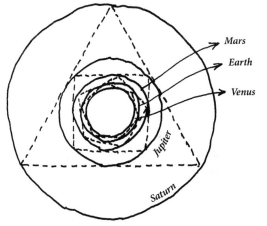

Fig. 4.18 *A first version of Kepler's system: a first wild idea!.* The five planetary orbits are separated by polygons of 3, 4, 5, and 6 sides, the triangle being inscribed in the Saturn orbit, the Jupiter orbit being inscribed in it, and so forth.

clined to choose the five perfect polyhedrons (Fig. 4.20) instead of regular polygons (Fig. 4.19).

Thus Kepler, guided by both intuition and the consideration of some "perfect" correspondence between the real world and the world of number and figures in the manner of the Pythagorians, thought of five polyhedrons, as there are only five of them. These were the five polyhedrons

Fig. 4.19 *The infinity of successive polygons.* The construction of Fig. 4.18 can be continued ad infinitum, the limiting polygon being a circle of a *finite* radius.

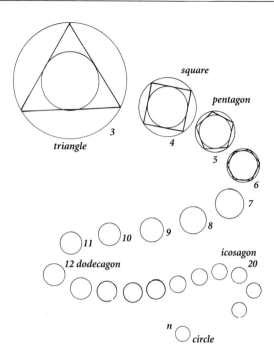

used by Plato, in quite a different context. Kepler constructed a complicated system explaining the relative distances of planets, as we see in the representation of his drawing (Fig. 4.21), using the polyhedrons in an order which seemed to him more in line with the Copernican distances. He did not follow either the number of faces, as he did in the case of the regular polygons, or the order followed by Plato (see Fig. 2.3).

Actually, these successive attempts (there were a few more) by Kepler were wrong. He also tried to link the distances of the planets to their periods of revolution, through a rather strange construction not explicable to our logic (Fig. 4.22). He also tried to establish simple ratios between them (2, 3, 4 ...).

Kepler also attempted to solve the problem by linking motions and distances to musical tones for each planet (the "harmony" of planetary spheres) (Fig. 4.23). This series of efforts led Kepler in due course to discover his third law, which is basic to Newtonian cosmology. We shall come back to this.

Note that, remarkably, Kepler did not realize earlier how the very Copernican ideas, which identified either the epicycle or deferent with the

Fig. 4.20 *The five regular polyhedrons.* The fact that there are only 5 regular polyhedrons derives from Euler's equation wherein the number of summits, S, plus the number of faces, F, minus the number of edges, E, equals two (S + F − E = 2), an equation which admits only five solutions in which F, S, and E are integers. The fact was known (by Plato, and later by Kepler) much before its mathematical demonstration.

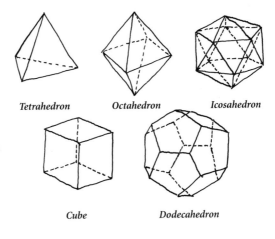

Tetrahedron   *Octahedron*   *Icosahedron*

Cube   *Dodecahedron*

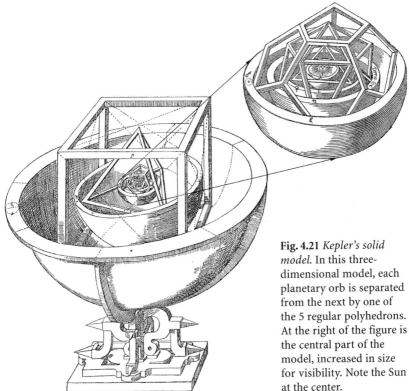

Fig. 4.21 *Kepler's solid model.* In this three-dimensional model, each planetary orb is separated from the next by one of the 5 regular polyhedrons. At the right of the figure is the central part of the model, increased in size for visibility. Note the Sun at the center.

Fig. 4.22 *Planetary distances: Kepler's second wild idea.* The distances of the planets to the Sun are plotted on the ordinate. The driving force, or, allegedly, the velocity of each planet on its orbit, is plotted on the abscissa. The stars are at point zero, while the Sun is represented by the segment AD. The relation $vr = constant$ would not be represented by a circle as it is here, but by a hyperbola (adapted from Kepler's drawing).

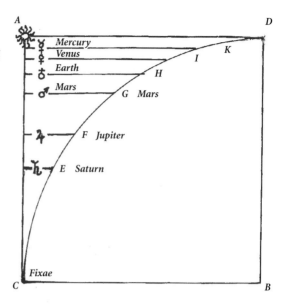

image of the Earth's orbit (same size, same time of revolution of the center of the epicycle along the deferent), made possible the deduction of the relative planetary distances from the observations of retrogradations and stations of planets

4.4.2
The Elliptical Motion

Kepler's greater innovation was to introduce the elliptical motion as a good way to model the planetary motions and to save the phenomena, using an Aristotelian methodology, but at the expense of circularity and uniformity of motions. This progress was accomplished by a very careful and tedious analysis of the apparent motion of Mars, as determined by Tycho.

Ptolemy had also worked on the motion of Mars, by a trial-and-error method (Fig. 4.24). He was able, in this way, to determine the location of the equans and the characteristics of the orbit. Kepler was more fortunate than Ptolemy, in that he was able to use Tycho's observations, the accuracy of which was greater than that of Ptolemy's observations (1 to 2' against about 10').

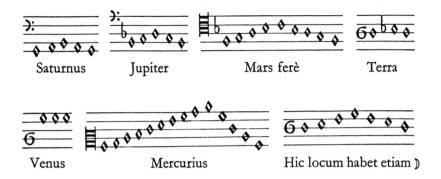

Fig. 4.23 *The harmony of spheres: another wild idea of Kepler's.* The musical interval of the harmony of each planet is a measurement of its eccentricity. The fundamental tone will be the velocity of the planet at its aphelion. A nice, but quite unlikely planetary choir!

In essence, Kepler, although a "believer" in many things, from the numerical harmony that was thought to rule the world, to the detailed Copernican mechanisms, was still primarily an empiricist. Although he certainly held preconceived ideas, he was always ready to change them in order to reach a better agreement with the observed data.

He first noted, along Copernican lines, that practically any motion, any trajectory, could be explained or at least described using a combination of circular motions – circular but not necessarily uniform. This gave him the idea of elaborating on the equans methodology of Ptolemy. As we have noted earlier (see p. 106), the Ptolemaic location of the equans is not arbitrary. Hence, there is a very clear-cut choice between only two possible motions, one truly uniform on the orbit around the mean Sun, the other uniform around the equans. This does not give enough flexibility to the solution. Therefore, Kepler looked for a new parameter and assumed that the equans was not necessarily symmetrical with respect to the center S_m (mean Sun) of the orbit. This is the *hypothesis vicaria*.

Kepler's efforts continued for several years before they could really be completed in a satisfactory way.

First, he determined the obliquity of the orbit of Mars on the ecliptic (1°50'), checked that the Sun is in the plane of this orbit (a fact absolutely not obvious in the Ptolemaic constructions), and verified that this inclination is constant over time.

Fig. 4.24 *Ptolemy's analysis of the motion of Mars.* If the motion, as seen from the equans, is uniform, the relation can be expressed: $\alpha_1/\alpha_2 \doteq t_{12}/t_{23}$. If the revolution time t_{rev} is known, and the durations t_{12}, t_{23} are known, then the angles α_1 and α_2 are known. Successive approximations, starting with an equans identical to the Earth, and located at the center C, lead to a longitude of the apogee 115°30' and to an eccentricity $CE/R = 0.200$. Compare with Fig. 4.30.

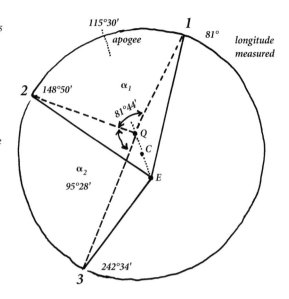

However, once an apparent motion is determined, in longitude, many different orbits can satisfy the basic condition of saving the phenomena. Assuming 1, 2, 3 ... 6 to be the successive directions of the points so observed (Fig. 4.25), at equidistant time intervals from the Earth, there is an infinity of possible orbits, if we do not impose simple laws of motion.

But what laws should we impose, now that we have given up the concept of uniformity-circularity (a concept, let us remember, quite contradicted by Ptolemy's systematic study)? What laws could now be introduced?

(i)   Uniform motions on a circle, with the Sun at the center, an Aristarchian notion. The motion, as seen from the Earth, will not appear uniform, but there is no reason why it should so appear. Two observations (points 1 and 2) are sufficient to define the orbit. There are two unknowns, the radius R of the orbit, the angular velocity v of the planet on its orbit or the period P. The model lacks flexibility (Figs. 4.25, curve A, and 4.26).

(ii) Circular motions around a center which is still the mean Sun but with uniformity of the angular motion around the equans. This is a heliocentric quasi-Ptolemaic view. The motion is not uniform around the Earth (Figs. 4.25, curve B, and 4.27). It coincides with

Fig. 4.25 *The basic indeterminacy of the orbit.* Points 1, 2, 3, … represent the direction of successive observations. Fitted to the various orbits possible or even impossible, they yield variations of velocity on the orbit that are represented at the bottom. Uniformity cannot be achieved without a centered circular motion (Such as A).

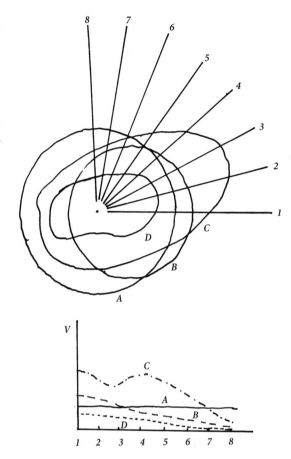

model (i) at two points 1 and 2 only, but from point 3 on, they differ substantially, as can be seen in Fig. 4.26. In this model, there are four unknowns, R, eccentricity e, location of the Earth with respect to the Sun (or longitude of aphelion), and angular velocity ϕ around the equans.

(iii) A Copernican view: Uniform and circular motions along an epicycle circle, of radius R_E, whose center describes the deferent circle, of radius R_D, in a uniform motion. This description (Fig. 4.28) coincides with description (ii) at three points, 1, 2, and 3. There are still, as in (ii), four unknowns to determine, the radius R_D of the deferent, the radius R_E of the epicycle, the velocity V of the epicycle center on the deferent, and the angular velocity ω of the planet on its

Fig. 4.26 *Kepler's attempt at an Aristarchian solution.* The orbit is circular. Its center is not the Earth. On the orbit, a uniform motion gives rise to observations from the Earth that do not correspond to uniform rotation (points 1, 2, 3 …). The points 1, 2, 3 … and the dotted arrows correspond to the dotted lines of Figs. 4.27 and 4.28.

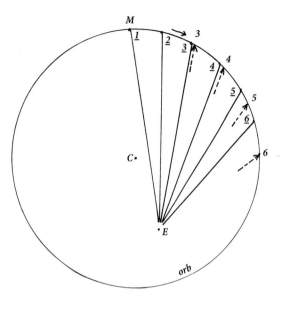

Fig. 4.27 *A heliocentric and Ptolemaic effort by Kepler.* The motion is uniform on the eccentric circle (points 1, 2, 3, 4 …), but not on the planet's orb (points 1, 2, 3 …), as seen from the Earth. Points 1, 2, 3, … correspond to the solid arrows, here as in Fig. 4.26.

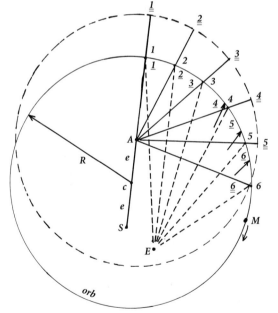

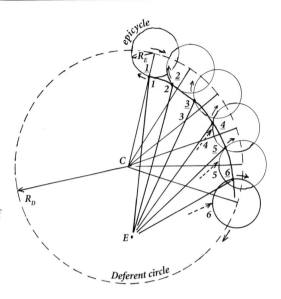

Fig. 4.28 *A Copernican effort by Kepler using the epicycle-deferent.* The motion of the epicycle's center on the deferent is uniform. The motion of the planet on the epicycle is uniform. The resulting orb is drawn, through the points successively occupied by the planet, 1, 2, 3 ... The dotted arrows, as in Fig. 4.26 and 27, represent the points 1, 2, 3 ... of the quasi-Ptolemaic attempt.

epicycle. These laws of motion may be applied. Whatever the details, there are discrepancies for points 4, 5, 6, and 7 and incompatibilities with the observations.

Kepler had several possibilities to explore. There was the possibility of getting rid of the uniform motion on the epicycle, but Fig. 4.29 shows quite clearly that this excessive freedom may lead to absurd consequences.

The *vicaria hypothesis* is another way and it has the virtue of limiting the number of necessary hypotheses. Kepler's first attempt (Fig. 4.30), made use of Tycho's Mars observations. By using the Mars oppositions (the Sun, the Earth, and Mars are then aligned), a relation among the four parameters is introduced. Hence, strictly speaking, one reduces to three the number of parameters necessary to determine the equans, in the quasi-Ptolemaic view. Adding the *vicaria hypothesis* increases the number again to four. Therefore, four points are necessary, or in other words, four oppositions of Mars. Fortunately, there is one opposition approximately every two years, so Kepler used the four oppositions of 1587, 1591, 1593 and 1595 to derive, by trial and error, and through geometrical constructions, the values $e'/R = 0.07232$ and $e/R = 0.11332$, R being the radius of the orbit.

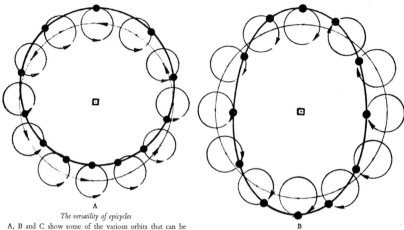

A

The versatility of epicycles

A, B and C show some of the various orbits that can be constructed by a combination of epicycles and deferents.

A shows how an epicycle can produce an effect equivalent to an eccentric. If the planet moves on the epicycle with such a speed that, half-way round the deferent, it has travelled once round the epicycle in the same direction, then the effect, as seen from the centre of the orbit, will be the same as if the planet had been moving along an eccentric circle

B shows how an elliptical orbit may be produced. In this case the path of the deferent is in an anti-clockwise direction, while the planet moves in a clockwise direction on the epicycle. This time, the planet travels half-way round the epicycle during the time it takes to go half-way round the deferent

Fig. 4.29 *The versatility of epicycles* (after Toulmin & Goodfield)

A, B, and C show some of the various orbits that can be constructed by a combination epicycles and deferents. (A) shows how an epicycle can produce an effect equivalent to an eccentric. If the planets moves on the epicycle with such a speed that, half-way round the deferent, it has traveled once round the epicycle in the same direction, then the effect, as seen from the centre of the orbit, will be the same as if the planet had been moving along an

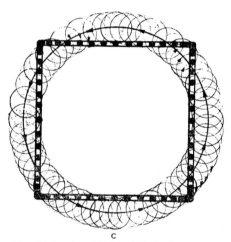

C

C demonstrates how, given a suitable choice of radii and speeds, one can get even an effectively square orbit. By appropriate variations in radii and speed, any kind of orbit likely to be met with in astronomy can be constructed

eccentric circle. (B) shows how an elliptical orbit may be produced. In this case, the path of the deferent is in an anti-clockwise direction, while the planet moves in a clockwise direction on the epicycle. This time, the planet travels half-way round the epicycle during the time it takes to go half-way round the deferent. (C) demonstrates how, *given a suitable choice of radii and speeds,* one can get even an effectively square orbit! By appropriate variations in radii and speed, any kind of orbit likely to be met in astronomy can be constructed.

At first this appeared to be a good representation. Twelve oppositions were represented with errors averaging less than 2°12″ in longitude. Unfortunately, using two further oppositions (1585, 1593) for the latitudes, Kepler found results quite different from the early ones, and obtained eccentricities of 0.08 and 0.099. As the average of these two values and $(e + e')/2R$ roughly coincided, Kepler decided to return to the Ptolemaic simplicity and to the equality $e = e'$, thus ultimately rejecting his own *vicaria hypothesis*.

Still, errors of about 8′ were found for various positions of the planet, different from the oppositions, hence some new attempts were necessary. The *vicaria hypothesis* had been abandoned and the Ptolemaic equans was not helpful. A very difficult, painful, but geometrically correct, work was undertaken. Kepler had the idea that the trajectory is not a circle, but an "oval." In order to better accomodate the data, without any preconceived idea about the shape of the orbit, Kepler derived the distance d of Mars to the Sun at different times, both from the circular hypothesis and, independently, from the observations. This is quite difficult and requires extensive trial and error work. The curve he then computed was not a circle, as motions turned out to be too slow far from the perihelion and too fast far from the aphelion if circular motion is maintained. The data are represented in Table 4.3.

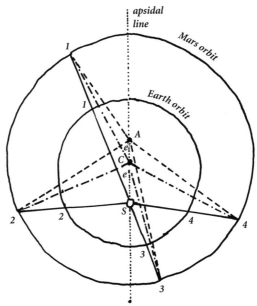

Fig. 4.30 *Kepler's Mars analysis.* Kepler used 4 oppositions 1, 2, 3, 4 and derived, by trial and error, the value of the ratios e/R and e'/R. Compare with Fig. 4.24.

Table 4.3 *Kepler's measurements of Mars' distance*

date	distance to aphelion	d (circular motion)	d' (from observed data)	Δd
31 Oct 1590	9°37'	1.66605	1.66255	0.00350
31 Dec 1590	36°43'	1.63883	1.63100	0.00783
25 Oct 1595	104°25'	1.48539	1.47750	0.00789

units : Astronomical Unit (average distance from Earth to Sun)

The curve that resulted was not a circle. As was often the case for Kepler as he meandered along his strange intellectual path, after some new trials, he found that the Δd max = 0.00858 was too large and should perhaps be divided by 2. He arrived at this idea after 22 different observations, yielding a figure for Δd of 0.00432, and finally, after some new trials 0.00429. At the same time, the eccentricity was determined to be 0.0926. Kepler noted that $0.00429 = (0.0926)^2/2$ or Δd max $= a - b = e^2/2$.

This is a known property of ellipses (Fig. 4.31), provided the eccentricity is not too large and this was common knowledge to the geometers of that period. Simply put, $(ea)^2 = a^2 - b^2$, by the very definition of the eccentricity of an ellipse. From this, one can derive: $e^2 = (a - b)(a + b)/a^2$ $\cong$ (if b does not differ too much from a) $2a(a - b)/a^2 = 2(a - b)/a$. If we remember that a is the unit in all these computations, the relation quoted above becomes clear.

Having noted this remarkable numerical coincidence in 1605 (by inspiration, not through any strict methodology), Kepler was finally satisfied. He had proven that the orbit is indeed an ellipse, not a circle.

4.4.3
The Elliptical Laws of Motion

Having noted that Mars' orbit is elliptical, Kepler immediately extended this statement to all planets, including the Earth. Kepler was lucky that he chose to begin with the Mars observations, since Mars' orbit has the greatest eccentricy of all the planets known in his time, save Mercury. Studying, therefore, the motions of Mars and the Earth along their orbits, Kepler arrived at his second law, regarding the areas swept by the orbiting planet in a given time interval. Since the Sun is the really important motor of the entire planetary system, it is necessary to consider the motion of line SP (Fig. 4.32) as essential. This is not a uniform motion. It

Fig. 4.31 *Circle and ellipse.* The figure defines the semi-major and semi-minor axes a and b, and the eccentricity e. The foci F_1 and F_2 are poorly located on this figure, for readability. Remember that an ellipse is the locus of the points M such that $F_1 M + F_2 M$ (constant). Eccentricity may be computed with the equation: $e^2 = 2a^2 - b^2$.

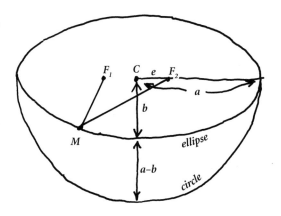

is immediately clear that the planet moves faster at the perihelion than at the aphelion.

This was actually already known by Ptolemy and was the very reason for introducing the uniform motion around the equans. A more detailed study led Kepler to an important result. The area swept in a given time interval Δt by the radius $r = SP$ is constant. In other words, the hatched areas of the figure are equal in surface, corresponding to the same time interval. This can be written as:

$$vr = \text{constant.}$$

4.4.4
The Distances of Planets to the Sun and Kepler's Third Law

The motion of a planet is faster when the planet is closer to the Sun. This idea led Kepler to study the possible relations between the revolution time (the period) P, and the semimajor axis a of the orbit, for all planets. As the velocity v on the orbit is of the order of a/P, v must be larger for a large a. If there is some relation between P and a, it must be that P grows with a, but at a slower rate. After some thinking, involving perhaps his ideas about dynamics (see p. 228 ff.), Kepler discovered, more or less empirically, that:

$$P^2 = Ka^3,$$

where K is a constant, unique for the whole planetary system.

This is Kepler's third law. Note that it seems to be the first physical law represented by a simple equation, and fully rigorous, although not demonstrated by Kepler. It was demonstrated indeed by Newton, in the frame of reference of the universal law of attraction and we should certainly note in this context, that all Newton's thinking was perhaps derived from, or at least justified by, this law.

4.4.5
Kepler's Concepts Regarding Dynamics

A very important idea in Kepler's thinking is that of the *solar dynamic action* on the planets.

Kepler believed that a *unique* dynamic approach (unlike the Aristotelian souls attributed to *each* planet) was necessary to account for all the planetary motions, as well as a direct intervention of the solar power, the Sun being, no doubt, the physical center of the system.

This was a new idea. Up to this point, the Sun had not necessarily been considered so important. Neither Ptolemy, Copernicus, or Tycho really thought so. This idea of a physical center being also the center of the motions is new and was essential, later on, to Newtonian thinking. As the French philosopher Jules Vuillemin noted, the whole problem is that, now, we have several "centers." C_m is the geometrical center of the motions of celestial spheres; C_v the geometrical center of the velocities (around which the planets turn uniformly); C_o the center of observations, i.e. the observer; C_r the "real" physical center; and C_p the center of the motions of the planets. Ptolemy, Copernicus, Tycho, and Kepler himself, all had different views as to the main questions that arose from this multiplicity of centers, as described in Table 4.4.

Actually, Kepler himself was critical of Copernicus. For example, he could not accept the "third motion" of the Earth, necessary to permit the Earth's axis to stay parallel to itself. Nor could Kepler accept the Copernican efforts to account for the trepidation, clearly a spurious effect (Fig. 4.33). Copernicus was overconfident in the observations of his predecessors, so far as Earth's obliquity and precession were concerned. Actually, the true variation is almost linear with time, unlike what Copernicus was inclined to believe.

Of course, even if Kepler's interpretation of Copernicus' views cleansed them from what was not essential to heliocentrism, still, Kepler, essentially replied "yes" to the last question of Table 4.4, and insisted

Table 4.4 *The center of planetary motions* (adapted from J. Vuillemin)

question	Ptolemy	Copernicus	Copernicus, according Kepler	Tycho
$C_v = C_m$? Is the motion of planets uniform w.r. to the sky?	no (cf. equans)	yes	no (it does not include the "third" motion of Earth)	no
$C_m \sim C_o$? Is the system geocentric?	yes	no	no	yes
$C_m = C_p$? Is the center the same for all planets?	yes (Earth)	yes (Sun)	yes	no
$C_m = C_r$? Is the center of motions a "real" center, a physical body, – Earth or Sun?	no (the equans is not "real")	no (the "mean Sun" is not "real")	no	no (Earth is not Sun!)

that the physical Sun was the source, the cause of motion if not its geometrical center.

Let us see how he did it.

Remember the Platonic-Aristotelian debate on the subject of the
"souls" of celestial bodies and later the ideas of Buridan and others on
the "impetus." After an initial period of interest in the "animal forces"
(angels?) acting separately upon each planet expressed in his early work,
the *Mysterium*, Kepler later looked for some *unique* driving force, a driving soul perhaps, but unique for the whole system of planets. Only the
Sun could be that unique force. In his first attempt to convey this idea
(in his *Astronomia Nova*), he suggested the analogy of a boat, pushed in
the current by the whirls produced by the pilot's paddling.

The action of the paddle makes the boat turn. The effect of the Sun on
the paddler forces the trajectory in the current to be eccentric (like the

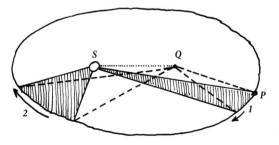

Fig. 4.32 *Kepler's second law.* The hatched areas are described in the same time intervals. They are equal in area.

Fig. 4.33 *The variation of the precession of the equinoxes and the obliquity of the ecliptic.* These curves, constructed from data going back to several centuries before the common era, represent the variation, as used by Copernicus, of the location of the point γ (vernal equinox), expressed in arcseconds per year, and of the obliquity of the ecliptic. The evolution of the curve representing the variations of γ is the so-called *trepidation*, which gave Copernicus a problem to solve. Actually, the old data are not as correct as Copernicus believed. Now, the obliquity is known to satisfy the equation: obliquity = 23°27'8.6" − 46.845" T − 0.0059" T^2 + 0.00181" T^3, where T is expressed in centuries or so, with $T = 0$ in 1900. This is practically a linear law, which results, in the year 100 BC, in a value of 23°42'51'.

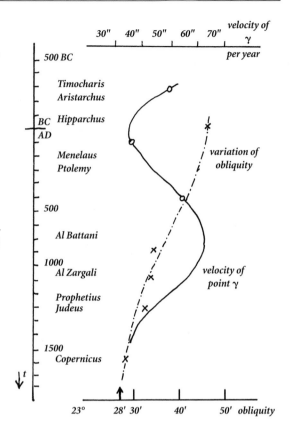

combined effect of wind and current in sailing). But the direction of the steering wheel at C would (top of Fig. 4.34) differ at two successive passages. It was not, in any case, clear how the Sun would influence the paddler.

Kepler turned to the *magnetic model*. In this model, there is no need to posit a current embedding the Sun and planets. Since the publication of William Gilbert's work in 1600, it was known that the Earth was a magnet. Assuming that other planets were also magnets and that they would be always oriented in the same way, one magnetic pole (*soli ami-*

ca, friend of the Sun) would be attracted to the Sun, the other one repelled (*discors* or hostile), as described in Fig. 4.35.

At A or E, there is no resulting attraction. When the planet moves, this motion, almost circular a priori, has to be kept eccentric precisely by this attraction which is manifest at B, C, and D, whereas at F, G, and H, the repulsion is dominant. At C and D, the axis is inclined by this effect, which explains the observed motion of the line of apsides. The flaws in this model are obvious, when one knows that the magnetic axis of the Earth is close to its rotational axis.

The magnetic model had, nonetheless, its historical importance. Is it possible that Newton was influenced by it to the alternate hypothesis that one attractive force was responsible for the motion of planets? This is actually not very likely: In Kepler's thought, there is the two-part notion of the motor along the trajectory and the action of the Sun responsible for this motor, although he does not explain how the Sun has this effect. This was not at all a part of the Newtonian view. It is possible, however, that Descartes' theory of twirls *(tourbillons)* was strongly influenced by Kepler's discussion.

Would the dynamic considerations of Kepler, as primitive as they were, have led him to his third law? It was logical to think of two possible simple laws. The first was to state that v (on the orbit), being proportional to $1/r$ (Kepler's second law), one could generalize this to a, and write $P \div a$. The second possibility was to state that v, as the driving force, being proportional to $1/a$, the period P is thus proportional to a^2, the circumference being itself proportional to a. Although the two ex-

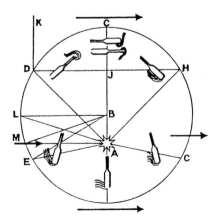

Fig. 4.34 *Kepler's interest in dynamics: the marine analogy.* The games of wind and paddling under the power of the Sun.

pressions are quite coherent, one implying a $1/a$ relation, the other a relation in $1/a^2$, taking into account the uncertainty about adaptability of the second law to all planets and considering all the observed data, Kepler finally proposed the law $P^2 \div a^3$. This solution was a middle ground between the two previous extreme suggestions. We may, however, be overrationalizing. It is quite possible that Kepler was guided only by some considerations of celestial harmonies.

4.4.6
Kepler's Other Studies

Kepler published many books, in many different fields. Before moving on, we must note at least the last of his astronomical books – the *Tabulae Rudolfinae*, published in 1627, built according to the new principles and using the best available observations. It was used for a long time after his death.

Despite his sucess, Kepler, like his master Tycho, was one step behind Copernicus with regard to several questions! In particular, he believed at one time that all the stars were included in a thin (2 German miles in thickness) starry sphere, at a distance compatible with the "harmony" of the Universe. It is possible that this really archaic view was more or less encouraged by the discussion of the idea (see the diagram of Thomas Digges, Fig. 4.36, p. 235, or the work of Giordano Bruno) according to which the Universe is full of stars, some far away (not bright), some close-by (and brighter). If the Universe were full of stars infinite in number, in a homogeneous distribution, then the sky would not be black at

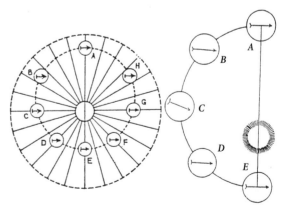

Fig. 4.35 *A further effort by Kepler to use dynamics: the magnetic analogy.* The arrow is supposed to represent the magnet-Earth. Each of the two figures is imaging two phases of the interpretation of Kepler's ideas.

night, as it is, but as bright everywhere as is the Sun itself. This idea was later formalized by Loÿs de Chézeaux and by Olbers. Under the name *Olbers' paradox*, it is among the main arguments of modern cosmology.

4.5
Galileo, Physicist and Observer

The great figure of Galileo is perhaps better known to the public at large for his difficulties with the Catholic church, his condemnation, and his renunciation of Copernican doctrine, than for his essential contributions to science. The reasons for his conflict with the Church are manifold. First, the onset of the Counter-Reformation after the Council of Trent and the increasing rigidity of the Catholic theologians, who reverted to the notion of "saving the Holy Scriptures" first, fostered an anti-scientific atmosphere. Second, Galileo himself was an aggressive individual, which made him many enemies. Third, the popular style of his writings, notably the *Dialogues*, written in the vernacular Italian, resulted in a wide readership, which allegedly mocked the ignorance of the pope. All of this led the Church to condemn at one stroke, Copernicus (in the *Index* of 1620), Kepler's Copernican treatises, and the devoted Copernican, Galileo. In addition, Galileo's atomism, also expressed in his *Dialogues*, may have been instrumental in turning the Church against him. Atomism was regarded as an argument against the reality of transsubstantiation, but this criticism is not mentioned in the text of the actual judgment made against him.

This episode was a shame for the Church, not a glory for Galileo. It had an important impact on the political and ideological evolution of the western world, but not so much upon the evolution of science. Soon after the condemnation of Galileo, many scientists, philosophers, even several priests, shifted to the Copernican point of view. Actually, Galileo, although he was a devoted Copernican, did not really contribute much to the conceptualization and the mathematics of the planetary system, as Copernicus, Kepler, and later Newton did. He did not even seem to know Kepler's laws, although he had a long correspondence with Kepler.

Still he was, basically, a great physicist, and, together with René Descartes and Francis Bacon, must be considered one of the fathers of modern physics.

4.5.1
Pre-Galilean Astronomical Observations

As we shall see in Chap. 6, Galileo was not the first to know and use the refracting properties of glass to build optical instruments. The tradition, however, and quite rightly in view of his observing achievements, considers him the first astronomer to efficiently use the refracting telescope to observe astronomical objects, the Sun, planets, Moon, and stars.

We will take a moment now to quickly review the evolution of instrumental astronomy and define the properties of any astronomical instrument, be it based on pure alignments of material devices, without glass, or on the refracting properties of glass as well.

An astronomical, or more properly, astrometrical instrument has to give us a precise idea of the location of objects on the sky, with reference to each other. Basically, one has to measure the angle between two directions on the sky, or between one direction and the horizontal plane, or between one direction and the vertical. Astrometry is a determination, a measurement, of angles. It must also give a precise location of these objects with respect to the Earth and compile a proper inventory of all these objects.

In addition, the astronomer needs to measure time, by reference to the local noon time, well determined.

Other observations, implying the quantity of light (photometry), its nature and its quality (spectrometry), are essentially made with *astrophysical* instruments. They came into use only much more recently, during the 18th century. We shall come back to them, never forgetting that, on the one hand, instrumental progress is necessary to any change in ideas, while on the other hand, the main focus of this book is to follow the development of ideas. This emphasis necessarily means that both the technical conditions for this development and its social, political, and religious context must be dealt with in an abbreviated fashion.

The ability to measure angles with sufficient accuracy is essentially a mechanical problem. Instruments must be able to resolve objects in their components. They must make angular determinations, but we should remember that all optical instruments, including the eye, have diffracting properties, i.e. they do not give a point-image of a point-source, but rather (in the first approximation) a dot, of a certain diameter, which reflects the aperture of the instrument (Fig. 4.37). The angular separation between two point-sources has to be larger than the radius of that dot

Fig. 4.36 *The Copernican World (according to Thomas Digges).* In this figure, the universe of stars is extended to infinity beyond the "last" sphere, which is only a lower boundary to the starry universe. Kepler stayed within his bounded description of the starry orb, as shown in Figs. 4.5a and 4.7.

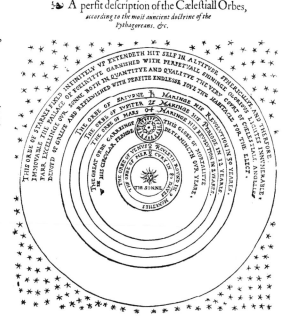

for both the point-sources (stars) to be seen. The angular radius of that dot, is, in visible light, about $\alpha = 10/L$, where L is the radius of the circular aperture of the instrument, in centimeters, when the angle α is expressed in minutes of arc. This angle α is the *resolved angle*. For the naked eye, L is the radius of the pupil of the eye, perhaps 0.3 cm at night. Hence, the minimum visible separation, whatever the mechanical properties of the instrument, is as high as 30'! But pointing the center of the dot may be done perhaps more accurately, even with the naked eye. A good observer, using large instruments, such as those used by Tycho, was able to reach a pointing accuracy of about 2'–3'.

An interesting consequence of diffraction is its limitation on the observation of faint objects. On the area of the dot, image of a star, its brightness is evenly distributed. If the star is faint, the spreading of light in the solid angle corresponding to the dot of radius 30' may be such as to make the star unobservable, even against the very faint light radiated by the night sky. If the spreading is limited to, say, only a 3' radius dot, the light is concentrated a hundred times better (one has "gained" a factor of 100, for a division by 10 of the radius of the diffraction dot) and the star, invisible to the naked eye, may become visible using even a

small instrument. The *resolved angle* determines the number of stars that
one can see.

The main instruments used in astrometry were becoming increasingly
large, as we have seen when describing Tycho's achievements. However,
astrolabes had a great vogue as far back as Ptolemy's time and were built
in several different forms up to Galileo and even later. In essence, the
best astrolabes are the "spherical" astrolabes, inspired by the *armillary
sphere* (Fig. 3.14b, p. 181 and Fig. 4.39, p. 241), which is a 3-D representation of the sky and of apparent motions within, provided its axis is
pointed towards the celestial pole. A planar astrolabe (Fig. 3.14c, p. 182)
is nothing but a convenient projection, on one vertical plane, of the armillary. The great circles, in quarter circles, used by Tycho and others,
are nothing but large planar astrolabes.

If is is relatively easy to understand how a large circle works, one is
amazed at the complexity of the astrolabe which was much more commonly in use than any other instrument of the time. The main function
of the astrolabe is to determine the local time, in daylight (measurement
of the height of the Sun above the horizon) and at night (the same for
bright stars). A more practical use was the orientation of a traveller (they
are rather light and transportable objects). Quite an interesting use of

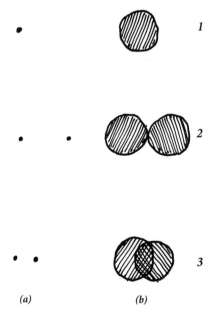

Fig. 4.37a,b *Effect of diffraction on the observa-
tions.* At the left (**a**), are
one or two point sources,
as they would be seen if
the aperture of the eye, or
of the instrument, did not
produce diffraction ef-
fects. But they do; a point
source is seen (**b**) as a dot
of a certain diameter. At
the right (**b**), the diffraction dot is represented in,
respectively, the case of:
(1) one source, (2) one
double source well sepa-
rated, and (3) one double
source where one cannot
distinguish the double
character, because of the
diffraction dot.

the astrolabe was, by considering it more or less (be it spherical or plane) as a model of the sky, to use it to compute various quantities, as a sort of abacus. Note that the word *abacus*, and many others used in medieval astronomy, like *almucantar* (a circle of the celestial sphere corresponding to the horizon), *azimuth*, and *zenith* are from the Arabic. The 9th century was a period which saw extensive progress in the construction of astrolabes, as described by al-Farghani and Albamazar. They were used in the Arabic-Spanish schools, in the Byzantine and Christian monasteries, and among Jewish scholars, from the 9th century to Galileo's time and even later, when Islamic muezzins used them for religious purposes. Most of the transportable astrolabes were plane astrolabes. This instrument has a ring and a loop, which allow it to hang vertically (Fig. 4.38a).

✳ A flat body circular in shape, the *mother,* is the main part of the instrument. It bears an axis around which many other parts are installed, as shown in Fig. 4.38b. Some of these parts are circular plates (*tympanons*), corresponding to various latitudes, and fixed with respect to the mother. The purpose of tympanons is in essence auxiliary to the computations and in practice they were often used by astrologers. Two essential parts are mobile. One is the *alidade,* which allows the placement of a certain diameter in the direction of a given star. The other, the *spider,* is a projected map of the sky, on which very precisely pointed needles point to the location (Fig. 4.38c) of some 20 to 30 very bright stars, where the zodiacal signs are indicated, and which can therefore be oriented with the pointing of the alidade towards the solar disk.   ✳

Without going into detail regarding the very astute geometry of this sky map, based on spherical triangles, we can at least indicate some of the various uses of the instrument: (i) Determination of the time, knowing the local latitude, the height h of the Sun or of a star (measured with the alidade) above the horizon, the date which in essence defines the declination of the Sun (its distance to the celestial equator), and the right ascension of the star, in the stellar case, (through the use of the spider). (ii) Determination of the location of the Sun with respect to the local meridian plane, knowing the latitude, the height of the Sun above the horizon, and the date. It is difficult with an astrolabe to reach to as low as about 10 minutes of time in accuracy. Without being astrolabes in the strict sense, the large quadrants used by Tycho Brahe were indeed an elaboration of the astrolabe, without its astrological additions and without its computing properties.

Fig. 4.38a-c *The plane astrolabe (after d' Hollander).* (**a**) Seen when used vertically hanging by the finger. For (**b**) and (**c**) see next pages.

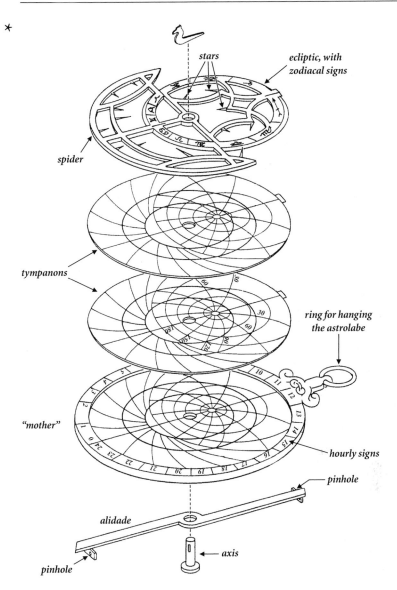

stars

ecliptic, with
zodiacal signs

spider

tympanons

ring for hanging
the astrolabe

"mother"

hourly signs

pinhole

alidade

axis

pinhole

Fig. 4.38b Reduced to its elements. The tympanons are mainly computing auxiliaries
and can be interchanged. They are to be located at a fixed position with respect to
the mother, which is slightly larger. The alidade is mobile in order to align any star
with the two pinholes which it bears.

Fig. 4.38c A drawing of the spider of the astrolabe of Abu Bekr (built circa 1216–17). The stars indicated on it, and numbered on this drawing, are, on the northern side of the zodiac: 1. ζ Ceti; 2. Aldebaran; 3. Rigel; 4. Betelgeuse; 5. Sirius; 6. Procyon; 7. Castor; 8. α Hydrae; 9. Regulus; 10. γ Corvi; 11. Spica; 12. Antares; 13. δ Capricorni; 14. ι Ceti. On the southern side of the zodiacal band they are: 15. Algol; 16. α Aurigae; 17. ι Ursae Majoris; 18. θ Ursae Majoris; 19. η Ursae Majoris; 20. Arcturus; 21. Margarita; 22. α Ophiuchi; 23. Vega; 24. Altaïr; 25. ε Delphini; 26. Deneb; 27. β Pegasi; 28. β Cassiopeiae – all very bright stars, visible from the Mediterranean area.   ✶

What time accuracy is necessary to properly make the best visual observations? A star located on the equator describes an entire sky, i.e. an arc of 360°, in 24 hours; hence a minute of time corresponds to 15 minutes of arc in angular displacement, at least for stars located near the equator of the sky. Incidentally, this correspondence shows why, in order not to be confused, standard abbreviations have been legalized: °, ', ", for degrees, minutes, seconds of arc; and: h, mn, s for the hour, minutes, and seconds of time. They are related, but different! To be able to properly exploit Tycho's measurements made at an accuracy of 2', one needs an accuracy in time of 8 s, in a sort of almost absolute time, the time in solar value, referring to a precise determination of the solar noon. This is indeed a high level of accuracy and this is not an easy task. It is under-

Fig. 4.39 *The armillary sphere, according to Petrus Apianus (Cosmographia, Antwerpen, 1584).* This is a representation of the sky, with its equator, the zodiacal path, the two tropics, and the polar axis around which the system moves, the meridian plane being fixed. The armillary sphere used a pinhole alidade to point at any direction in the sky. (Observatoire de Paris)

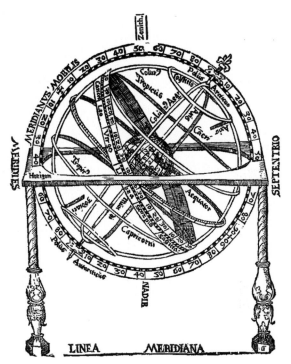

standable, therefore, why it was so difficult to properly describe the relative motions of the planets with respect to the Sun or to the fixed stars.

This is why the instruments built to measure time intervals are of such importance in astronomy. Until the time of Galileo, astronomers had only a few devices to use. Of course, sundials, and clepsydrae, were somewhat obsolete. Accurate sundials were used to determine the solar noon. The study of the passage of the Sun through the meridian was done more efficiently with accurate mechanical devices, meridian circles, analogous to the quadrants described above, but maintained in the meridian plane. To "keep" the time, the medieval observers used clocks moved by the fall of weights and regulated by dented wheels (Fig. 4.40). Wheel escapement clocks using a spring instead of a weight appeared in the 15th century and were more accurate. They are described in Galileo's *Dialogues*. Galileo had the idea to regulate them better by using pendular motions as regulators, but it was only Huygens in 1657 who really made the first pendulum clocks.

Fig. 4.40 *The mechanism of a clock, as it appeared in the 15th and 16th centuries.* (Observatoire de Paris)

4.5.2
Galileo in 1610: The Telescope and the Telescopic World

Real optics, using glass as a refracting medium, was in use before Galileo. The Venetian opticians, besides making art glass, constructed lenses of all types to improve human vision. Spectacles had been known for centuries, even if they were not perfect. Roger Bacon had claimed that he was able to combine lenses enabling him to see distant objects as if they were close by. Thomas Digges (1546?–1595) in England, Giambattista della Porta in Italy (1535–1615), and H. Lippershey (1587–1619) in Holland, used similar combinations in the late 16th and early 17th centuries. Galileo heard about their work, or perhaps, in the hands of some traveller, even saw these devices, which were displayed as an amusement at the fairs. In any case, he worked empirically on the idea, Kepler having earlier arrived at the theory of such combinations. Galileo took great care in selecting the glasses that he used to combine. By combining one convex lens and one concave one in a tube, he could obtain first a magnification of a factor of three. Later, by improving the choice of lenses, he reached a factor of 20 and even 30. Fig. 4.41 shows the principle of the Galileo telescope.

As clearly seen in Fig. 4.41, the telescope is not inverting the images of the objects it observes. This is why this combination had been used at the fairs. As inversion does not matter much when looking at the heavenly objects, much more efficient instruments were later built. In any case, as it was, Galileo made good use of the telescope. He was not, truly enough, the only one to do so. In England, Thomas Harriot (1560–1621), and in Germany, Simon Marius (1570–1624) did so too. Unlike them, however, Galileo observed constantly with remarkable obstinacy, with great skill, and with the full realization of the implications of his discoveries. His discoveries were actually published in the *Sidereus Nuncius* (1610) and later, and their implications as confirmations of the Copernican system were noted. Although they cannot be considered real proof, they did support the intuition that the Earth is an ordinary planet, with a satellite, our Moon, just like other planets, which may also have satellites, often called "moons."

Galileo's discoveries, made with a proper use of the telescope, were indeed of major importance, whatever their implications for the Copernican system.

(i) The Moon appeared to have mountains, and Galileo, from the study of the variations of their shadows, determined their height to be on the order of some miles, a rather good value. This confirmed Tycho's

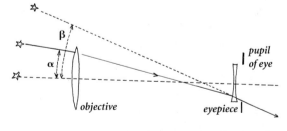

Fig. 4.41 *The Galilean telescope.* The telescope is essentially made out of an objective, a converging lens of a "large" size (a few cm at the time of Galileo) and a diverging lens in the form of an eyepiece which one looks into. The eyepiece must be mounted such that its distance to the objective is smaller than the focal length of the objective. The eyepiece can be moved along the optical axis, for focusing. A double star, the apparent separation of which seen with the naked eye would be α, is, when seen through the telescope, $\beta > \alpha$. The ratio β / α is the *magnification* of the telescope.

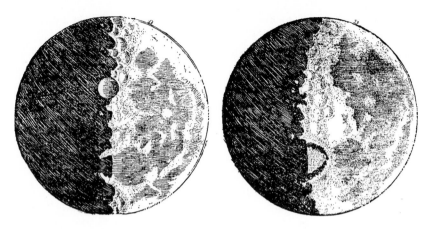

Fig. 4.42 *Two of Galileo's drawings of the Moon* (from the *Sidereus Nuncius*), showing the "libration" (apparent oscillation of the Moon around a North-South axis).

disproof of the Aristotelian idea of the perfection of the heavenly bodies of the *astral world* (Fig. 4.42).

(ii) The stars appeared very numerous. Their number was multiplied, using Galileo's telescope, by an order of magnitude, due precisely to the reduction of the diffraction dot. The diameter of the eye is about six mm at night. The aperture of the telescope, or the diameter of Galileo's first objectives in 1610 being only about four centimeters, the "gain" of the telescope is about a factor of 50, increasing the contrast between the stars and background light (see above, pp. 235–236). In the Pleiades, for example, six, perhaps seven, stars were known. Galileo found 36 stars instead. The Milky Way was indeed a great accumulation of stars, not a cloudy feature, as it was thought to be earlier. This was similar to Digges' vision of the starry world. It was also related to the mystical concepts of Giordano Bruno, who considered the Universe infinite in size, infinitely populated by stars, all surrounded, like the Sun, by a family of planets. Bruno, who extrapolated the Copernican model to other stars, basically understood the Sun to be a star like other stars. Bruno was dead by the time of Galileo's observations, but his Copernicanism was given a very strong a posteriori support by Galileo.

(iii) The Jupiter System. Galileo looked at planets, Jupiter and Saturn, as well as Venus and Mars. Around Saturn he found two patchy figures, which disappeared at times. He did not dare to announce his discovery of the ring of Saturn, except in a cryptogram sent to Kepler, but he did

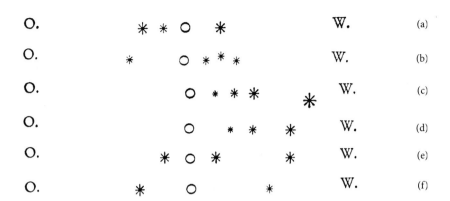

Fig. 4.43 *Jupiter and its satellites: seven successive observations* (a, b, ... f) made at intervals of several days in 1610. The configuration, always different, of the four satellites can be easily explained by assuming they are in a constant rotation around the planet and that the Earth (i.e. the observer) is practically located in the plane in which these satellites are rotating (compiled here from the *Sidereus Nuncius*). (Observatoire de Paris)

indeed discover three moons (called satellites by Kepler) around Jupiter on January 7th, 1610 and soon he found a fourth one (Fig. 4.43).

He called them *Medicean planets* in honor of his patron, or as we would say, sponsor, Cosimo di Medici. Soon, he embarked on a study of the motion of these Jovian satellites. They turned around Jupiter, just as planets turn around the Sun. This was given as an argument for Copernicanism, because of the obvious similitude between the Moon around the Earth and the satellites around Jupiter. Beyond that and suggesting a starting point for a theory of the dynamics of heavenly bodies, the fact that around Jupiter there is a miniature planetary system, just as there is a large planetary system around the Sun, led to the idea that massive objects tend to force smaller objects into orbits around them. The problem of how this force works was the object of theoretical thinking during the 17th century, from Galileo to Bacon, to Descartes and, finally, to Newton.

(iv) Venus of course was also observed frequently by Galileo. Its phases, very similar to those of the Moon, were observed (Fig. 4.44).

These appearances are quite easy to understand from the Copernican point of view (Fig. 4.45, adapted from Kuhn), and are contradictory to the Ptolemaic system, as shown in the figure. In the latter, at *opposition*, one sees the dark side of Venus, hence its brightness should be greatly reduced. Of course, this is not a proof of Copernican theory, as long as

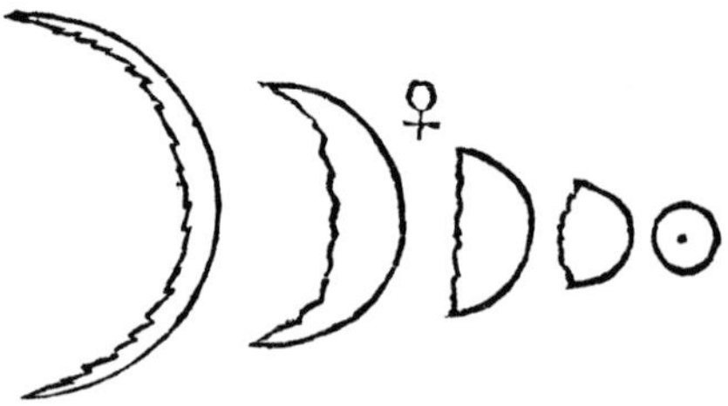

Fig. 4.44 *Galileo's observation of the surface of Venus.* It is clear that the width of the crescent is decreasing, or that the planet is increasingly crescent-shaped, while the apparent size of it is increasing (taken from *Il Saggiatore*). (Observatoire de Paris)

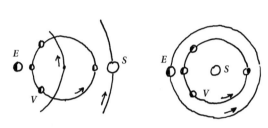

Fig. 4.45 *Venus' phases in the two world systems.* The phases of Venus, as conceived in the Ptolemaic system, at the left, and in the Copernican system, at the right (the Tychonian system is equivalent to the Copernican in this respect). According to Ptolemy, a terrestrial observer should never see more than a thin crescent of the illuminated side of the planet. According to Copernicus (or Tycho), the observer should see almost the entire illuminated side of Venus, just after or before the passage of Venus behind the Sun. This was an impossible occurrence in the Ptolemaic view. A comparison with Fig. 4.44 shows that the Ptolemaic system must be excluded, as it is unable to account for the observations.

there is no true measurement of this brightness, in our modern terms, only vague estimates.

(v) The Sunspots. Finally, Galileo studied the Sun and observed the sunspots at the end of 1610 (Fig. 4.46).

Several astronomers before Galileo, or at just about the same time, made similar observations, which did not require very elaborate instrumentation. A *camera oscura*, a large piece of cardboard with a small hole in its center, what we call a "pinhole," was all that was required to see the spots, if not to draw them accurately. Most of these observers did not attribute the observed features to the Sun, but to passages of Mercury in front of our luminary, or to other objects. This was, in fact, Kepler's understanding of the spots. Still Harriot, John Fabricius, Ch. Scheiner, s.j., and his assistant Cysat, probably came to the conclusion, at the same time as Galileo or even before, that the spots were solar.

Galileo's demonstration of the solar nature of sunspots was, however, quite convincing. In essence, they did not move in a uniform way, but appeared to turn as if they were at the same distance from the center of

Fig. 4.46 *Sunspots* (from Galileo's *Macchie Solari*)

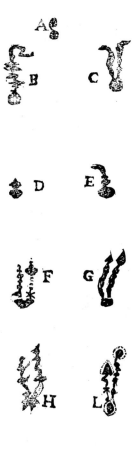

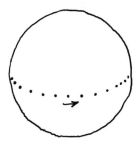

Fig. 4.47 *Solar rotation.* This schematic view shows
how a point located on the solar equator appears on
the apparent solar surface at successive days. A sunspot goes from one side to the other in about 14 days.
The period of the rotation, in an absolute system of
coordinates (i.e. corrected for the motion of the Earth
on its orbit around the Sun) is equal to 25.4 days at
the equator, and about 30 days near the poles: The
Sun does not turn like a solid body, but like a sphere
of gas. It displays a differential rotation, as scientists
learned in the beginning of the 20th century.

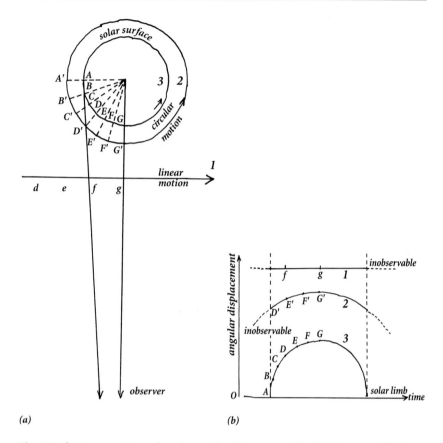

(a)   (b)

Fig. 4.48a,b *A sunspot is on the solar surface.* Let us consider (**a**) three possible locations for the "object" which we see as a sunspot. It may be moving along a straight line like a bird (1), or it may be moving around the Sun like Mercury (2), or it may be moving with the Sun, on its surface (3). At regular time intervals, the "object" would be: (1) at points a, b, c, ..., f, g; (2) at points A', B', ..., F', G'; (3) at points A, B, C, ..., F, G, as shown on the left. On the right (**b**), the actual angular motion of the "object" in each case is represented. The apparent angular displacement on the disk between two successive positions would be quite different in the three cases. In (1), they would be equal to each other; in (2), they would be shorter at the limb but not much shorter than near the center; in (3), the difference between limb and center is more accentuated. The observations are in strict accordance with case number 3.

the Sun as is the solar surface. The geometry of the problem is indeed quite simple (Fig. 4.47 and in detail in Fig. 4.48).

This discovery, like the 1572 and 1604 supernovae and like the mountains on the Moon, was proof of the imperfection of the sidereal, astral world, another blow to a strict Aristotelianism. The fact that the Sun rotates in about 27 days was also discovered by the same study of the motion of the spots.

The solar rotation was an additional argument in favor of the idea that the Earth is turning on its axis. We can imagine what Kepler's reaction to Galileo's discovery would have been, since Kepler thought that the Sun revolved in only three days. Note that, had his third law been applied to a distance to the Sun equal to the solar radius, it would have resulted in a period of rotation of neither 27 nor three, but about 14 days!

Very soon, other characteristics of the solar spots (the existence within a given spot of a dark *umbra,* and a somewhat less dark *penumbra,* the brighter *faculae* surrounding the spots) were discovered either by Galileo or by Scheiner, himself an excellent and obstinate observer.

One could say that, through the remarkable observations of Galileo, a new era in astronomy began, the era of instrumental astronomy. This was the real beginning of the studies of lunar physics, the solar system, solar physics, galactic structure and the physical exploration of the depths of the Universe.

Still, I do not think that these discoveries brought so much to Copernicanism. They could well have been accommodated by the Tychonian system, and, as important as they were, they did not at the time bring much to our understanding of the organization of the Universe. Galileo made other astronomical discoveries, for example that of the Moon's libration. However, although he advanced the progress of our knowledge of the Universe and astronomy, notably through his very popular books, which were easy to read and to understand, his work did not deeply alter our views of the World system. The really important development which caused learned society to progressively abandon Ptolemaic ideas was the practical success of Kepler's *Rudolphine Tables.* This development was accompanied by a progressive tendency within the Society of Jesus, whose members grasped rather quickly that religious dogma would have to make room for scientific discoveries if it wished to survive, even if this broadmindedness took place at the expense of the strictest interpretation of sacred texts.

4.5.3
Galileo's Mechanics

Much more important to the progress of science were Galileo's achievements as a physicist, as one of the creators of modern mechanics, of dynamics, a science which changed entirely during this period.

First of all, we must bear in mind that the past history in this field is a
course of continuous progress, from Antiquity to the mediaeval world,
and then to the Renaissance thinkers.

Aristotle thought that "natural" motion was uniform and circular, but
at that time no mathematical apparatus was linked with kinematics. How
could a motion be described? and, a still more difficult question, how
could some cause be assigned to this motion and how could the action of
this cause be described in mathematical terms? A way of thinking of motion in mathematical terms simply did not exist, apart from the very
simple considerations of uniformity and circularity. Geometry progressed, however, as did algebra, notably under Arabic influence. The
Merton school (around 1330–1350) of Heytesbury and his pupils and, in
France, Nicole Oresme (around 1360) had started to represent motion on
a graph. Uniform velocity, but also uniformly accelerated motions could
be described in such a way. The very idea of analytical integration and
derivation existed in germ form in these attempts. The notion of acceleration was introduced as the rate of change of velocity.

Galileo showed, experimentally, that a falling body was subject to a
constant acceleration. This was an experimental way to study mechanics
as well as to make astronomical observations.

Describing the kinematics of motion is one thing. Understanding its
dynamics is another. What causes the motion of moving bodies? Buridan
had reintroduced the point of view of John Philopon, followed by Averroes, and spoke again of the *impetus*. Whereas Aristotle thought that an
arrow in motion is continuously pushed by the air surrounding it, this
air progressively filling the vacuum left behind the arrow, Buridan argued that the air opposes the motion instead of helping the bodies move.
For him, the impetus was proportional to the velocity (v) of the body
and to the quantity of matter (the body's mass m). In our modern language, this notion is close to that of momentum, or quantity of motion,
$m\,v$. In addition, weight acts upon a body, at least on Earth, where this
notion has some meaning, but the heavenly bodies do not experience
weight, or gravity (at least, so it was thought at that time). So, impetus

acts along the circular path, forcing the motion of heavenly bodies to be circular.

In 1585, Benedetti used the impetus theory to explain how a sling could throw a stone along a straight line.

Without some force coming from air or elsewhere, planets, as well as stones, would actually follow "their own track" (circular). This definition, in very simple terms, is the basis of the principle of *inertia*. Galileo, as well as Benedetti, defended this concept of some inertia in motion, inertia meaning essentially that once the body is in motion, it will passively follow its natural course of motion. However, experimenting on balls rolling on inclined planes led Galileo to the conclusion that impetus, a property of motion, perhaps even a measure of motion, was not its cause, since a ball was eventually brought to a stop by friction on the plane and also by the resistance of the air. In other words, inertia is opposed by some acting force (which can accelerate or slow down the motion), impetus being only the measure of the real resulting motion.

In essence, Galileo asked the question: What force can be said to cause motion? (except that he did not use the word "force" with the meaning the word has now).

The "straight line" as a natural inertial motion was not yet discovered. The main conclusion of all these experiments was, expressed in modern language, that on the circular orbit (and on Earth, motion on a plane is a circular motion around the Earth, like that of a boat), one must apply a relation of the type:

$$F = \pm\, m\, \gamma \,,$$

where γ measures the acceleration, i.e. the rate of change of velocity with time, along the trajectory (the sign + in accelerated cases, the sign – in the case of deceleration). The study of the motion of bodies, the (perhaps imaginary) experiment from Pisa's inclined tower, the laws of falling bodies, all became the subject matter of Galileo's experiments, which paved the way for the further development of dynamics and modern mechanics. We shall come back to this subject when describing the Newtonian system.

It is obvious, at this point, that a massive intrusion of mathematical formulation must now develop. Galileo and Kepler were both ardent propagandists for the mathematical description of the Universe. Whereas Copernicus thought that mathematics was more or less a plaything for mathematicians, Galileo said clearly (in the *Saggiatore*) that "philosophy is written in the open book in front of us, the Universe; one can understand it only on the condition that one uses the language in which it is written; and this language is the mathematical language, in which letters are triangles, circles, etc..., without which any human understanding of it is quite impossible, without which one wanders in an obscure labyrinth." His point of view could not be more clear.

However, one thing must perhaps be clarified. Galileo had seen many things, observed the sky with great skill and had a true genius for realizing the importance of good and continuous observations. He introduced the need for a mathematical formulation of physics. However, one may be tempted to give him more credit than he deserves, in some fields at least. I have said that he did not contribute much to our understanding of the structure of the Universe. Likewise, although he made many experiments and set the basis for the principle of inertia, he did not present it in a satisfactory form, to be used in the mathematical description of motion. In fact, mathematics was not yet sufficiently advanced, but this period is certainly the dawn of a systematization of all that had been accomplished previously. It is the outstanding contribution of Galileo that he risked his life, pushing with such tremendous energy, against winds, tides and priests, towards the splendid achievement of Newtonian physics and astronomy.

CHAPTER 5

Dynamics Enters Astronomy:
From Galileo to Newton

5.1
Galilean Dynamics

We have seen how Galileo, the physicist, described the free fall of bodies and suggested the possibility of the existence of inertia. Clearly, through his experiments, Galileo was one of the founders of modern mechanics. However, the situation was far from simple. Galileo did indeed succeed in describing motion in a way which approached a mathematical description of the kinetic reality. Nonetheless, Galileo, in his effort to arrive at a complete explanation, to actually build a dynamic model of the observed kinematics, was often unsuccessful. His basic principle of a proportionality between the velocity of the moving body and the path already covered was obviously wrong, although it was almost true in a small number of cases, such as that of the almost uniform motion of slowly falling bodies. The truth, expressed in our mathematical language, is that in the uniformly accelerated motion (γ = constant), the velocity v is equal to $K (t-t_0)$ hence the acceleration γ (the derivative of v with respect to time t) is constant, equal to K. Galileo would have written $v = K (x-x_0)$; hence $\gamma = Kv$ (v being the derivative of space with respect to time t), which is clearly wrong! Note that the constant K is, in the case of falling bodies, the acceleration of weight. If one attributes this acceleration to a force f, this force is equal to $m\gamma$, where m represents the inertial mass of the quantity of matter put in an accelerated motion. This is the way Galileo's experiments were expressed, not by him but later, notably by Newton.

Let us, as a typical example of some of his misconceived ideas, give an account of the theory of tidal motions according to Galileo, and presented as a central proof of the Copernican system in the fourth day of the *Dialogues*. To quote Galileo: "There are many who refer the tides to the Moon, saying that it has a particular dominion over the waters; lately a certain prelate has published a little tract wherein he says that the Moon, wandering through the sky, attracts and draws up toward itself a heap of water which goes along following it, so that the high sea is always in that part which lies under the Moon." It may be interesting to note that the "prelate" was Marco-Antonio de Dominis, Bishop of Ris, in Illyria (now in Croatia), a prelate who was indeed condemned by the Church (like Giordano Bruno who was burned at the stake) and whose works were burned at the same place as he himself was later cremated! Galileo completely rejected de Dominis' idea, because it contradicted some "obvious" aspects of tidal motion. For example, the Moon crosses the Mediterranean sky every day from East to West, but the tides are observable in Venice but to no great degree elsewhere, such as in Corfu. To attribute the tides to the Moon would be a miracle and Galileo did not believe in miracles, or in action at a distance. He even wondered why such a distinguished mind as Kepler "despite his open and acute mind ... has nevertheless lent his ear and his assent to the Moon's dominion over the waters" and to such "puerilities." To Galileo, the motions of oceans and sea waters were so dominated by local conditions that the dominion of the Moon was an absurd myth.

So what was Galileo's theory?

For him, the tides were closely linked to the motion of the Earth (one of the Copernican "conquests") around the Sun and around its axis (Fig. 5.1).

At night, the Earth counteracts the motion attributed to its revolution around the Sun. During the day, it follows it. As the Earth's motion has a dragging effect on the waters (as upon anything), and as the waters have some inertia, a tide results, but a once-a-day tide! The facts contradict Galileo's theory since, even in Venice, high tide occurs twice a day. It is important to note that de Dominis and Kepler's intuitions, linked only to the Moon's attraction, also gave rise to once-a-day high tides!

However, the Greeks (Poseidonius for example) already knew about many aspects of tidal motions that are absent from the Galilean description. Pope Urban VIII, still a friend of Galileo's in 1630 when the completed book was submitted for his approval, tried to convince Galileo that it was wrong to publish his tidal theory, which was even, so it seems, the initial title of the most important of his "Dialogues", later published as *Dialogue on the Two Main Systems of the World*. Galileo was as stubborn as he was brilliant. He followed the Pope's suggestions, but only reluctantly, so proud was he of his misguided tidal theory.

Fig. 5.1a,b *The Galilean theory of tides.* **(a)** The theory of tides according de Dominis, who, like Kepler, noted the lunar effect on the tides. **(b)** According to Galileo, the low tides occur on the dark side of the Earth where the rotation of the Earth is opposite to its revolution around the Sun. On the sunny side of the Earth, we have high tides because of the reverse effect.

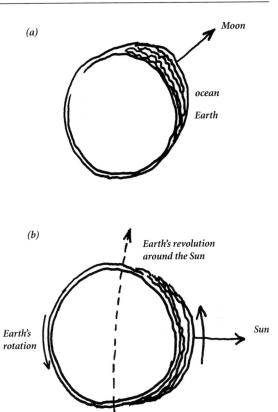

We, nevertheless, will have to wait for Newton to give a correct and complete theory of tides, which will take into account the proper dynamics of the motions in general.

5.2
Francis Bacon in England

In 1620, Francis Bacon published the *Novum Organum,* which took advantage of the astronomical progress being made in central and northern Europe and the mechanics developed primarily in Italy. England was not at that time a very peaceful country. Unlike Galileo's Italy, it was not dominated by the reigning power of the Jesuits. The struggles in England extended beyond the Catholic community led by Mary Stuart, to the newly created Church of England of Henry VIII and Elizabeth I. This heterodoxy may have given more freedom of speech to English thinkers, as is often claimed, but this is only partly true. Bacon, for example, born in 1561, was initially protected by Elizabeth and later by James I, but was ultimately convicted of corruption in 1621 and died sadly in 1626. As everyone knows, many commentators attributed some of Shakespeare's plays to Bacon. Although this is probably not true, he nevertheless wrote several books and had a great influence on his contemporaries both in England and on the continent. His primary contribution must be considered the efficient and somewhat prophetic vigor with which he advocated the strength of the experimental method. This was by no means original. Gilbert, Galileo, and Kepler, almost his contemporaries (see Table 4.1), contributed much more to the progress of science by actually spending a large amount of time experimenting, measuring, and using the measured data. The primary emphasis was given to the observed phenomena and to physical experimentation. It was only in this way that so-called modern science really began, dominated by the progressive acquisition of better and more numerous experimental and observed data. Perhaps, it would be more accurate to say that modernism in science was actually defined by this development, more than by the evolution of ideas or concepts. Both were so strongly interconnected, however, that it is often difficult to study them separately. Whatever Bacon's small contribution to science itself may be, we will credit him with expressing the new principles (really Aristotelian principles!) in a very systematic and convincing way. He was a good propagandist and an intelligent philosopher, but not a scientist, and still less a mathematician.

The influence of Bacon's ideas in England during his lifetime was considerable. After the victory over the Spanish Armada of Philip II, England was on the threshold of its world power, mostly on the seas. Science developed to a great extent in the service of the navy, as well as for the sake of knowledge itself. The title of Bacon's posthumous book, *The New Atlantis*, speaks for itself.

Under the influence of Baconian philosophy, and often against the establishment, a few wise men gave a new impetus to the scientific world. One was John Wilkins (1614–1672) in Oxford (where the old tradition was still vital). The Royal Society, created in 1660, from individual initiatives, became a very active nucleus of science. It was an organization characterized by an extreme openness of viewpoints. Its members were actively tolerant, although in general tended towards a strong belief in the Creator of the World. Among the famous members of the Royal Society, we should mention Boyle (the experimentalist who, before Mariotte in France, discovered the law of compressibility of gases), and Robert Hooke, astronomer and expert in mechanics. The Royal Society was open to illustrious foreign scientists. Its correspondents were none other than Hevelius, Huygens, Spinoza, Malpighi, Auzout and Cassini. This group of scientists did perhaps more for the triumph of human reason over irrational bigotry than any other group in the world. It is natural that we will see in this group the rising influence and the chairmanship for quite a time of Isaac Newton, who still dominates today's science by virtue of his work, so short and concise, but so very far-reaching.

5.3
The French School: Descartes and His Contemporaries

During the Middle Ages, the French universities produced a large number of good scholars. However, the bitterness and violence of the religious wars had considerably weakened the country. Isolated philosophers like Montaigne, or epic poets like Du Bartas, were indeed more interested in human nature than in the outer world, the "natural" world.

However, Henry IV, the French king, inspired perhaps by the creation of the Sacred College of the Jesuits in Rome, encouraged the French Jesuits to create schools and promote scientific studies. Perhaps under their influence, a few good spirits emerged from these schools, such as René Descartes (1596–1650). As early as 1611, at the Collège de la Flèche in western France, he studied the Galilean discoveries, as well as the Aristotelian dogmas, from his Jesuit teachers. An excellent student in philosophy, as well as in physics and mathematics, he published several well-known books and had a very lasting influence. We shall concentrate on his manifold contribution to natural philosophy. For Descartes, tradition had little weight; at least he disputed its undue influence, and he wanted to think for himself.

5.3.1
Cartesian Mechanics: Huygens and Centrifugal Force

Descartes' contribution to mechanics is essential. He asserted that the "natural" motion of a body is a straight line, not a circle, and he claimed that deviations from the straight line (hence circular motions) have to be the result of the action of some force. This is a systematization of the inertia principle, more or less implied in the works of Galileo, of course, but also of Beekmann, Cavalieri, and Gassendi. The innovation of Descartes was his geometrical conceptualization of the inertial straight line. Descartes thought in geometrical terms, hence he considered motions in a three-dimensional space, a Euclidian one of course, in which he defined the so-called "Cartesian" coordinates. He looked for a force that would be able to keep a ship on the ocean moving on a spherical path. This force was the weight of the ship. Descartes introduced several other mechanical principles. Notably, in studying collisions, he arrived at a version of what we now call the law of conservation of momentum or of conservation of quantity of motion.

The Cartesian view was based on metaphysical thinking, not the result of experiments. However, experimental developments followed soon after. The influence of Father Marin Mersenne, a Jesuit, friend, and contemporary of Descartes at the Collège de La Flèche, where they were both students, was considerable. Mersenne generated a vast network of scientific ideas and suggestions through a very elaborate, lengthy, and voluminous correspondence with everyone of importance in the Europe of scholars and scientists. In southern France, Nicolas de Peiresc had been one of the first users of Galileo's telescope.

Fig. 5.2 *Centrifugal force.*
If a given kinematic system has a "center," of certain privileged properties, its effect is as follows: a certain weight is attracted by it, but this force or attraction is compensated by a centrifugal force linked to the rotation of the body in its equilibrium path around the center, similar to that of the motion of a ship around the Earth. The centrifugal force equilibrates the weight. It is an inertial force. Without the weight, the body would move along a straight line.

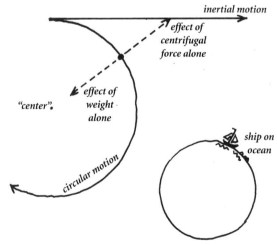

A remarkable scientist of that time, influenced by Mersenne, and critical of Descartes, especially the latter's views on collisions, was Christiaan Huygens (1629–1695), a Dutchman. We will return shortly to his astronomical and mechanical inventions (p. 280). His reflections and experiments led him to the discovery of *centrifugal* force. This force could create acceleration or compensate for another acceleration, thereby opposing the weight (in a motion around the Earth), or the effect of material binding (hence of a binding force) (Fig. 5.2).

The prefix *centri-* is obviously essential here. It clarifies Descartes' principle in that it shows that a linear motion, in the vicinity of the Earth where objects have weight, results from the compensation for the weight by a centrifugal or inertial force! The straight line is the natural motion. In a vacuum there is no weight, no weight to counterbalance, hence no centrifugal force, unless rotation is imposed by some force. In a strict vacuum, there is no imposed force and no center. This debate continues in the late 20th century, as the question of a "center" to the Universe immediately gives rise to the problem of the existence, or absence, of an absolute system of reference in time and space or space-time.

This idea of centrifugal force spread quickly within the astronomical community and penetrated the methodology of celestial mechanics. Giovanni Alfonso Borelli (1608–1679) applied Huygens' formula of centrifugal force to the circular motion of planets and concluded that this motion is due to the attraction of the Sun (Fig. 5.3). Obviously, to us, this solar force is similar in nature to the effect of weight in the case of a body on Earth (a ship for example). This similarity was not, however, fully formulated by Borelli. Descartes, Hooke, and Newton had to continue Borelli's work.

5.3.2
Descartes' Theory of Vortices and Some Reflections on Space and Time

One of the main points which has fueled intellectual conflict since the time of Galileo has been the question of force in a vacuum. If there is some force acting on celestial bodies, how can that force be exerted in a vacuous medium? How can anything act at a distance, without material contact? How can this action be instantaneous, given the great distance between celestial bodies?

Remember Galileo was also worried about this problem. His response is explicit in his theory of tides. He suggested some kind of local drag

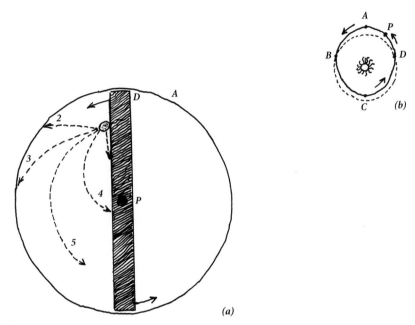

(a)

Fig. 5.3a,b *The planetary theory of Borelli (adapted from Kuhn).* (a) The model is a basin full of water A, in which a cork C is floating. The cork C and the center of the bar P are magnetized, attracted to each other. Normally, left to itself, C would move towards P (1). A bar D is turning in the basin. If this bar were violently moving, giving an impulse to C, C would move along a straight line (2). The bar, in a rapid motion (3), pushes the cork in a spiralling motion, from its initial position to the outside of the basin. In a slow motion, the magnetic attraction dominates (4), and the bar D pushes the cork towards the center of the basin. At a convenient intermediate velocity (5), the cork moves along a circle – a Copernican orbit, in essence. (b) Borelli's derivation of an elliptical orbit. When the planet moves along the circular orbit (in discontinuous lines), the centrifugal force compensates exactly for the *anima motrix* (the moving attracting force). If the planet is put in position A (by some random effect?), it will be slower, as in Fig. 5.3a where bar D is moving more slowly. It will, therefore, describe a spiral towards the Sun, like the cork in motion (4). On the other hand, in position C, the anima motrix is greater, the motion faster, hence the motion of the planet, like the cork in motion (3), will describe a spiral away from the Sun. Borelli assumed (without much of a demonstration!) that this constitutes an ellipse. It is an eccentric, in any case, and probably an "oval" (remember Kepler's doubts).

acting on the moving bodies. One could say that Galileo's tidal theory is very close to the vortices theory of Descartes, which we shall now describe. It was indeed in order to arrive at a solution similar to Galileo's by a clear denial of the possibility of action at a distance, acting presumably without material vectors of this action, that Descartes developed his widely discussed theory of vortices. This theory dominated cosmological thought in France till at least 1730, whereas in England Newtonian cosmology, a drastically different solution, was widely accepted long before. The battle raged up to the middle of the 18th century. Even then, new observational and experimental data had to be gathered in order to resolve the conflict. The theory of vortices had one grave defect. It was in essence a qualitative theory.

Throughout antiquity and the Middle Ages, philosophers were preoccupied with the problem of whether or not the Universe extended to infinity.

Leucippus and Democritus were both atomists, in so far as the structure of matter is concerned, as was Epicurus at a later date, and much later again Bruno and Galileo. In fact, as we have said, this attitude was not entirely irrelevant to Galileo's condemnation by the Church. An atomist also, Descartes envisioned an infinite Universe composed of many smaller systems – many suns, many planets in motion around them, a system indeed quite similar to Bruno's. This was a construction of the macrocosm as an aggregate of elementary bodies, similar to the microcosm. It was, in fact, the reverse of the notion of the medieval astrologers and medical alchemists, who went from the structure of the Universe as they saw it, to the microcosmic world and the structure of the human body.

Descartes posited an infinite Universe, without Empyreum! Again, this concept about the geometry of the Universe was not new, but it was systematized by Descartes. Nicola da Cusa had already expressed similar ideas for purely metaphysical reasons. For him, a limited sphere was not compatible with the omnipotence and the glory of God. Copernicus, who suggested that the starry vault was essentially fixed and deep, brought back the idea of an infinite Universe. Digges, as we have said (p. 235) came to the same idea, at about the same time if not earlier. On the other hand, Kepler rejected this idea (because the sky did not look bright at night, a prefiguration of Olbers' paradox), and reconsidered a very thin starry sphere. Bruno, in essence, reconciled the idea of infinity and the world of appearances, for reasons which had little to do with observational astronomy. Infinity, linked with atomism, was also associated with the vacuum. Since matter was now discontinuous, the vacuum could be anywhere, could fill any space, so-to-speak, between celestial bodies.

All the principles of Cartesian philosophy derive from a certain duality between atomism and Copernicanism, in an infinitely extended fra-

Fig. 5.4 *The birth of a vor-*
tex (in French: *tourbillon*)
according to Descartes. A
particle would follow an
inertial track, along a
straight line, until collisions with similar parti-
cles cause the incident
particle to deviate. Successive collisions create the
conditions for a vortex.

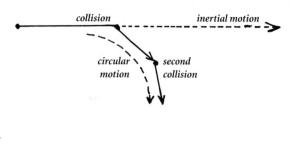

mework, an Euclidian space. It is within this frame of thinking that
straight line motions were introduced, as the only strictly inertial motions, far from any Sun. An important aspect of this world view is Descartes' conception of the vacuum between stars or planets. At every
point, this vacuum has a corpuscular structure (a renovation of the
aether, perhaps!). The collisions between these corpuscles constitute the
motor by which all things move. Vortices are formed, since the only
stable motions are indeed, and cannot be otherwise, circular (or elliptical). Each vortex becomes a new solar system, in which the motions result from the combined effects of inertia and collisions. Collisions perfectly equilibrate the centrifugal tendency given to each particle by the
centrifugal inertia (Fig. 5.4). These circular motions, at the eyes of the
cyclonic system of vortices, produce a continuous vibration, which itself
generates waves, i.e. light.

We shall not describe in all its details Descartes' very elaborate theory
of vortices (Figs. 5.5, 5.6, and 5.7). The theory attempts to explain everything – tides, motions of projectiles and weight itself, which is linked to
collisions between particles.

The main aspect of the theory, as we see it now, is that it enables us to
conceive of non-instantaneous action at a distance. Since the origin of
the forces keeping the planetary motions stable against inertia must be
located in the Sun, the vacuum must be filled so that forces are transmitted point to point progressively from the inside of a whirl or a vortex
to the outskirts of the system.

This concept was superseded by the Newtonian idea of instantaneous
action at a distance, gravitation. However, modern physicists, reintrodu-
cing the essential cohesion of the distribution of matter and the very
structure of space (Einstein), and the modern view of aether (Dirac), are

Fig. 5.5a,b
(a) *Descartes' description of the solar system* (from *Le monde ou le traité de la lumière*). Points S, E, A, and ε, are the centers of vortices, contiguous to each other. Rapid motions make these centers luminous. These are the stars. S is the Sun. Around it, the black points are the planets of the solar system. The body C (near the top) is dragged from vortex to vortex. It is a comet, moving in the outer regions of the vortices, from point I (at the upper right) to point R (at the upper left) where the motion is too slow to keep the vortices moving in circles. For (b) see next page.

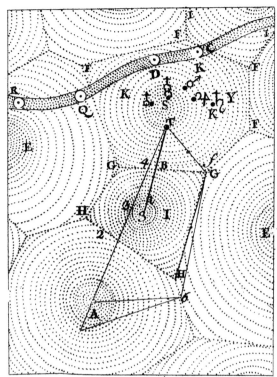

in a way returning to a Cartesian point of view. Contemporary physics, of course, has the benefit of observations impossible in Descartes' time. Again, we see here the pendulum swinging from one philosophy to another. Galileo and Descartes were Aristotelians, sticking to the observed facts and more or less excluding any *deus ex machina*. On the other hand, Newton's impeccable description represents a Platonic point of view, where some idealistic perspective permeates the idea of instantaneous action at a distance, similar to ancient and medieval notions of the influence of the heavens on the affairs of men.

Descartes' theory was too qualitative to be very convincing (except in France!). It left open many questions for Newton to ponder, as expressed by the great epic poet Pope, in a significantly religious tone:
"Nature, and Nature's laws lay hid in night;
God said: 'Let Newton be!' and all was light."

Fig. 5.5b *Descartes' description of cometary motion* (from *Principes de la Philosophie*, 1657). A comet, born in a small vortex, unable to produce a real star, follows the lines of least attraction by successive vortices. When it crosses one vortex (here, vortex Y near the top of the figure), it is because it has strong inertia, and a high initial velocity.

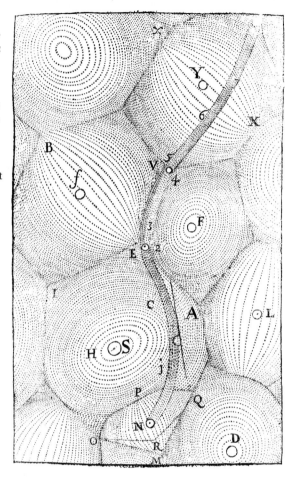

Newton had to take into account the discoveries made previously, and often quantitatively expressed. There is no essential difference between the astral and physical worlds (Tycho vs Aristotle). There is no special role for the Earth, essentially different from other planets (Copernicus vs Ptolemy). The motion of planets follows very precise laws (Kepler). The concepts of the dynamics of motion are emerging (Galileo, Hooke, Gassendi). Astronomical optics is born and applied to the celestial realm (Galileo), leading to new methodologies for checking the theories and world systems. All of this was indeed the garden in which Newton was left to pick the best flowers.

Fig. 5.6 *The rotation of the Earth* (from *Principes de la Philosophie,* 1647). The particles of the vortex A push the Earth towards Z. They have a tendency to avoid the Earth, to move in a linear fashion towards B, instead of D. Hence, they force the planet to turn in the direction ABCD.

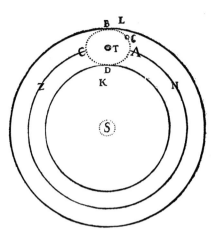

Fig. 5.7 *The Cartesian concept of the Earth's attracting effect* (from *Principes de la Philosophie,* 1647). T is the Earth's center. The Moon is at B. EFGH is the Earth. Curves 1234 is the water, 5678 the air. The space within ABCD is filled by the matter of one vortex, as described in Fig. 5.6 (ABCD). The motions within it of the particles of the vortex keep the air and the water from escaping from the Earth. A rock will fall and be dragged at the same time by the Earth rotation (between 1234 and 5678).

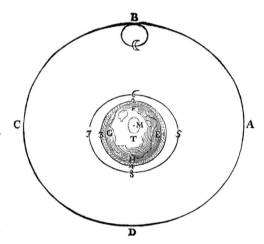

Nonetheless, the Cartesian attempts have attracted the attention of many thinkers to the very nature of space and time.

Descartes, basically involved in geometry, did not conceive of the time dependence of events. He considered space the "locus" of motion, but did not go much further in his assessment of the properties of space, essentially linked to the system of vortices, hence more or less absolute. For him, matter is to be more or less identified with space, as it was for Einstein three centuries later.

5.3.3
Objections: Henry More and Space; Isaac Barrow and Time

These opinions were often considered weaknesses of the Cartesian world view. Henry More (1614–1687), one of the brightest Cambridge scholars, a very deep student of Cartesian theory, did not accept this identification between space and matter. In other words, space is an entirely immaterial entity for More. This is a characteristic it has in common with spiritual entities. On the other hand, for Descartes, matter was divided into vortices of corpuscles. More was a very religious person. For him, the Cartesian ideas were marked by some impossible materialism. One cannot, as Descartes attempted, explain phenomena by their mechanisms. One can only explain their immediate causes. Although More accepted the system of vortices, he claimed that it was not sufficient, and that some "Spirit of Nature" (i.e. God) must be the ruler of the World system. More's arguments belong more to metaphysics than to astronomy. No undisputed fact is really advanced in support of his ideas. It is noteworthy, however, that More's thinking is not entirely foreign to the Newtonian views. It is easier to put God in Newton's system than in Descartes'.

More looked at the nature of space. At about the same time, Isaac Barrow (1630–1677), a Professor at Cambridge, a learned theologian, clever mathematician, and skilled optician, went so far as to leave his Chair (a Chair of Mathematics) to none other than Isaac Newton, because he thought that he could not adopt Euclid as a Bible. Barrow certainly had a strong influence on the Newtonian concept of time. For Descartes, duration is a property inherent to each body. For Barrow, time has to be an absolute notion, the duration of each thing being its perseverance, as such, in its state, in its motion. Each thing has to be located with respect to other things, hence with respect to an absolute frame of reference of time. Barrow's analysis is extremely fine and subtle.

Some modern commentators have gone so far as to suggest that the Cambridge scholars, More and Barrow, were in some way building modern space-time at Trinity College. I do not think so. The More-Barrow construction was indeed close to the concepts of time and space as they were understood before the General Theory of Relativity. The concept of space-time, as it is now understood from relativistic thought, cannot be dissociated from the distribution of matter and, in spite of the weaknesses of the Cartesian view, is closer to it than to the Newtonian.

As we have stated, space, according to Euclid, with Cartesian coordinates, is infinite. Descartes wanted to fill it with "matter," or with material vortices. Time, in Barrow's view, is eternal. We can put along side these ideas the famous statement of Blaise Pascal: "The eternal silence of these infinite spaces frightens me." ("Le silence éternel des espaces infinis m'effraie.") It is interesting to see here the various tendencies of the 17th century, joining each other in statements that so strongly violate the creationist letter of the Scriptures. Although the mood of the time was a return to some form of Platonism, implying some Creation, the old Thomism is also present, perhaps close to Jansenist thinking. The concept of Creation is of a metaphysical nature and does not contradict the mathematical Euclidian-Cartesian concept of infinite time and infinite space.

5.4
Newton and Universal Gravitation

As we have already said many times, astronomical observations, continuous, and regular in their progress, are the necessary background for the emergence of new theoretical syntheses of the old observations and the new. The 17th century, after Galileo, was a period of great advances in instrumentation and many discoveries. On the basis of new observations, the 18th and 19th centuries saw the triumph of Newtonian mechanics, until questioned, of course, by new experiments, such as those of Michelson and his collaborators. At the dawn of the 20th century, although we are only starting to fully realize it a hundred years later, the balance swung back again from a Platonic point of view to something closer to an Aristotelian perspective. In contemporary science, those reluctant to fully accept the Einsteinian Universe are now returning to either a neo-Platonism (the so-called big bang), or a neo-Pythagorianism (the physics of the Grand Unification, or the beauty of superstrings and the like).

But let us stay with Newton for the time being.

As legend would have it, Newton used an apple tree instead of a telescope. This is of course not true, but it is a highly symbolic representation of the fact that Newton's work was essentially a musical pause after a period of observational progress and a synthesis of the progressive evolution of ideas before him. Again, it is difficult to conceive of Newtonianism as a revolution. It is more appropriate to view it as key point in the formulation of a progressive evolution.

5.4.1
The *Principia*

The idea of attraction was not new, nor was that of inertial motion. Nevertheless, although Copernicus and then Kepler attributed the cohesion and sphericity of celestial bodies and the tidal motions to attraction (gravity) between neighboring material objects, they still considered the orbital motions of planets as essentially of a different nature. Galileo thought that inertia acted in a circular way, which meant that the permanence of the circular motions of planets did not need to be explained. Descartes thought of inertia as a linear motion and he proposed an aetheral fluid to drag the planets in their motion. Huygens tried (in 1669) to establish some relation between gravity and orbital motion, in an obvious reaction against the qualitative views of Descartes. Finally, Borelli (in 1663) and Huygens (in 1673) succeeded, by successive steps, in building a rather complete theory of centrifugal force.

The predominant idea of the time was already that an attraction directed to the center of the celestial orbits works upon celestial bodies, opposing the effects of inertia. The question remained as to how this attraction would be a function of distance. Of course, it would decrease with the distance, but how? Newton not only answered this question, but he expressed it in mathematical terms. He constructed a coherent explanation of Kepler's laws, then completely accepted everywhere, through the extensive distribution of the Rudolphine Tables, so efficient in astronomical (and astrological!) computations.

Newton's *Principia* was first published in Latin, in 1687, with a substantial preface by Newton, and a grand ode by Sir Edmund Halley. It was printed by the famous Samuel Pepys, published again in 1713, and a third time in 1725–6, with a new preface written by the old Sir Isaac. The first translation in English is credited to Andrew Moote in 1729.

This book, unlike Galileo's *Dialogues*, or Descartes' *Le Monde ou le traité de la lumière*, is a work of applied mathematics, with propositions, or postulata, theorems, lemmae and scholia. It is as dry as it could be. Several geometrical properties of ellipses are included, but often without demonstration. The book has the accuracy of mathematical language and the strength of rigor and coherence. It is noteworthy and significant, that Newton never quotes his predecessors, which fact certainly helped later scholars to forget about them!

The *Principia* begins with definitions, such as those of the "quantity of matter," the "quantity of motion," the "vis inertia" or inertia, the "impressed force," the centripetal force and its quantities, "accelerative quantity," and "motive quantity." Having defined his terms, Newton was able to express the laws of motion as follows:

Law 1. *Every body continues in its state of rest, or of uniform motion in a right line, unless it is compelled to change that state by forces impressed upon it.* (This law is already essentially present in Galileo's thinking.)

Law 2. *The change of motion is proportional to the motive force impressed and is made in the direction of the right line in which that force is impressed.* (A principle of dynamics)

Law 3. *To every action there is always opposed an equal reaction; or, the mutual actions of the two bodies upon each other are always equal and directed to contrary parts.* (The so-called "principle of action and reaction")

These laws are completed by several corollaries and scholia. From this apparatus the laws governing the motion of bodies were first demonstrated in a series of lemmae, scholia, and propositions or theorems.

5.4.2
Kepler's Law of Equal Areas

The first proposition derived by Newton was the area law (Kepler's second law), not as an empirical fact, but as a theorem deriving from the propositions as follows. Here, we will adhere to Pannekoek's derivation. It is in essence the same as Newton's, but expressed in modern terms and somewhat elaborated for clarity.

Kepler's Proposition: *The areas which revolving bodies describe by radii drawn to an immovable center of force lie in the same immovable plane and are proportional to the times in which they are described.*

Newton's original figure is reproduced in Fig. 5.8. The Sun, at S, exerts a centripetal force. In a given time interval ΔT, if this force were not effective, the body would be moved from a point A to a point B, in an inertial motion. In B, we represent the integrated action during the time ΔT of the centripetal force, as if the position of the body would be changed

Fig. 5.8 *Kepler's law of*
equal areas as demon-
strated in Newton's Princi-
pia. This original figure
from the *Principia* is made
more explicit in a modern
way in Fig. 5.9.

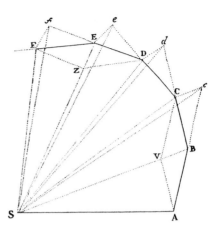

by an additional centripetal motion BV. So instead of continuing along
Bc, during the next time interval ΔT, the body will move along BC.

Geometrically (Fig. 5.9), we see that AB = Bc, as ΔT is uniquely chosen.
In addition, BV = Cc, and both BV and SB are parallel to Cc. The triangles SBC and SBc have, therefore, the same base SB and the same height
$h = $ cH, as Cc is parallel to SB. Next, AB = Bc. Thus, the two triangles SAB
and SBc are equal, as they have the same base AB and Bc and the same
height $k = $ SK. Therefore, SBC = SBc = SAB.

The equality SBC = SAB demonstrates Kepler's second law, the so-
called law of equal areas. To return to Fig. 5.8, one can, therefore, extend
what we have said to the triangles SAB, SBC, SCD, SDE, SEF, all swept in

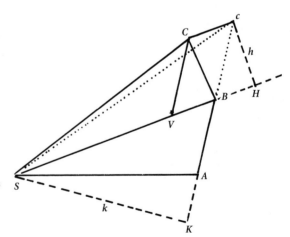

Fig. 5.9 *Kepler's law of*
equal areas. The two areas
SAB and SBC, described
in the same time interval
ΔT, are equal (see text).

the same time ΔT. If this time interval is reduced to infinitesimal value δT, so that the force can be said to be acting continuously, then the law still holds and the trajectory, instead of being a polygonal curve, is a curved orbit.

5.4.3
Newton's Law of Gravitation: The Inverse Square of Distance

Newton used Fig. 5.10 to demonstrate that *if* motion is elliptical, *if* the force PS is centripetal, and *if* the "center" is the focus of the ellipse, then the force f is equal to K/r^2, where the constant K is equal to $2a/b^2$, a and b being the semi-axes of the ellipse.

Newton's demonstration requires a knowledge of the geometrical properties of ellipses. Regrettably, modern readers are not as familiar with these properties as were Newton's readers (or Pascal's, or even Apollonius' readers). We will begin, therefore, by enumerating the properties considered by Newton.

The semi-axes of the ellipse being a and b, the foci S and H, as in Newton's book, we can assert, without demonstration here, the following purely geometrical propositions: (1) $PS + PH = 2a$; (2) $EP = a$; (see Fig. 5.10) (3) $h/a = b/b'$; (4) $d/e = a/a'$ (see Fig. 5.11). The demonstrations are given, for the curious reader, in the appendix to this chapter. It should be pointed out that Newton was always very concise in reducing his demonstrations to the strictest essentials, assuming, thereby, a considerable knowledge on the part of his readers concerning the geometrical properties of ellipses and other "conical" curves (parabolae, hyperbolae).

Let us now introduce motion, i.e. the time element of the orbit. We shall use an amendment of the Pannekoek derivation, which is close to

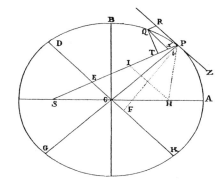

Fig. 5.10 *The inverse square law, according to Newton's Principia*. The original figure from the *Principia* is shown in a slightly different way in Fig. 5.11.

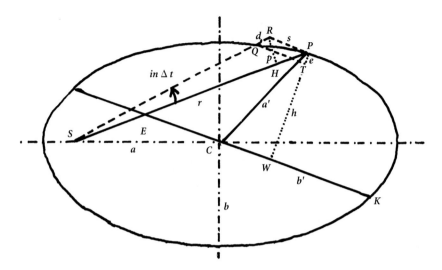

Fig. 5.11 *The inverse square law.* The figure is adapted from Pannekoek. See the text for the demonstration. See also the appendix to this chapter for a more correct and complete demonstration.

Newton's but perhaps easier to understand. Let us assume that in its motion, after passing P, the planet is slightly deviated from the elliptical orbit and subject only to the centrifugal inertia, following, therefore, a tangent to the ellipse. The planet would then reach point R after a time interval Δt. The deviation due to the centripetal force would be d. This deviation, which is in essence the "fall" of the planet under its "weight" in time Δt, would be $d = f(\Delta t)^2$, following the law of the acceleration of falling bodies observed by Galileo. But, because of Kepler's second law, Δt is proportional to the surface of the triangle RPS, if we consider the arc RP as sufficiently small. The proportionality constant may be considered equal to unity. Then, $f = d/(\text{area (RPS)})^2$.

If (Fig. 5.11) we now draw RH $= p$ normal to PS, the area of PRS is equal to SP $\times p/2$ or letting SP $= r$, to $rp/2$. Let us also draw the normal PW $= h$ on EC, the axis of the ellipse parallel to the tangent PR. The triangles PWE and PRH are "similar" to each other. As the tangent to the ellipse in P, PR is parallel to the axis EC, by construction. Therefore, $p/h = s/\text{EP} = s/a$ (geometry). Then, we can write $rp/2 = rhs/2a = rbs/2b'$ (using the geometrical properties described above), and finally

$$f = \frac{d}{(rsb/2b)^2}.$$

The deviation d can also be computed: $d = a \, (s/b')^2/2$, using again the ellipse's properties. Then the expression of f is easy to derive:

$$f = \frac{2a/b^2}{r^2} \, .$$

This law is valid at any point on the orbit, as it no longer contains any of the quantities specific to the choice of point P. It is the very famous law of universal attraction, inverse to the square of distance. The proportionality constant is, of course, a function of the very particular choice of units we have made. We must comment here on the value of the constant in question. Obviously, we have used special units when expressing the Galilean acceleration and when using the surface of a triangle as the square of a time interval. This choice of units is, in essence, the story of the apple. In other words, the attraction of a body of a mass m by the Earth is $f = m\gamma$. This f is the weight K/r^2, where r is the Earth's radius. The attraction of the Earth of mass M by the Sun is $f = M\gamma = K/R^2$, where R is the distance between the Earth and the Sun, *with the same constant K* (weight and universal attraction being the same things). In order to actually have the same constant K, it must be proportional to the two masses involved, that of the attracting body and that of the attracted body (the law of action and reaction!). Hence, we may write the equation as follows, where G is the so-called *Cavendish constant* because it was actually measured by Cavendish much later:

$$f = G \, \frac{mM}{r^2} \, .$$

Whatever the value of the "constant" G, we note that the mass of an object may be defined by Galileo's law $f = m\gamma$, i. e. by the acceleration a certain force exerts on a body of that mass. This is the *inertial mass*. The *gravitational mass* could have been defined by the law of gravitation, positing $G = 1$. But, equating a priori the gravitational mass with the inertial mass is one of the unexpressed hypotheses, hidden behind Newton's system and universally accepted, although it might be interesting to question it.

Fig. 5.12 *One of Newton's problems with dynamics, according to the Principia.* If the center of force is the center of the ellipse, what is the relationship between $CP = \varrho$ and the attracting force? In this case, the force is inversely proportional to the distance CP to the center.

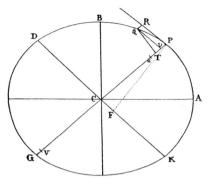

5.4.4
The Systematization of Mechanics

It is interesting to note that Newton asked other similar questions, quite irrelevant to the case of planets. For example, he wondered (Fig. 5.12) how to express centripetal force in the case of an elliptical motion, if the center of force were not the focus but the center of the ellipse. In that case, the force is proportional to the inverse of the distance:

$$CP = \varrho; \quad \text{or} \quad f = 1 / \varrho.$$

But the observations did not confirm the reality of such a motion. The Sun (Kepler) is actually at the focus of the ellipse. The role of the Sun is essential, witness its observed location at the focus of the elliptical orbit.

From the inverse square law, Newton's law, one can easily deduce Kepler's third law. It is even possible to demonstrate that it can be found only under that condition.

We will not go into more detail regarding the *Principia*. It is a long book, difficult to read, and often difficult to understand, since most of the demonstrations, as rigorous as they may be, are not expressed in the same way as we would express them today.

As an overview, it is worth noting that Book I of the *Principia*, dealing with many more problems under the general title of "Motions of Bodies," offers a very satisfactory explanation of the relative motion of two bodies, each of them "attracted" by the other with a force proportional to the inverse square of distance, the proportionality constant being only a function of the geometrical characteristics of the orbit for two bodies known a priori. It describes the motion of the two bodies around the center of gravity of the system, a notion introduced by Newton and typically Newtonian.

In Book II, devoted to the motions of bodies in a resisting medium, frictional forces, i.e. the resistance of the medium to the motion of the bodies within, are introduced and defined. Newton provides us in essence with the very basis of modern hydrostatics (after Archimedes!) and hydrodynamics.

Book III is devoted to Newton's "system of the World." It starts with a few "rules" of logic. Rule 1: "We are to admit no more causes of natural things than such as are both true and sufficient to explain their appearances" is indeed nothing else but good old Occam's razor, applied to an Aristotelian σώζειν τὰ φαινόμενα concept. This is followed by rule 2: "Therefore, to the same natural effects, we must as far as possible assign the same causes;" rule 3: "The quality of bodies, which admits neither intensification, nor remission of degrees, and which are found to belong to all bodies within the reach of our experiments, are to be esteemed the universal qualities of all bodies whatsoever;" and rule 4: "In experimental philosophy, we are to look upon propositions inferred by general induction from phenomena as accurate or very nearly true, notwithstanding any contrary hypotheses that may be imagined, till such time as other phenomena occur, by which they may either be made more accurate, or liable to exceptions." This was Newton's logic, very close indeed to our own, a truly scientific method!

Then Newton enumerated carefully the phenomena he wanted to describe, most of them emerging from the observational work of several 17th century astronomers, such as Cassini, Boulliau, Huygens etc. He was interested in the relation between the periods of the Jovian satellites and their distance to Jupiter. He even used the word "satellite" as well as "planets" for these circumjovian objects. From there, he wanted to extend the theory to Saturn's satellites and investigate the circumsolar motion of the five planets using Kepler's second and third laws and Kepler's laws as they pertained to satellites – all observed facts!

From observation came theorems, propositions, lemmae, and corollaries. Newton constructed a very strict treatise of celestial mechanics, the first one based on the power of calculus. It included the theory of tides, the beginning of a perturbation theory (the effect of the Sun's attraction on the Moon's motions), and the detailed study of the motion of

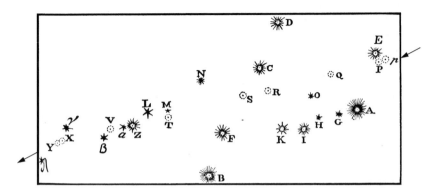

Fig. 5.13 *Newton's observations of the 1680 comet, in February and March, 1681.* Most of the observations were made by Flamsteed, and corrected by Halley. Some of the observations, the ones in this figure, are by Newton himself. In this diagram, taken from the *Principia*, except for the arrows at the right and the left, A, B, C, D, O, Z, α, β, γ, δ, are field stars. Points p, P, Q, R, S, T, V, X, Y represent the comet's successive positions. At that time, the tail of the comet was not visible.

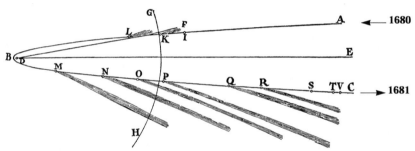

Fig. 5.14 *The orbit of the Comet of 1680, as derived by Newton.* The points here do not correspond to the observations of Fig. 5.13, but derive from much more complete observations done elsewhere in the world (London, Coburg in Saxony, Rome, Avignon, La Flèche, Cambridge, Boston, Jamaica, Virginia, Padua, Venice, Ballasore-East Indies, and Nuremberg, from November 1680 to March 1681). D is the Sun. ABC the orbit HG the Earth's orbit. S, T, V, represent Newton's observations as in Fig. 5.13. M to R are Flamsteed's observations. The first observations (point I) are by Gottfried Kirch, in Coburg.

comets (assumed to be parabolic and with orbits determined by three observations) (Figs. 5.13 and 14).

Newton also predicted the flattening of the Earth at its poles. In a *General Scholium*, he objected to the theory of vortices on the basis of Kepler's third law. In a non-mathematical appendix, intended for the reading of the layman, and called "The system of the world," he described

Fig. 5.15 *Launching, a satellite, as seen by Newton.* How to launch a satellite around the Earth, from some high point above sea level, according to the *Principia*.

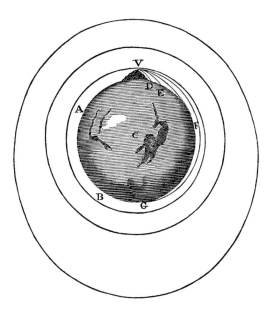

satellization around the Earth (Fig. 5.15), among many other things, a real prefiguration of the most advanced modern celestial dynamics.

5.4.5
The Refutation of the Cartesian Theory of Vortices

Newton devoted Book II to an analysis of the motions of bodies immersed in a resistant fluid material. He concluded that "the planets are not carried around in corporal vortices." In essence, the argument was based on Kepler's third law, about which he speaks ironically. "Let philosophers then see how the phenomenon of the 3/2-th power can be accounted for by vortices." Note incidentally that, while Newton quoted some of Descartes' followers (Cassini for example), he never quoted Descartes himself, presumably thinking of him as a mere philosopher! Regarding vortices, Newton considered this a law in power 2, assuming vortices to be of uniform nature and the motion within them uniformly circular. This was based on the assumption of a strong viscosity. The demonstration is difficult and rests upon various hypotheses, some expressed and some not expressed. Many scientists remained skeptical, including Huygens and the Paris astronomers, led by Jean-Dominique Cassini (Fig. 5.16).

Fig. 5.16 *A caricature of the Newton-Cassini debate.* This figure is taken from the well-known appendix to the *Principia*, written by Florian Cajori, in the late 1920's, and published in 1934 after his death.

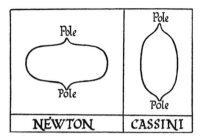

One argument seemed to promote their dubious response. Newton computed the oblateness of the Earth as 1/230, but geodetic measurements gave Cassini an opposite result. He determined that 1° of the arc of a meridian circle is equal to 57098 toises south of Paris, and 56970 toises north of Paris. A toise was equal to 1.949 meters. If true, this would tend to indicate an elongation of the Earth towards the poles. Descartes had not really predicted this result, but it was certainly in clear contradistinction to Newton's prediction. Hence, several scientists, notably the French geodesists, Cassini, and Huygens maintained their doubts with respect to Newton's theory.

Newton himself had no concern about the controversy surrounding his theory. He was looking for an operational method of doing practical mechanics and celestial mechanics and he had discovered it. Whatever deeper cause was acting behind the scenes was in essence irrelevant, although he was a firm believer in God as the primary cause. The attractive force K/r^2 accounted for everything. Universal gravitation became a universal law. Period! The doubts, the controversies and the reemergence of the legitimacy of Descartes' efforts came back much later.

5.5
The Triumph of Newton

As is was later noted by Laplace, the great advantage of Newton's theory, particularly as compared to Descartes', was its ability to predict quantitatively many observed or observable phenomena. In the language of the modern epistemologist Karl Popper, we would say that Newton's theory was "verifiable," or its reverse, "falsifiable," which was not the character of Descartes' system. The condition that a theory be falsifiable is considered essential by Popper.

5.5.1
The Development of Instrumental Astronomy

Since Newton, for at least three centuries, astronomy has progressed
through advances in instrumentation. New ideas have emerged from a
scrutiny of observations and new explorations have been undertaken.
Until now (with the exception of advances made under the influence of
Einstein's theories), all of this progress has fallen within the solid frame
of reference of Newtonian physics. Even Einstein's general theory of relativity, although it cast gravitation in a completely different light, has not
altered the value of Newton's laws as tools enabling anyone to perform
quantitative mechanics, in particular quantitative celestial mechanics. In
fact, Newton's theory may be conceived of as an excellent first approximation of Einstein's physics and it can be demonstrated to be valid in a
very large number of cases, where general relativity becomes necessary
only either at very large velocities or very large time intervals, or again
in very large scale phenomena.

Observational astronomy was given a new start with the first of Gali-
leo's telescopic discoveries in 1610. It was Kepler, however, (1611 in
Dioptrice) who was the first to formulate a theory of image formation.
He showed how a real image could be obtained with two curved lenses
(Fig. 5.17). The image was "reversed" (upside down), so was not practical
for use in everyday life, but astronomy was another matter.

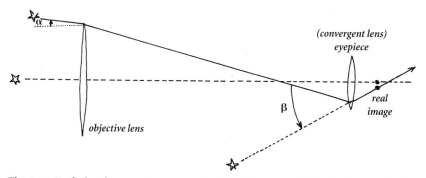

Fig. 5.17 *Kepler's telescope.* Compare with Fig. 4.41, where Galileo's telescope is described. The eyepiece is a converging lens, contrary to that in the Galilean telescope.
In this mounting, the distance of the eyepiece to the objective is larger than the focal
length of the objective, by a distance close to the focal length of the eyepiece. The
magnification can be very high. The focal length of the objective can be very long.
This results in a real image which improves the possibility for studying it.

In Holland, the Janssens (Zacharias and his son Hans) are said to have constructed such a long telescope in 1619, but it is likely that it was not a telescope of the Kepler type. In 1608, 1630, and 1645, Fontana, Scheiner, and Schyrrle seem to have constructed similar instruments. Kepler's construction had a larger field than the Janssens'. From 1640 on, it was the Kepler telescope which was used by every astronomer. It was improved further by micrometer wires (Auzout, 1667) put at the focus which allowed greater accuracy in pointing. Gascoigne (1640) also made such a device, but it was discovered only later in his unpublished papers. Galileo reached an enlargement of a factor of 30, in the best cases. The limitation was due to the fact that he used short focuses, hence rather curved lenses, producing optical aberrations which were indeed the true limitation to the increase of the power of the instrument. In order to avoid these perturbing aberrations, Huygens constructed a much longer instrument with no aberrations to limit the gain resulting from a great enlargement, several times that of Galileo. He actually built a 12-foot telescope, with a 57-mm aperture and another one 23-feet long. With it, he discovered a satellite of Saturn (used by Newton in his theorizing) and later Saturn's rings (Figs. 5.18 and 5.19).

Huygens came to France in 1667 after the foundation of the Paris Observatory. He joined a school of very gifted astronomers (Picard, Cassini,

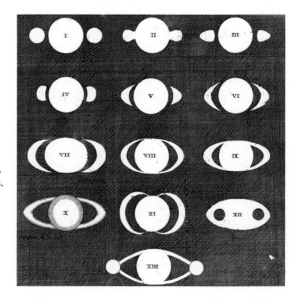

Fig. 5.18 *A few observations of Saturn.* Whereas Galileo had just barely seen "something," the rings were clearly visible, but often as two satellites, by various later observers. This figure is taken from Huygens' *Systema Saturnium.* It reproduces the observations of Huygens' predecessors including those of Galileo (top left figure).

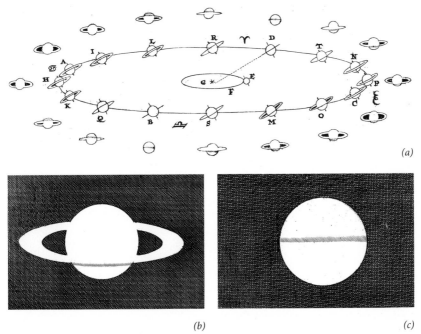

(b)   (c)

Fig. 5.19a–c *Saturn's ring, after Huygens' Systema Saturnium.* (**a**) The explanation of the variable appearance of Saturn, as seen from the Earth, during Saturn's path around the Sun. (**b**) Saturn at the maximum of the visibility of its ring. (**c**) Saturn, when the Earth is in the plane of the ring. Note that one sees on these drawings only one ring. It was Cassini who discovered that the ring is divided in two main parts by the so-called *Cassini's division.*

Römer, Lacaille, Auzout) and made many new discoveries, including four new satellites of Saturn. Huygens' new type of eyepiece was instrumental in this progress, as was the use of a long focus.

The study, in Paris, of the motion of Jupiter's satellites, by the Dane Ole Römer led to his determination that the velocity of light was not infinite. Indeed, until that time, most scholars did not even question that problem. A majority of those who referred to it thought the velocity of light was infinite. Descartes himself thought that light propagated instantaneously. At that time, of course, no link was made between the behavior of light and that of gravitation. The General Theory of Relativity gave the question a completely different meaning, but that was not before the early 20th century.

Fig. 5.20 *The observation of the velocity of light by Ole Römer, from Römer's drawings in the* Journal des Savants, *1676. At A the Sun, at EFGHLK, the Earth, at B, Jupiter, at DC one of its moons.*

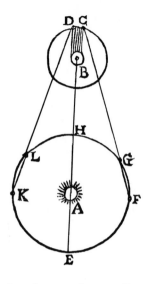

Römer's data (Fig. 5.20) were, in a sense, paving the way for Bradley who in 1725 constructed a proof of the motion of the Earth around the Sun using the phenomenon of aberration. This was without doubt a very important observation.

The time lag determined between phenomena (Fig. 5.21), eclipses of Jovian satellites by their planet, or their maximum apparent distances to the planet observed at six month intervals, corresponds to twice the Earth-Sun distance. At E2, the occultations of satellites by Jupiter are delayed by a time $2Rc$, where c is the velocity of light. This perceived delay

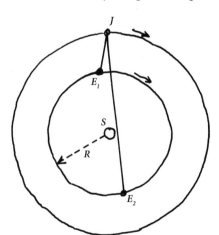

Fig. 5.21 *The determination of the velocity of light by Römer. See the explanations in the text.*

is extrapolated from the observations made when the Earth was at E1, six months or so before. The delay actually was measured as more than ten minutes. The actual measurements showed Römer that the light took 11 minutes or so to travel from the Sun to the Earth (now we know it to be 8 minutes). This method yielded a rather good evaluation of the velocity of light. It certainly showed that propagation is not instantaneous!

Greenwich Observatory was founded in 1676, under Flamsteed, only a few years after the Paris Observatory, more or less under Cassini, in 1666. A very stimulating competition ensued, much to the benefit of science!

Observations of planets (by Morin, Cassini, and others), continued up to the 20th century, as well as observations of the Moon (by Hevelius, and Langren). Catalogues of stars of magnitudes 7 and 8 were published by Hevelius, Hooke, Halley, and Picard.

The Earth was also the object of intense study. As already mentioned, at the Paris Observatory, Cassini asked Picard to determine the length of an arc of meridian in France, a rather limited one actually. This was a key point in the dispute between Cassini and Newton.

In essence, up to contemporary times, Newton's ideas predominated. The World system was based on the concept of universal gravitation. The data were accumulating; the extent of the explored Universe was expanding with the use of better instruments and improved methods, but science was still characterized by Newtonian-Platonic thinking. One described the Universe; one knew that it had been created by God; one did not look at the way it had evolved, or indeed if it had evolved. If some evolution were invoked (such as that of the solar system by Kant or by Laplace), it was in the frame of reference of an eternal Newtonian Universe.

For a more extended listing of the main actors in this ongoing and progressive drama and their main instrumental achievements and results, see Chap. 6.

5.5.2
The Flattening of the Earth

Some of these achievements, those which helped Newtonian theory to triumph over any objections, must be mentioned here. The first was the measurement of the oblateness or flatness of the Earth, a measurement which was decisive in promoting universal acceptance of Newton's system. The methodology was to measure arcs or meridians near the equator and as close to the pole as possible. Cassini's above mentioned argument against Newton was not convincing. At that time, people did not present their results with any estimation of errors and this type of measurement is difficult. It necessitates measures of angles in difficult conditions, in a network of triangles, and intermediary determinations. Therefore, in 1735, the Académie Royale des Sciences in Paris appointed a first expedition to the Equator. Bouguer and La Condamine went to Equador, then to Peru. They measured an arc of 3° latitude in the plain of Quito, but they were unable to measure, as they were instructed to, a longitudinal arc. In 1736, Maupertuis and Clairaut went to Lapland, again under the auspices of the Académie des Sciences. They measured an arc of 0°57'. The length of the degree was then 56753 toises near the equator, as against 57438 in Lapland, and 57057 in France. The accuracy was then unquestioned. The measurements were a great improvement over those previously made by Picard or Cassini. In this way, the reality of the flattening of the Earth was proven. Cassini had been wrong and Newton had been right. One should mention, however, that the flattening of 1/114 deduced from the comparison France-Lapland was too large. The other one, Peru-France, of 1/279, was much closer to Newton's value. In France at any rate, the proponents of Newtonianism, notably Voltaire, noted this as a triumph.

5.5.3
The Return of Halley's Comet

The second important new element of the discussion was the return of Halley's comet. As we have seen, Newton had assumed parabolic trajectories for some comets to be observed. A parabola, however, is nothing but an infinitely elongated ellipse, so to speak. Kepler conceived of long straight lines ending at, or in, the Sun; Cassini thought of very oblique (with respect to the plane of the ecliptic) circular orbits; Borelli had assumed a parabola, in 1664 before Newton; and, in 1680, Dörfel, a Saxon priest, gave the same interpretation to the large comets then visible. Newton had amply demonstrated the possibility of parabolic orbit, as a consequence of the inverse square law. Following Newton's method, Edmund Halley computed orbits for 24 comets so far observed in different epochs. Some of them had almost identical orbits, strangely enough, in particular the three observed in 1531, 1607, and 1682, which is to say at an interval of 75–76 years. He suggested that it could be a single comet, moving on a very elongated orbit, with a period of 75–76 years. Hence, he predicted its next return for 1758 and discussed the possibility of identifying it with the 1466 and 1378 comets as well.

In 1758, the date predicted for the return of the famous comet, Sir Edmund Halley (1656–1742) was dead, but the problem of his cometary orbit was still alive. Clairaut (1713–1765), a skillful scientist, optician, and astronomer, but most of all an excellent mathematician and determined Newtonian, had refined the perturbation theory only sketched by Newton. This allowed him to solve the three-body problem as a problem of successive approximations. Euler and Bernouilli (in Basel), Clairaut and d'Alembert, and finally Lagrange and Laplace had indeed brilliantly used Newton's ideas to develop celestial mechanics (and general mechanics). Clairaut used his perturbation theory to compute the orbit of the comet at its return, knowing the period had to be 76 years in 1531–1607 and less than 75 in 1607–1682, assuming of course that the intuition of Halley was correct. A step-by-step computation was necessary, long and tedious, following the comet in its approach to Jupiter and Saturn. It was undertaken, under the supervision of Clairaut, by Jérôme de Lalande and Nicole-Reine Lepaute, one of the first women in astronomy. They concluded that the date of the nearest approach to the Sun would be April 1759 (with an explicit margin of ± one month). Actually, the comet was discovered in December 1758 by Palitsch, an amateur astronomer in Saxony. The comet reached its perihelion in March 1759, thus splendidly confirming Clairaut, Lalande and Lepaute's predictions.

5.5.4
The Aberration of Light

A third essential development was the discovery of aberration by James Bradley (1693–1762) in 1728. At one and the same time, it proved the reality of the Copernican view of the Earth orbiting around the Sun and the finite velocity of light, as measured by Römer. In the early 18th century, it was possible to determine the positions of stars on the starry vault by their declination and right ascension, with an accuracy of less than 2" or 3" of arc, after corrections for air refraction. Then, strangely enough, the fine instruments used in Kew near London by Molyneux and after him by Bradley, showed that all stars were oscillating around a mean location in a period of one year, essentially following an elliptical path. These ellipses were, in their major axes, all equal to each other.

Bradley actually only really measured the major axis. Remembering Kepler's thinking on this subject, one might think that the ellipses were parallactic, due to the finite distance of the star, but since they all had the same major axis, such was clearly not the case. Rather, as the velocity of light was known, Bradley interpreted this effect as resulting purely from the velocity V of the Earth on its orbit combined with the velocity of light c, 10000 times greater (in gross figures) than the Earth's velocity, as shown in Fig. 5.22. Bradley called this effect *the aberration of light rays*. Although indirect (inferred from the velocity of Earth on its orbit), this was the first real "proof" that Copernicus was right, that the Earth was indeed moving with respect to the "fixed stars." Had the velocity of light been infinite, however, this could never have served as a proof. The size of the apparent elliptical aberration is such that the semimajor axis of the ellipse is equal to V/c. V is of the order of 30 km/s, c is 300000 km/s, the angle α (Fig. 5.22) is of the order of 1/10000 of a radian, about 20" of arc.

Later, this effect was compared to the effect of rain falling vertically on the windows of a car moving horizontally, with velocity of the same order of magnitude (Fig. 5.23), but we should not push this type of analogy too far. Sometimes it may be misleading, notably when considering the theorem of addition of velocities, as shown by the Michelson and Morley experiments that we will describe below on pp. 391, 419, 424.

It is important to note that these were not the only significant advances of the period. Herschel's discoveries pertaining to galactic structure and the idea of island-universes constituted a glimpse into the later cosmological period of astronomy, but they were not directly related to the solar system and to the Copernican concepts. We will return to these discoveries and ideas. For the moment, however, following the course taken so far, mainly the description of the solar system, its motions, its dynamics, and universal gravitation which is the cornerstone of celestial mechanics, the next logical step was improved accuracy in angular determination of stellar positions.

5.5.5
Annual Stellar Parallaxes

Between 1830 and 1840, three astronomers, Bessel, Struve, in Dorpat (later Tartü, in Estonia), and Henderson, in the Cape Colony, measured the first annual stellar parallaxes ever known, thus enabling a determination

Fig. 5.22 *The aberration of light.*
During the Earth's quasi-circular
motion around the Sun, the velocity
of the Earth is near 30 km/s. The
velocity of light is near 300 000 km/s.
The apparent direction of the star
is displaced from its true direction,
hence the star appears to describe
on the sky a quasi-circular orbit of
semi-angular diameter of about 20″.
If the star is close to the plane of
the Earth's orbit, the apparent
orbit appears very elongated.
Its semi-angular diameter is
the same, about 20″.

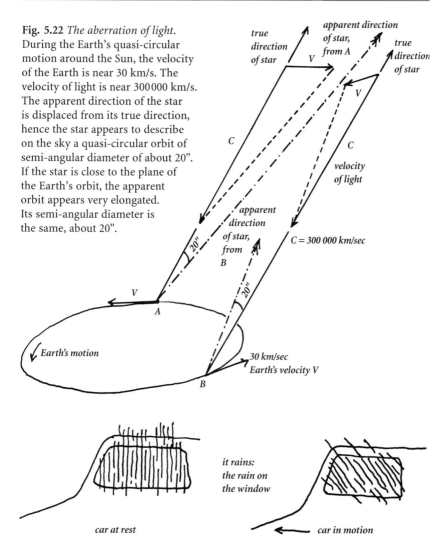

Fig. 5.23 *An analogy to aberration.* The windows of a car, during a vertical rain,
when the car is idle (left) or in motion (right).The apparent motion results from the
combination of the fall velocity of the rain and the velocity of the car.

of the distances to the stars (Fig. 5.24). The distance to α Centauri makes
it the closest neighbor of the Sun, until new discoveries disclose stellar
objects at a closer distance! We are now at the end of the 20th century
and nothing has altered this privileged position, except perhaps the dis-

Fig. 5.24a,b *The measurement of annual parallaxes of stars.* **(a)** *The motion of a nearby star S, with respect to distant stars, as seen from the Earth E which rotates around the Sun (successive positions* E_1 *and* E_2, *at times separated by about 6 months here). The Earth orbit is quasi-circular and is represented here in perspective. The annual parallax is the angle* ϖ, *under which, from the star, one sees the radius of the Earth's orbit (more precisely, the semimajor axis of this orbit, or the astronomical unit of distance). For* **(b)** *see next page.*

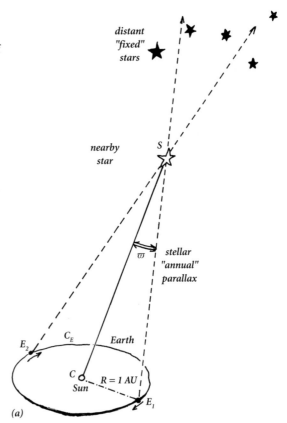

(a)

covery of the triple nature of the star, the component C, or Proxima Centauri, being the closest to us. The distance of a star of which the parallax is one second of arc is, by definition, a parsec (pc). One parsec is equal to 3.26 light-years. The distances found by the observers were thus as follows, in various units:

	Parallax		Distance		
	ϖ	in pcs	in light-years	in A.U.	in 10^{13} km
61 Cygni	0".348	2.87	9.3	590 000	8.9
Vega	0".26	3.84	12.5	790 000	12
α Centauri	0".76	1.32	4.31	270 000	4.0

Fig. 5.24b The motion
of the same star S, as seen
during the year from the
Earth. It describes an orbit
in the sky. Points S_1 and
S_2, correspond to the location of the Earth in E_1 and
E_2, as shown on Fig. 24a.
The orbit of S, around its
center O, is identical to
that of the Earth, and parallel to it. The distance of
the star from the Earth
makes the apparent orbit
seem smaller. If the star S
is located near the plane
of the ecliptic, the apparent parallactic orbit has a
very elongated shape.

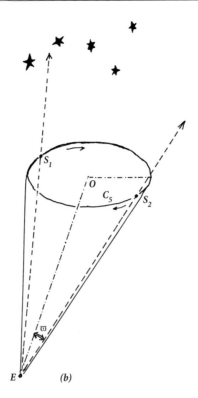

(b)

This then was the first conclusive proof, without any conceivable appeal, without the need of knowing that light had a finite velocity, of the
actual revolution of the Earth around the Sun, the frame of reference
being the fixed stars. Errors were made by Henderson, Struve and Bessel,
but they were of secondary importance. Modern data are somewhat different from the original ones. They are as follows:

	Parallax ϖ	in pcs	Distance in light-years	in A.U.	in 10^{13} km
61 Cygni	0".294	3.40	11.1	701 000	10.5
Vega (α Lyrae)	0".123	8.1	26	1 670 000	25
α Centauri C (Proxima)	0".762	1.31	4.27	270 000	4.04
α Centauri A,B	0".745	1.34	4.37	276 000	4.13

Note that the aberration ellipse has nothing to do with the parallactic ellipse, although they are both described in one year. In one case, the semimajor axis is 20", less than 1" in the other ... not the same order of magnitude. Whereas all stars are affected by the aberration of the elliptical annual motion with respect to the celestial coordinates linked to the Earth's motion, the annual parallax is measured with respect to the "fixed stars", presumably much more distant than the nearby stars for which the parallax is detectable.

5.5.6
The Fundamental Distances in the Solar System

In the preceding section, we have used the value of the velocity of light c, in order to transform parsecs into light-years. But the value of c, as deduced from the study of the motion of Jupiter's satellites, is a function of the actual measure of the astronomical unit of distance (in abbreviation A.U.), i.e. the average distance from the Earth to the Sun. So we need to determine well the value of 1 A.U., which we have used in the last three columns of the tables above in anticipation of correspondences which were actually not available in Struve's time. This is done by measuring the diurnal parallax of the Sun.

Even though the distances of the Sun and Moon were relatively known, at least approximately, the determination of the diurnal parallax of the Sun and of the planets was a major step in the progress of the knowledge of the solar system and of stellar distances as well. This investigation occupied an important part of astronomical activity in the 17th and 18th centuries.

To measure the diurnal parallax (Fig. 5.25), it is necessary to simultaneously observe the same object on the background of fixed stars from two different points on Earth.

Richer (1630–1696) in Cayenne and other observers (Picard, Cassini) in Paris (1671–1673) determined the distance of Mars in terms of terrestrial units of length and Cassini deduced from this measurement the diurnal parallax of the Sun. (The ratio of the Mars distance to the Sun-Earth distance is deduced from Kepler's third law.) He found 9".5. Incidentally, when Richer was in Cayenne, he deduced from pendulum experiments the effect of the weight of a mass on acceleration and found it to be less than in Paris. Newton correctly used that argument in support of his case for a flattening of the Earth. Later, using methods similar to

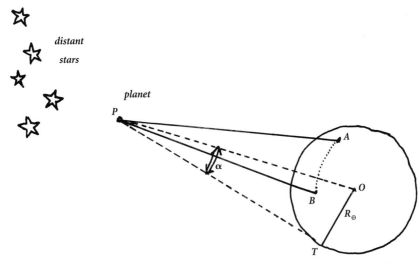

Fig. 5.25 *The diurnal parallax of a planet.* From two points on Earth, A and B, one observes simultaneously the position of a planet P on the background of distant objects (including the Sun, if needed, see next figure.) Knowing the positions of points A and B permits us to determine the angle α (diurnal parallax) under which, from the planet P, one sees the radius R of the Earth at the equator. This difficult computation requires a precise knowledge of the shape of the Earth.

those of Richer and Cassini, LaCaille (in the Cape Colony) in cooperation with Lalande (who observed from Berlin), and Bradley (who observed from Kew), determined the distances of Mars, Venus, and the Moon to the Earth and corrected their data by taking into account the measured flattening of the Earth, as determined by the expeditions of La Condamine and Maupertuis. Using Venus' orbit and Kepler's law, a parallax of 10" could then be arrived at. This was close to Cassini's result, but accuracy was still low.

From LaCaille and Lalande's determinations of diurnal parallaxes of the Moon and the planets, it became clear that a still better method (Fig. 5.26) to determine the solar parallax would be to use passages of Venus in front of the Sun, in June 1761 and 1769. Such transits occur in pairs eight years apart, every 105 years (in round figures). The study of the time duration of the passage during the crossing would provide an even more accurate determination of the distance of the Sun than the previous methods.

Fig. 5.26 *Determination of the Sun's parallax.* The distance of Venus is determined by its parallax, using either distant stars (as explained in Fig. 5.25), or the passage of the planet in front of the Sun, as described in Fig. 5.27. The parallax of the Sun is determined from the relation: $a_{\text{Venus}}/P_{\text{Venus}}^{2/3} = a_{\text{Earth}}/P_{\text{Earth}}^{2/3}$ where a are the semimajor axes of, respectively, Venus and Earth's orbit, and P the respective periods of revolution (Kepler's third law).

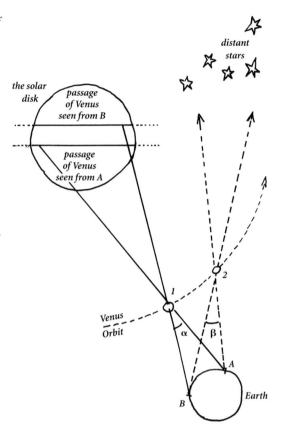

An international series of expeditions was, therefore, organized. Lalande, for the Paris Academy of Sciences, was the coordinator (Fig. 5.27). In 1761, observations were made at Tobolsk, St. Helena, The Cape, India, and in many European stations. On an even larger scale (now including Siberia, California, Tahiti, Hudson's Bay, and Madras), the same was done in 1769. The American Philosophical Society took part in this operation, as did many other learned societies in the world. It was the first entry of America into the astronomical community, since the splendid accomplishments of pre-Colombian history. The value determined for the solar diurnal parallax by Lalande was 8".72. Later, this value was corrected, by Encke in particular, but Lalande's value is still very near the currently accepted value of 8".79418.

Fig. 5.27 *The two passages of Venus in front of the Sun (1761–1769).* In each of the two cases, four lines indicate four observing sites: In 1761, 1. Rodrigues island; 2. Paris; 3. Tobolsk; 4. Tahiti. In 1769, 1. Tahiti; 2. Batavia; 3. Vardö; 4. Paris. The time indications refer to the Paris time. The passage was visible from well-located sites for about 5 to 6 hours. Reductions were performed primarily by Lalande. Drawing possibly by Pannekoek, whose book served as a reference here.

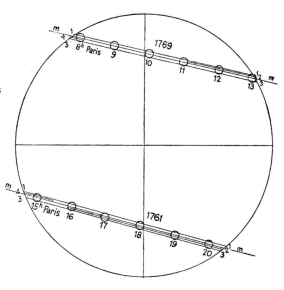

Intense astronomical activity (see Chap. 6 for more detail), from Galileo to the period of modern telescopes, has permitted more thorough exploration of the solar system and the stellar system, the discovery of the existence of our Galaxy, then of the external galaxies, the determination of the rotation of our Galaxy, its shape, the rotation of the more distant galaxies, their groupings, and their shapes. We have gained an understanding of the solar spectrum features and the stellar spectra. We have diagnosed the physical characteristics of millions of stars and galaxies, of the matter between them, and of their chemical composition. Finally, in the tradition of Tycho, we have studied solar, stellar, and galactic evolution, and our success in understanding this has been an indication of the validity of the physics without which astronomers could not work. The universality of the laws of physics, at least in the already observed Universe, has paved the way for modern cosmology to develop a new physics in order to understand some of the still unexplained facts of observation (see Chapter 8, Appendix II).

Newtonian astronomy kept a place for Creation and for God. Should we take this problem into consideration, or assiduously ignore it, as did the Thomists? Newton retained some rather simple ideas pertaining to the relative motions of bodies and the systems of reference with respect to which motions are measured. Newtonian physics rested upon the instantaneous propagation of gravitation in a vacuum and on a very oversimplified notion of the nature of light. Reflections on these points and further reflections following new experiments led modern cosmologies (notice the use of the plural!) to emerge, with difficulty and in a rather disorderly fashion. The aim of these cosmologies has been to describe and arrive at a coherent understanding of all that is or will be observable and how it evolves.

5.5.7
Newton and God

Newton's laws of motion and the law of universal gravitation, which is their expression in the realm of dynamics, are a description of practically all celestial motions that we observe. This description is, as we have demonstrated, more complete and at the same time more condensed than any other description. In a sense, the question "how?" is replaced in Newton's thought by the question "why?" True enough, Newton had already answered the question "how?" in a more elaborate fashion than that question had been answered earlier by Kepler or, still better, by Kepler's predecessors; and Newton's reply was an ultimate reply. Therefore, the shift from the "how" to the "why" became a logical imperative in Newton's thought. Why are orbits elliptical? Why do the planets follow Kepler's laws? Furthermore, these "whys" suggest a new "why." Why do the laws of universal gravitation act as they do? So Newton's reply to Kepler is only a better form of "how." A deeper "why" remains to be answered. Is there an answer in some contemplation of Creation? God? Eternity? Have we returned to Plato or Aristotle? This is a problem that follows from a concern with the *immediate* causes, to the exclusion of the *primary* causes.

At the end of the *Principia,* in the form of a five-page *General Scholium,* Newton wrote an essay to suggest a new "why" to be added to his own "how," as the final punctuation to end the Big Book. "This most beautiful system of the Sun, planets, and comets could only proceed from the counsel and dominion of an intelligent and powerful being" … "This being governs all things, not at the soul of the world," (a Platonic idea) "but a Lord over all." (The Judaeo-Christian tradition.) "He is omnipresent not virtually only, but also substantially; for virtue cannot subsist without substance." … "Hitherto we have explained the phenomena of the heavens and of our sea by the power of gravity, but we have not yet assigned a cause of its power" ... "But hitherto I have not been able to discover the cause of these properties of gravity for phenomena, and I frame no hypotheses" ... "and hypotheses, whether metaphysical or physical, have no place in experimental philosophy" ... "To us it is enough that gravity does really exist." Finally, Newton adds, without any real explanation, a paragraph concerning a most subtle "spirit" which pervades and remains hidden in all gross bodies, responsible for attraction, gravity, coherence, electricity, light, and heat. He can neither explain this in a few words, nor provide "that sufficiency of experiments which is required to an accurate determination and demonstration of the laws by which this electric and elastic spirit operates." In such language, Newton ends his book, the *Principia*.

We see then contradictory attitudes (as in almost any other scientist) in Newton's position:

- A quasi-positivist attitude (see Thomas Aquinus, or Occam). Hypotheses, whether metaphysical or physical, cannot be framed without proper bases ("hypotheses non fingo").
- A theistic belief, deeply rooted and barely commented upon.
- And in a few lines, a vague allusion to something other than gravity; a more subtle "spirit," encompassing more phenomena, possibly a bridge to future work, in particular to Newton's alchemical attempts.

The *Principia* was indeed a serious and purely scientific book, but Newton revealed in its pages a mystical tendency which preoccupied him from his fifties to the end of his life.

5.6

* **Appendix**

Written in collaboration with Prof. Daniel Pecker

In order to show its rigor and its degree of geometrical perfection, we shall reproduce here (§ 5.6.2) Newton's own demonstration of the law of universal attraction, almost exactly in most cases as Newton presented it and with his own notations (except for the symbolism of fractions).[14] We must note that all Newton's contemporaries, as well as Newton himself of course, were familiar with the geometry of conical sections, or conics, a field of study dominated by Apollonius of Perga (262 BC–190 BC) since ancient times. We shall print and designate the propositions, theorems, and properties (A, B, C ...) in boldface in order to be able to quote them more easily. Our own comments, intended to clarify the original demonstrations for modern readers generally not familiar with the properties of conics, are printed in italics. We will use the same figures as Newton's (except in Fig. 5.28 and 5.37), but we will reproduce only the significant parts of them, as the need arises. Note that our notations differ somewhat from Newton's. Our F_1 is his S and our F_2 is his H, but generally they follow Newton's.

Fig. 5.28

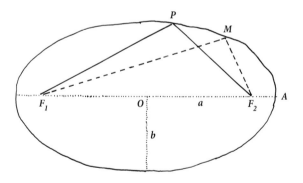

---

[14] Newton used A:B. We will use A/B which is the usual contemporary form.

Fig. 5.29

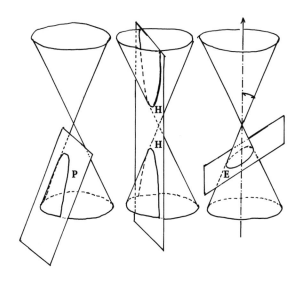

5.6.1
Geometrical Preliminaries

1. An ellipse is *defined* as the locus of points P, such that the sum of the distances $PF_1 + PF_2$ to the two foci F_1 and F_2 is given and constant (Fig. 5.28). Let (C) be the center of the ellipse. Let a and b be, respectively, the semimajor axis and the semiminor axis of the ellipse.

2. A basic *property* of ellipses, known since Greek times, was applied extensively, notably by Apollonius, and we shall use it as well. An ellipse is a section of a cone (or of a cylinder) based on a circle, constructed out of a plane, making with the plane which contains the circle a given angle ω. This explains the derivation of the term *conical curves* often applied to ellipses, parabolas and hyperbolas since antiquity (Fig. 5.29). If the cone is a circular cylinder, *the ellipse can be projected only as a circle* (Fig. 5.30). This basic property is important because *the projection does not alter some properties (the so-called "projective properties") of the projected figure, whatever it is.*
In particular: (this is **property A**) **the ratio of two parallel segments is identical to the ratio of their projections.**

Fig. 5.30

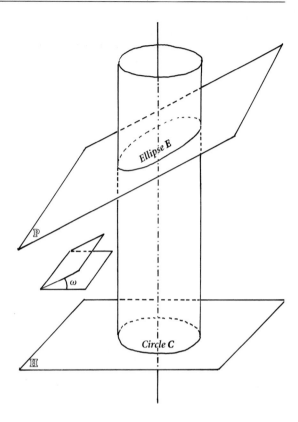

Fig. 5.31

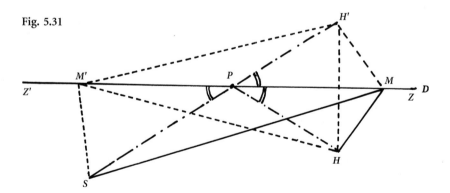

Fig. 5.32

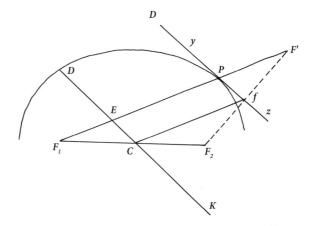

3. Using the definition and some of the the properties which we have or will demonstrate, notably the projective properties of ellipses, we will justify Newton's basic demonstration of the inverse square law of attraction.

(Property B) The sum PF₁ + PF₂ is equal to 2a.

When P occupies the position A, for example, in Fig. 5.28, the $\Sigma = PF_1 + PF_2$ is equal to $(AC + CF_1) + (AC - CF_2)$. But $CF_1 = CF_2$. Therefore, the sum is equal to $2\,AC = 2a$. As this sum is a constant, by definition, the equation $\Sigma = PF_1 + PF_2 = 2a$ is always true.

4. **(Property C) EP (in Fig. 5.10, p. 271 or 5.37, p. 304) is equal to CA = a.**
(a) Let D be a straight line z'z, S and H any two points located on the same side of **D** (Fig. 5.31). Let M be any point on **D**. The sum $\Sigma(M) = HM + HS$ is minimum when M is located at a certain point P. Where is P located? To know this, let us construct H', symmetrical to H with respect to **D**. By symmetry, $H'M = HM$ and $\Sigma(M) = H'M + MS$. Obviously this sum is minimum if P is located on the line H'S, as in Fig. 5.31.

(b) When M moves from P to infinity on the line **D** (towards z), $\Sigma(M)$ takes all values between $\Sigma(P)$ and infinity. This is true also when M moves from P to infinity towards z'. Therefore, for any point on the line **D** other than P, there is another point M' on the line **D**, on the opposing side of P, such that $HM' + M'S = HM + MS$.

(c) If we look at Fig. 5.31, we see that there is an obvious angular property (applied by Newton without demonstration). Clearly, HPz = zPH' (symmetry); and zPH' = z'PS (opposed by the summit of the angle). Therefore, **the angle zPH is equal to the angle z'PS. This is property D.**

(d) Let us now take Newton's ellipse (Fig. 5.32), and a line **D** (line zPy) tangent to the ellipse. The sum $\Sigma(P) = F_1P + F_2P$ has a minimum value compared to all sums $\Sigma(M)$, where M is any point on the line **D**. If this were not the case, according to what has been demonstrated above in (b), we would have two points P corresponding to the same $\Sigma(P)$. This is contradictory to the fact that **D** is tangential to the ellipse at P. The two points coincide with each other. **Property C** demonstrated above in (c) shows that angles yPF$_1$ and zPF$_2$ are equal. It also shows that point F', symmetrical to F$_2$ with respect to the tangent **D**, is located on the segment F$_1$P. This shows us, incidentally, that if the ellipse is the locus of the points P such that the sum $\Sigma = PF_1 + PF_2 = 2a$, the "outside" (frontier not included) of the ellipse, hence all points M on the tangent zPy at P to the ellipse, is characterized by the property $\Sigma_+ = MF_1 + MF_2 > 2a$. The "inside" (frontier not included) is characterized by the property $\Sigma_- < 2a$.

Fig. 5.33a

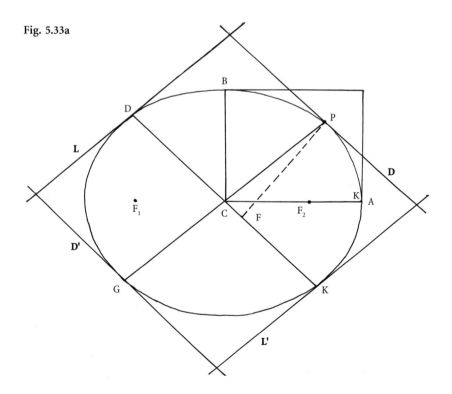

Fig. 5.33b

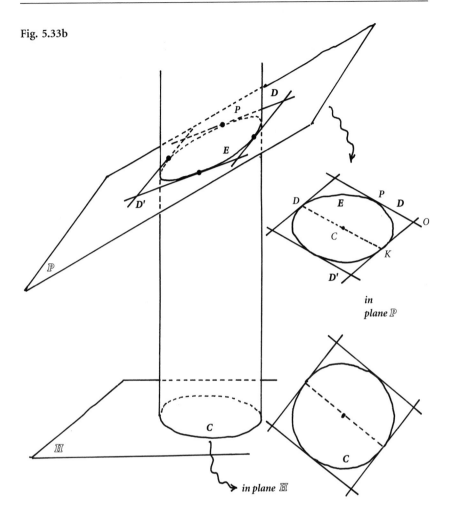

*in
plane* ℙ

in plane ℍ

(e) Now, we can come back to the ellipse of figure 5.32. The line DK is, by construction, parallel to **D**. Let us call f the point located on D in the middle of segment F_2F'. C is by definition the middle of F_1F_2. F_1F' is equal to $F_1P + PF_2$ (see Fig. 5.31, where we replace F_1F' by SH', F_2P by HP = H'P), hence to $2a$. In addition, the two triangles F_1F_2F' and CF_2f are homothetic, each side of the first one having twice the length of the corresponding side of the second. Therefore Cf, which corresponds to F_1F', is equal to CA= a. In the parallelogram CfPE, if Cf is equal to a, so is EP. EP = CA= a. Q.E.D.

5. **(Property E) Apollonius' theorem and its formulation by Newton:
CA.CB = CD.PF.**

(a) Let us consider (Fig. 5.33a) the tangent **D** to the ellipse at P, and the tangent **D'** to the ellipse at point G, symmetrical to P with respect to C. Obviously, **D** and **D'** are parallel to each other. Let us consider the diameter PCG, i.e. the so-called *conjugated* diameter to the tangent **D**, and the tangents **L** and **L'** at D and K to this ellipse. We have built a parallelogram. Let us show that this parallelogram has a constant area.

(b) Let us consider the ellipse as the section by a plane $\mathbb{P}$ of a cylinder of which the base is a circle **C** located on a plane $\mathbb{H}$ (Fig. 5.33b). Let us consider the projection of **D** and **D'** and of **L** and **L'**. Tangential properties are conserved by projection. Therefore, the parallelogram described above in 1 is projected as a square circumscribed to the circle **C**. All parallelograms circumscribed to the ellipse project as squares; but all these squares are identical. Therefore, all these parallelograms have the same area. So have the "quarters" of these parallelograms, such as PCKO. If PT is the normal constructed from point P to the straight line DK, then the area of such "quarters" is always CD.PF. The projective property shows that is equal to CA.CB = CD.PF = $a.b$.

(c) The next question that arises is: *How is this area linked to the area of the ellipse?* The ratio of the area of the parallelogram to the area of the ellipse is equal to the ratio of the area of the square to the area of the circle, because all areas in the plane $\mathbb{P}$, when projected on the plane $\mathbb{H}$, are

Fig. 5.34

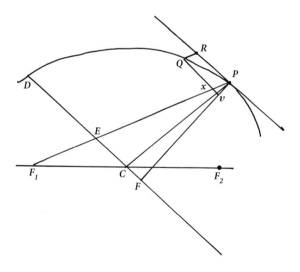

Fig. 5.35

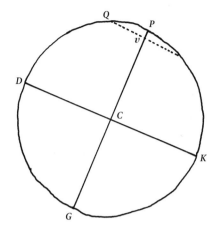

reduced by the same ratio. Let the radius of the circle be R. Clearly, the area of the square is $(2R)^2$, the area of the circle is πR^2. The ratio in question is $4/\pi$, which is unchanged in the projection. *It is the ratio of the area of the parallelogram to the area of the ellipse.* The area of the parallelogram is therefore $4R^2 a/b$.

These properties were demonstrated by Apollonius in antiquity!

6. (Property F) $(Gv.Pv)/Qv^2 = PC^2/CD^2$. Let us come back to Newton's ellipse of Fig. 5.37, although it is not quite clear in Newton's drawing. In its motion (hereafter, we shall consider only its geometrical properties), the mobile located initially at P follows the ellipse. A few geometrical considerations can be taken into account. Actually, during that time, it reaches Q, on the ellipse, because of the attraction. The pro-

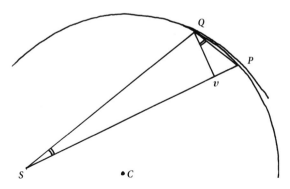

Fig. 5.36

jection of QR on PF, is Pv, in Newton's notations, and the projection of QR on PE is Px. The triangles PEF and Pxv are homothetic.

Let us consider the equality (which we will now demonstrate) $(Gv.Pv)/Qv^2 = PC^2/CD^2$. If we invert the terms, then: $(Gv.Pv)/(PC^2) = (Qv^2)/(CD^2)$. All the segments in each of the two terms are parallel. Therefore, these two ratios are conserved by projection and the property is valid if valid on the projected circle (Fig. 5.35).

But on that circle, $PC = CD$ is the radius of the circle. Therefore, *on the circle,* $Gv.Pv = Qv^2$. These two terms are indeed two different expressions of the "power of point v" with respect to the projected circle. They are therefore equal to each other. The equality of the projected terms is demonstrated, as is the equality on the projection, i.e. on the ellipse.

For those not familiar with the notion of "power of a point with respect to a circle," the property $Gv.Pv = Gv^2$ can be demonstrated in another way. Let us (Fig. 5.36) consider in the projected circle the triangles GQv, vQP, and GQP, the three being right triangles, similar to each other. The corresponding sides are expressed as $Pv/Qv = Qv/vG$. *Q. E. D.!*

5.6.2
Newton's text commented

We come now to Newton's text itself. Newton's text is between broad margins. Ours is normally set. Note that, if the attraction of the mass *M* located at P by the mass located at the focus F_1 (S in Newton's notation) did not occur, the massive point P would follow the tangent D. Over a "small" period of time Δt, it would reach point R.

Fig. 5.37

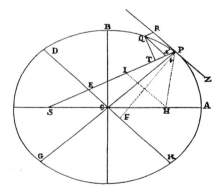

Fig. 5.38

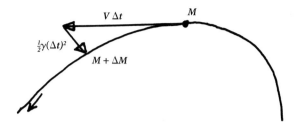

Proposition XI. Problem VI[15]. *If a body revolves in an ellipse, it is required to find the law of the centripetal force tending to the focus of the ellipse.*
Let S be the focus of the ellipse[16]. Draw SP cutting the diameter DK of the ellipse at E, and the ordinate Qv in x and complete the parallelogram QxPR. It is evident that EP is equal to the greater semiaxis AC[17]. For drawing HI from the other focus H of the ellipse parallel to EC[18], because CS and CH are equal[19], ES EI will also be equal[20]; so that EP is the half-sum of PS + PI, that is [because of the parallel HI, PR, and the equal angles IPR and HPZ] of PS, PH which taken together are equal to the whole axis $2AC$[21]. Draw QT perpendicular to SP and putting L as the *principal latus rectum* of the ellipse (or for $2 BC^2/AC$)[22], we shall have: $(L.Qr)/(L.Pv) = QR/Pv =$ [*because triangles* PEC *and* Pxv *are similar*] $= PE/PC = AC/PC$ [*because of property C*], also, $(L.Pv)/(Gv.Pv) = L/Gv$ [*obvious simplification*], and $Gv.Pv/Qv^2 = PC^2/CD^2$ [*because of property F*].

---

[15] This shows, incidently, that Newton had previously established some, if not all, of the bases of his demonstration.

[16] We use in our drawings F_1 and F_2 for the foci S and H.

[17] This is demonstrated above in (D). Newton gives another demonstration here.

[18] HI = F_2I in our notation; H = F_2 in our notation

[19] CS = CF_1 in our notation; CH = CF_2 in our notation

[20] ES = EF_1 in our notation

[21] PS = PF_1 in our notation; HI = F_2I in our notation; HPZ = F_2PZ in our notation (this equality is demonstrated by *property C*); PS = PF_1 in our notation; PH = PF_2 in our notation.

[22] SP = F_1P in our notation.

7. To continue (see Fig. 5.34), we must come back to the physical meaning of R and Q. We assume that Dt is as small as possible. Let us, in other words, make P and Q coincide. PR and PQ are infinitesimal displacements. Some very small quantities can be considered almost equal, as equal as they can be not taking terms of higher degree into account (speaking in modern language). Then Qv = Qx. They are two "infinitely" small quantities, of which the ratio tends towards unity when Q tends to P. In the triangle QxP,vx is negligible with respect to Qx or QP or PR Now, we can come back to Newton's text.

> When the points P and Q coincide, Qv^2 = Qx^2, and (Qx^2) or (Qv^2/QT^2) = (EP^2)/(PF^2) [*in the similar triangles* QTx *and* PFE] = (CA^2)/(PF^2) [*because of property C*] and (Gv.Pv)/ Qv^2 = (CD^2/CB^2) [*because of property E*]. Multiplying together corresponding terms of the four proportions and simplifying[23], we have:
>
> (L.QR) / QT^2 = (AC.L.PC^2.CD^2) / (PC.Gv.CD^2.CB^2) = 2PC / Gv since AC.L = 2BC^2. But the points Q and P coincide[24], 2PC and Gv are equal. And therefore, the quantities L.QR and QT^2, proportional to these, will be also equal. Let those equals[25] be multiplied by SP^2/QR and L.SP^2 will become equal to (SP^2.QT^2)/QR.

How can we arrive at Newton's law from this demonstration? If we reconsider Galileo's experiments about motion (Fig. 5.38), we see that along the trajectory of the mobile M, the velocity is V, the displacement is VDt and the "weight" or the acceleration imposed by the attractive body is, according Galileo's experiment, ($\gamma \Delta t^2$)/2 and oriented towards the attracting center. To return to Newton's conclusion:

> And therefore, the centripetal force [*expressed as RQ, whenever the time is expressed as* Δt] is inversely as L.SP^2, that is inversely as the square of the distance SP. Q.E.D.

---

[23] i.e. (L.Qr) / (L.Pv); (L.Pv)/(Gv.Pv)/Qv^2; and Qv^2/QT^2, written in the above in our notation.

[24] for Dt "infinitely" small.

[25] i.e. equal quantities.

Remembering Kepler's law of equal areas (Kepler's second law) demonstrated above, p. 269, we can write in Newton's drawing:

SP.QT= twice the area swept in time $\Delta t = K\ \Delta t$. Using this relation, we can see now that $L.PS^2 = K^2\ (\Delta t)^2/QR$. In other words, what Newton tells us, is that the acceleration γ is equal to $2\ K^2/L.PS^2$. It is another form of Newton's "law of attraction in $1/r^2$" (where $PS = r$, by definition of r).   *

CHAPTER 6

From Pre-Galilean Astronomy
to the Hubble Space Telescope and Beyond ...

Written in collaboration with Dr. Simone Dumont

Technical progress advanced the quality and the extent of observed data, as well as the volume of the observed phenomena. Theoretical concepts were developed to account for these observations. Hence, the depth of ideas and their evolution over a period of four centuries was considerable. As we have said, observations are not the essential subject of this book, devoted as it is precisely to ideas, and more specifically to ideas about the Universe as a whole, about the system of the Universe (or the "World," as it was called in pre-modern times). However, we should always keep in mind that the ideas stem from the observed data, themselves a consequence of improvements in techniques and instrumentation.

We shall, therefore, cover the observations accumulated by astronomers over the centuries, but we shall limit ourselves to the building of a few thematic paragraphs, tables, and diagrams, each oriented more or less towards a consideration of the progress accomplished, hence in a rough chronological order whenever feasible. Roughly speaking, we shall create a table for each of the identifiable technical domains, indicating some of the milestones of pre-galilean times, which we have fully covered in Chaps. 1–4, and covering also the main discoveries made within the Newtonian framework (Chap. 5).

We shall attempt to cover the material completely for the 18th and 19th centuries; but the acceleration of research in the 20th century, and its more collective nature in the space era, will be treated somewhat differently, schematically and in a purely graphical form (Figs. 6.1a–c, 6.2–4) without quoting systematically most of the astronomers in the field.

6.1
Improvements in Techniques and Instruments

Over the last three centuries, research in the field of instrumental and observational astronomy, which was conducted initially by individuals interested in several aspects of astronomy, became, for practical reasons, more and more specialized and collective. We shall try to divide the field into some subfields, as follows:

1.1 Size and nature of instruments (objectives); measurements of time
1.2 Development of ocular instruments (spectrography, photography, photoelectricity, photometry, polarization, etc.)
1.3 Basic physical data important for astronomy
1.4 Aperture of spectrum (infrared, ultraviolet, radio, high energy radiations (the detection of which generally implies space research equipment)
1.5 Astrophysics of particles (cosmic rays, solar wind, solar activity, neutrino astronomy)
1.6 Mathematical techniques and theoretical tools for astronomy

6.1.1
Size and nature of instruments (objectives); Measurements of time

6.1.1.a
Non-optical Instruments

Obviously, we are concerned with the whole series of instruments developed and used before Galilean times: first, of course, the time-measuring devices, gnomons, merkhets, polos, and others; then, the instruments oriented towards the measurement of the positions of stars, staffs, sextants, theodolites, meridians, equatorial systems, and quadrants. The best were mostly elaborated by Tycho, and aimed at making angular determination more and more precise (hence they were of larger and larger size) and more and more reliable, notably with respect to the Earth itself. They were, therefore, fixed with respect to fixed axes, on fixed walls for example.

What follows is a list some of the main contributors to the development of non-optical instruments. Italics letters are used for those primarily responsible for progress in the field concerned in each paragraph.

Ptolemy, Claude (ca 110 AD, possibly in Thebaid, at Ptolemais of Hermias – ca 168, in Alexandria) – Astrolabe. Description of instruments necessary for an observatory: armillary spheres, mobile quadrant, equatorial instruments, astrolabe, celestial globes (with mobile poles), dioptres for the measurement of apparent diameters.

Ptolemy reached a few minutes of arc in accuracy.

Menelaus, of Alexandria (second century BC) is one of the authors often quoted by Ptolemy for his trigonometric methods (and for some significant observations).

Mansur (Abu Dja 'Far 'Abd Allah al) reigned, as an Abbassid caliph (714–775), and initiated a period of rich Arabian astronomy, following Ptolemy's methods. Not to be confused with the following, Mansur Abu Nasr, much more well-known, and a real astronomer more than two centuries later. Under the caliph's authority, and under his successors, an active school developed in Baghdad. Many translators were at work, notably the tradition of Hunayn Ibn Ishaq (808–873, a philosopher and a medical doctor, translator of many Greek philosophers and physicians). Ptolemy was later translated and commented upon by his followers, such as al-Farghani Abu'l'Abbas Ahmad Ibn Muhammad Ibn Kathir (Alfraganus, ?–after 861), author of an Arabic digest of the *Almagest*, later translated into Latin in 1135 by Jean de la Luna.

Al-Battani (Abou Abd Allah Mohammed Ibn Gabir Ibn Sinan (al Raqqi, al Harrani, before 858–929)) was an excellent observer. From his own observations, he computed the distances of planets in the Ptolemaic system, as did Ibn Rostegh at about the same time. Maimonides, not much later, in spite of his acutely critical mind, was an admirer of these measurements and determinations.

Mansur (Ibn Ali ibn'Iraq; Abu Nasr, 950? at Gilzan, Persia – ca 1036, Ghazna, in what is now Afghanistan) was close to the Sultan till his own death, with his student al-Biruni. Measured obliquity of ecliptic; trigonometric research; construction of an astrolabe; used the method of Ibn al-Sabbah to study the solar position. Seemingly not greatly influenced by Ptolemy's thinking, he nevertheless undertook the task of checking Ptolemy's data, oriented apparently more towards observations than towards a model of the World. Large and lasting influence in the Arabic world.

Al-Biruni (973–1050 or later). Treatise on the sextant. Elements of Ptolemaic astronomy. Also an astronomer of lasting influence.

Lest we forget the development during this period of the Arabic astronomical concepts which we have mentioned earlier, we must remind the reader of such prominent figures as Thabit ibn Qurra (836–901) (see p. 143). Let us not forget either that the development of astronomy continued in the Western world, often strongly under the influence of the Arab translators of Greek treatises, including Ptolemy's.

Gerbert of Aurillac (ca 945–1003) served as Pope Sylvester II, from 999 to 1003, Rome, astrolabes.

Nasir Al-din al-Tusi (or Abu Djafar Muhamed Ben Hassan al Thusi or Mohammad ibn Muhammad ibn al-Hasan. 1201, Tas, Persia – 1274, Baghdad), Persian – Many publications. Instruments: mural quadrant (4.3 m), azimuthal circle, armillary sphere (built by Mu'ayyad al-Din al-Urdi in 1261–62). Observatory in Maragha from 1259. Of the large instruments, his were the best up to the time of Tycho. 12 years of observations. Probably the greatest astronomer of the Moslem world.

Richard of Wallingford (ca 1292–1336), curate at St-Albans from 1326, was one of the greatest creators of instruments in the West: *albion* (meaning "all by one") to determine planetary positions; *rectangulus,* a sort of armillary sphere. Large astronomical clock, apparently the first entirely mechanical clock.

Ulugh Beg (1394–1449), in Turkestan, was the grandson of Tamerlane. Samarkand Observatory in Uzbekistan, astrolabes, sextants, large *Fahkri sextant* (radius 40.04 m) of unusual accuracy of 2–5″, comparable probably to Tycho Brahe's in the West. Similar concepts were used in India later.

In the meanwhile, Nuremberg artisans (with Regiomontanus and Bernard Walther) built precise instruments (instruments almost as good as those of Nasir Al-Din and even Ulugh Beg):

Walther, Bernhard (1430–1505) was a student and collaborator of Regiomontanus (although he was older!). There is a famous story about Walther that Albrecht Dürer bought him a house with a small observatory. He used a Jacob's staff.

Regiomontanus (Müller, Johann) was called Regiomontanus because he was born in Königsberg (1436–1476). Many publications, including translation of Ptolemy. Many observations of good quality. Reform of calendar. Advisor to Pope Sixtus IV on calendar reform. An astronomer of lasting influence in the West.

Levi ben Gerson (1288, in Bagnols, Southern France –1344) used the Jacob's staff, to measure angles, and the *camera oscura* to observe the

Sun. Many observations in 1321–1339. He rejected Ptolemy's systems (which disagreed with some of his own observations and with some aspects of the Aristotelian doctrine!) and proposed a system of 48 spheres.

Nonius (Nunez Salaciense, Pedro, 1502–1578) was known for the construction of a dividing instrument, the *nonius*), in order to measure small arcs of circle with good accuracy. Somewhat similar to the *vernier* but of a lesser quality.

Apian, Peter (or Apianus, Petrus, also Bienewitz, or Bennewitz, Peter. 1495, Leisnig, Germany –1552) – Writings about instrumentation. Observed five comets. He noted that the tail is in the opposite direction from that of the Sun. His son Philipp continued his writings on instruments.

Wilhelm IV, Landgrave of Hesse (1532–1592), ordered the construction of an astronomical clock (1560–61), "das Wilhelmsuhr." Observatory in Kassel with the first revolving roof (1561). Stellar catalogue of 179 stars. Determination of the latitude of Kassel with an accuracy of about 10".

Tycho Brahe (1546, Skone, Denmark –1601, Prague) constructed a 19-foot quadrant in Augsburg graduated in minutes (1566). Great celestial globe, 5 feet in diameter at Observatory in Hveen, in what was then the Danish Sound. Built Uraniborg (1576) and Stjerneborg (1584) in Hveen. Large instruments of the mural type, azimuthal semi-circle, Ptolemaic rulers, brass sextant, azimuthal quadrant, parallactic rulers, including his famous meridian instrument, the large mural quadrant of about 6 feet in radius, of which the arc was divided in minutes and each minute into 6 divisions of 10" each. Tycho reached an accuracy of 1'–2', as against a few minutes of arc previously, due to the size of his instruments, the care of their construction and orientation, and the method of graduation by transversal lines. Later (1599), he used a small observatory with smaller instruments in Benatek, or Benatky (near Prague). Tycho was, undoubtedly, after Ulugh Beg, the greatest observer of pre-galilean times (see above, p. 208 ff.), certainly the one who most affected the trends in ideas about the system of the World.

Bürgi (Bürg, Burgi, Borgis, Byrgis), Joost (Jobst? Justus? 1552, Liechtenstein –1632, Kassel) was a watchmaker for Wilhelm IV and Rudolf II in Prague. Discovery of the pendulum clock, the pendulum being a regulator, used in order to time the observations, in Kassel Observatory. Assisted Kepler in computing.

6.1.1.b
Optical instruments

(α) The precursors.
Optical astronomy is rooted in the elementary attempts to study refraction and reflection centuries before. Euclid and Ptolemy, were, in a sense, precursors. But, again, one finds builders of instruments in both the Arabic and the Christian world.

Alhazen (Ibn al-Haytham, Abu Ali Al-Hasan ibn Al-Hasan), the Arab scholar of the 10–11th centuries (965, in Iraq –1039, in Cairo), experimented with light transmission in different media. Two of his treatises on optics (the others have been lost) were later translated into Latin and deeply influenced occidental European scholars thereafter.

Grosseteste, Robert (ca 1168–1253). – Optics, calendar. He stressed the use of lenses as a way to bring distant objects "nearer."

Bacon, Roger (1214–1294), in Oxford, studied the structure of the eye, spectacles, reflection, refraction by lenses (plane-convex). Possibly responsible for first combination of lenses. Did he know the telescope? Perhaps! Many writings on optics, and calendar reform.

Witello (ca 1230, Poland – ca 1275), a contemporary of Bacon, strongly influenced by the optics of Ptolemy and of Alhazen, performed studies of perspective, reflection by mirrors, etc. He did not apparently build instruments.

Spina, Alessandro della (Pisa? –1313), and Armati, Salvino degli, also studied reflection and refraction, in Florence. They are sometimes said to have been the inventors of reading glasses.

Digges, Leonard (or Diggs, ca 1520 – ca 1559), was a mathematician, perhaps originator of some optical advances, such as the *pantometric mirrors*, and the development of spectacles. Not to be confused with his son Thomas (see below).

Digges, Thomas (1546?–1595) was a mathematician. He (perhaps) was familiar with the telescope for terrestrial use. But, more essentially, he defended the Copernican point of view with great vigor. He observed stars, such as the supernova of 1572, and made measurements, second only in accuracy to those of Tycho.

Dominis, Marko Antonije, (1560, Rab –1626) wrote books on optics, lenses, telescopes. Better known for his early defense of a Copernican point of view.

Kepler, Johannes (1571–1630) conceived of a refracting telescope.

(β) The constructors.

Porta, Giambattista della (1535–1615) was a mathematician and a philosopher. In 1593, his book on refraction described the basis of the properties of lenses. He probably invented the combination of lenses. He did not invent the *camera oscura,* but used it, adding a concave lens in the aperture. He combined concave and convex lenses. Probably, he did not invent the telescope, strictly speaking; but undoubtedly, he paved the way for the invention of the telescope and for its astronomical use.

Lipperhey (or Lippershey), *Hans* (1587–1619) was the more likely inventor of the telescope (perhaps simultaneously with two other Dutchmen). In 1608, he solicited from M.de Nassau a patent for a terrestrial telescope.

Metius, Jacob (?–1628, born in Alkmaar) requested a patent a few weeks after Lipperhey (1608). His brother Adriaen (1571–1635) worked with Tycho at Hveen and taught astronomy at Frasseker.

Galilei, Galileo (1564–1642) has made his mark on history. His telescope was an association of one converging lens (objective: plane-convex) and one diverging lens (ocular: plane-concave), with a power of 3 at first, but later 30. Used in many discoveries, from 1609 (the date at which he arrived in Venice with a telescope of power 9) which greatly expanded the field of modern observational astronomy. (See above pp. 233 ff. and below p. 334). At the time of Galileo, and later, many dozens of telescopes of the Galilean type were indeed built and distributed all over the western world.

Harriot, or Hariot, Thomas (1560–1621) used a telescope for the study of the Moon, then for a systematic survey of the sky. He had heard of the findings of Galileo and observed at almost the same time, but with less persistence. Noted for his observations of the solar spots. Some of his findings were published only long after his death.

Fabricius, David (1564–1617) discovered in 1596 the variability of the brightness of Mira Ceti, but not its periodicity. He was the assistant of Tycho in Bohemia (1601).

Marius, or Mayr, or Mayer, Simon (1573–1624) was the first to publish observations of the Andromeda nebula, known previously, for example, by al-Biruni. Tables of the motion of Jupiter's four medicean satellites, which he named, on the advice of Kepler.

Scheiner, Christoph, s. j. (1573–1650) built a telescope in 1611, and devoted his instrument (helioscope) to solar observations, from about the same date. Described for the first time a "Keplerian" telescope, with two convex lenses. Sunspot studies (see later, § 6.2).

Fabricius, Johann (1587–1615 ?), son of David, made telescopic observations (1611) of the Sun and of sunspots in Wittenberg (see later, Chap. 7).

Hevelius, Johannes (Johann Hevel, or Heweliusza, Hewelcke. Danzig, 1611–1687) built his own instruments and organized his observatory (Sternenberg). His second wife, Elisabetha, was very involved in his work, and in the publication post mortem of his writings. She was probably the first woman astronomer of modern times. He used a telescope of 6 feet in length and another one later of 12 feet in length with a power of 50. He built a new helioscope and sundials.

Picard, Jean (1620–1682) made systematic use of a regular time service at the newly constructed Observatory of Paris (architect: Claude Perrault, 1671). He mounted a telescope on a quadrant. This innovation greatly expanded the possibilities of astronomical observations (see below, p. 339). He used Auzout's micrometer in collaboration with Auzout himself (see p. 323).

Cassini, Giovanni Domenico (*Jean-Dominique*, or "Cassini I", 1625–1712) was the founder of an astronomical dynasty. Very large instruments, an objective by Campani first, of 17 feet in length. He used objectives up to 136 feet in focal length, without a tube, as did Huygens (see next entry). First general superintendent (not Director, but resident) of the Paris Observatory.

Huygens, Christiaan (or Hugenius, 1629–1695) was a very great physicist. He constructed telescopes of very high power, of long, or even very long focus, without a material tube. With his brother, he developed techniques for polishing lenses of high quality (1655). Invention of the pendulum clock (1656, Bürgi's discovery was forgotten by that time) (see p. 313). Observation of the first satellite of Saturn, Titan (1655), of the ring around Saturn (1656), of the period of Mars, of the Orion nebula. Use of a new micrometer to determine apparent diameter of planets.

Campani, Giuseppe (1635–1715) built a clock that was presented to Pope Alexander VII. He invented a composite lens eyepiece, and built a telescope with four lenses, three of them being part of the eyepiece. He constructed polishing machines and made lenses of good quality for more than 50 years, mostly for the use of G.-D. Cassini.

Cassegrain, Laurent (ca 1629–1693)[26] published his first paper in 1672, when he was in Chartres. Reflecting telescope with a hole in the middle of the mirror for the direct observation of the prime focus behind the telescope, putting in front of the prime focus a small convex secondary mirror.

Gregory, James (1638–1675) Scotland. First description of reflecting telescope 1663 (hence before Newton and Cassegrain).

Newton, Isaac (1642–1727). Reflecting telescope (constructed in 1668). A prism is used to send the light on the side of the tube of the instrument.

Flamsteed, John (1646–1720) was the first Astronomer Royal of England. The Observatory in Greenwich was built on his advice. Refined methods to construct star catalogues (using clocks, micrometers, a large mural arc of circle of 140°). First catalogue based on optical astronomy (2935 stars) in 1712 (published by Halley, against Flamsteed's advice, republished in 1725).

Castel, Charles, abbé de Saint-Pierre (1658–1743). Caen, around 1675, developed a method to determine longitudes at sea.

Graham, George (ca 1674–1751) was a successful and active instrument maker. He never built instruments for himself but for others, notably, in 1725, an eight-foot quadrant for Edmond Halley.

Hadley, John (1682–1744) built a large telescope of the Newtonian type in 1720, with a diameter of 6 inches (15 cm) and a focal lens of 63 inches (1.6 m). In 1719, he also used parabolic mirrors and built several other reflecting telescopes.

Molyneux, Samuel (1689–1728) built a telescope in Kew, almost vertical (towards the star γ Draconis). He worked with Bradley; ordered a large zenithal sector of 24 feet in radius from George Graham; also used a smaller, more convenient sector, installed at Bradley's home, with which continuous observations over more than one year allowed him to discover and determine stellar aberration (1723–1727). Molyneux later had a successful political career.

Bradley, James (1693–1762) discovered stellar aberration (with Molyneux, 1727). His zenith sector telescope was used in 1732 in Wansted for stellar positional absolute astronomy of several stars, not only γ Draconis! He quickly became famous throughout Europe and succeeded Halley (1742) as Astronomer Royal. Halley had succeeded Flamsteed in 1719.

---

[26] This unpublished date has been kindly communicated to the authors by Françoise Launay.

Harrison, John (1693–1776) made improvements in chronometers and timekeepers with new escapement systems for use in navigation to determine longitude.

Dollond, John (1706–1761) improved the optical combinations, with a view to reducing the chromatism.

Bird, John (1709–1773 England), built several instruments, in particular a movable forty-inch radius quadrant, cast in brass, for Bradley (Greenwich Observatory), but also for St.-Petersburg, Paris, Cadiz.

Short, James (1710–1768). Made several telescopes, notably a 21.5-inch mirror telescope.

Clairaut, Alexis-Claude (1713–1765) conceived in 1761–62 of combined lenses of different refracting indices, in order to build objectives of refracting telescopes (see below, p. 343).

Le Monnier, Pierre-Charles (1715–1799) constructed the great meridian on the grounds of the church Saint-Sulpice in Paris, which can be used as a sundial, together with a lens located at the top of one of the large doors of the church and several instruments for the Paris Observatory, notably the first transit instrument.

Mayer, Johann Tobias (1723–1762) contributed to the correction of instrumental errors of astronomical instruments. Cartographic methods and use of a repeating circle.

Ramsden, Jesse (1735–1800) invented several optical and mechanical combinations (such as theodolites).

Herschel, F. Wilhelm (called *William* in England, 1738–1822). Grinding and polishing of large bronze mirrors (which unfortunately tarnish quickly), development of eyepieces, construction of large reflecting telescopes, from 30-foot focal length and 3-foot diameter (1781) to 20-foot focal length and 18-inch diameter (his favorite instrument, 1783). Invention of the Herschelian arrangement (or front view) which allowed a gain in brilliancy (or contrast). Construction of the large telescope of 40 feet in length, with a 48-inch diameter mirror (see later, pp. 344). Probably the most important astronomer of the end of the 18th century.

Zach, Franz Xavier von (Pest, 1754–1832) was adviser to Duke Ernst II of Saxe-Coburg. He directed the Observatory built in Seeberg, near Gotha. Several observations of the Sun, stellar catalogues. He is responsible for having formed an association of 18 astronomers, in various locations, in order to systematically observe the sky, divided into zones. This network resulted in the discovery of small planets or asteroids.

Arago, Dominique *François* Jean (1786–1853), a leading astronomer of his time, suggested, built, and encouraged the building of many instruments. He had a substantial influence on the younger generation of astronomers.

Gambey, Henri Prudence (1787–1847). François Arago drew attention to his work. He built several instruments, theodolites, heliostats (for Fresnel), etc. for French astronomers

Lassel, William (1799–1880) mounted his telescopes on an equatorial mount, with a fork mount, a clever idea, that neither Herschel nor Lord Rosse ever dared to use. His larger instrument was built at Malta in 1859, with a diameter of 4 feet (122 cm), and a focal length of 37 feet (11.4 m).

Parsons, William (third Earl of Rosse, better known as *Lord Rosse* 1800–1867), constructed in 1842–45, at Parsonstown in Ireland, a reflecting telescope 54 feet in length with a mirror of 72-inches in diameter, mounted on a meridian mount. His polishing machines were steam-driven. He used an alloy of four parts copper and one part tin.

Nasmyth, James (1808–1890), Scottish engineer, invented optical combinations allowing observations of a star, through a refracting telescope, from a fixed location of the observer.

Foucault, Jean-Bernard *Léon* (1819–1868), one of the best physicists of his time, published in 1857 a treatise concerning modern techniques for silvering glass to make mirrors for reflecting telescopes. He invented a classical method to test the quality of an optical surface using a razor blade. He built several instruments, including his famous siderostat and a photometer.

Draper, Henry (1837–1882), in the U. S., used mirrors in silvered glass (1864).

The progress in ground-based observations from the end of the 19th century to the present has been tremendous. Let us note in short: *Large refractors* of more than 1 m diameter and long focus: Yerkes 1897, 102 cm, 40 inch; Lick 1888, 91 cm, 36 inch; Meudon 1896, 83 (and 62) cm, 32.7 inch; Nice 1887, 76 cm, 30 inch, to stop the list in 1900. *Large reflecting telescopes*: Mount Wilson; Hooker telescope 1917, a large 101 inch telescope (257 cm); Palomar Mountain 1942, 5 m, actually used systematically only after 1948; Zelentchuk, in the Caucasus (former Soviet Union) 1974, 6 m; Keck telescope, in Hawaii, 1993 (two composite mirrors of 10 m diameter each); VLT in Chile, installation well advanced in 1996, four telescopes of 8 m diameter each, each movable with respect to the

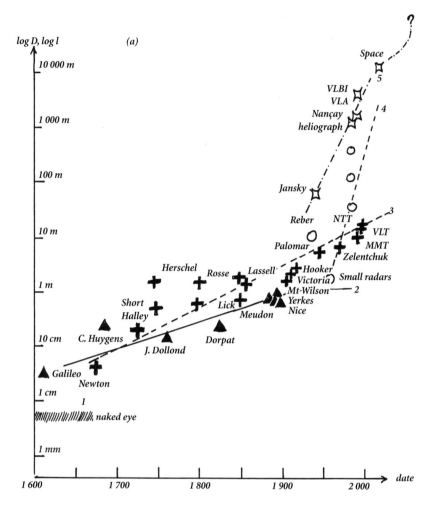

Fig. 6.1a–c *The evolution of the aperture size of the light collectors.* Black triangles indicate refractors (uninterrupted line, 2); crosses indicate mirror telescopes (dotted line, 3); circles indicate single-dish radio telescopes (broken line, 4); quasi-square white symbols indicate radio-interferometers (mixed line, 5). Abbreviations are VLA: Very Large Array; VLBI: Very Long Baseline Interferometry; MMT: Multiple Mirror Telescope; VLT: Very Large Telescope; NTT: New Technology Telescope. As the scale is logarithmic, the tremendous and rapid increase of the potential is clear. For (**b**) and (**c**) see next pages.

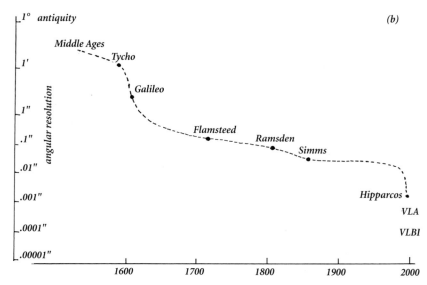

Fig. 6.1b *The evolution of angular resolving power in the visible domain since Antiquity* (see also black circles in Fig. 6.1c). The scale here is in fractions of seconds, not in fractions of radians, as in part c. Remember that: 1 radian = 57.29577°, 10^{-2} rad = 34.3775', 10^{-5} rad = 2.06264", 10^{-10} rad = 0.00002", 1° = 0.0147453 rad, 1' = 0.000291 rad, 1" = 0.00000485 rad.

others. *Large Schmidt telescopes* (having a huge field of view due to very elaborate optics), in Palomar 1948, 122 cm; Oak Ridge, 152 cm. But also solar telescopes such as "Themis," 1996, in the Canary Islands.

Figure 6.1a shows the progress in size of instrumentation of various types and Fig. 6.1b the parallel progress in angular resolution in the visible. Fig. 6.1c shows this development at all wave-lengths.

The measure of time is based both on astronomical observations of the Sun (sundials, gnomons, etc.) and stars (merkhets, meridian circles) and on instruments based on some physical principle such as hourglasses. Actually, the meridian circle was used up to 1845 or so to regulate pendulum clocks. We must, therefore, mention the meridian circles built since the Renaissance, using lenses, by Cassini (in Bologna and Paris) and the one in the Church of Sta Maria degli Angeli, which was used to regulate the clocks of Rome.

But, for short durations, clocks were used and the determination of time was based for a long time on pendulum motion, used as a regulator and scale. Quartz vibrations, using piezo-electricity, were developed and their accuracy permitted the determination of variations in the motion

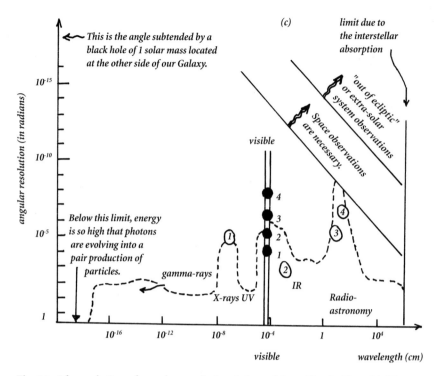

Fig. 6.1c *The evolution of angular resolution (adapted from Martin Harwit).* The resolving power (in a dotted line, the maximum resolution reached by astronomical techniques) is a function of both the diameter of the collector, and the wavelength. The atmospheric tremor, however, damages the quality of images seen from the ground unless mathematical methods of restoring this quality using the known properties of the atmosphere are used. Space instrumentation avoids that difficulty, irrespective of the size of the instrument. Still, it is necessary to go outside of the ecliptic to achieve better definition. Far from the solar system, the conditions may be still better. Symbols: In white circles: 1. Einstein X-ray Telescope; 2. IRAS, the Infrared Astronomical Satellite; 3. VLA radio interferometer; 4. SETI (search for extraterrestrial intelligence) array of radioastronomical dishes. In black circles corresponding to observations in visible light: 1. naked-eye observations of Antiquity; 2. Tycho Brahe's naked-eye observations; 3. Hubble Space Telescope, Hipparcos Space Telescope; 4. "next generation" optical telescopes (combining space techniques, adaptive optics, and corrections for image distortion).

of rotation of the Earth around its axis. Now, atomic time (using the fundamental vibrations of an atom of cesium) is much safer as a basis for formulating a definition of time than astronomical time; the relative accuracy obtained in the measurement of the duration of an astronomical phenomenon is in the best case about 10^{-14}. All modern systems of units are based on atomic time (TA), as opposed to UT (universal time, based on the motion of the Earth with respect to the stellar vault), or ST (solar time, based on the motion of the Earth around the Sun).[27]

6.1.2
Development of Ocular Instruments and Techniques

6.1.2.a
Micrometers

Roberval, Gilles Personne de (1602–1675), together with Auzout and Picard, used the micrometer Auzout-Picard (see later) for systematic observing efforts.

William Gascoigne (1612–1644), England, constructed (1640) the first micrometer with crossed wires. Angles of a few seconds could be measured. He was also the first to apply collimation to telescopes. His inventions were forgotten, because of his untimely death at war, until 1841 when a letter of his to William Oughtred was printed by G. Rigaud.

Picard, Jean (1620–1682) worked on the Auzout micrometer. The perfection of the instrument resulted from the use of a lattice of fine wires by Cornelio Malvasia. Auzout and Picard used two parallel hairs separated by a variable distance, revealing the apparent size of the image. Systematic observations beginning in 1666.

Auzout, Adrien (Rouen, 1622–1691) made improvements to the movable-wire micrometer (1666), actually forgotten since the untimely death of Gascoigne. He advocated the replacement of open sights by telescopic sights.

Huygens, Christiaan (1629–1695), using his discovery of the focal point, constructed the first micrometer used for the determination of planetary diameters.

---

[27] This is a great over-simplification. Many successive time-scales and units of time intervals have been in use.

The 20th century has seen many objective developments in ocular observation, with no permanent association with any particular scientist. Note the photometric zenithal tube, the objective prism astrolabe, and more generally the entire development of the automation of data.

6.1.2.b
Spectroscopy, Spectrography, Photography

Newton, Isaac (1643–1727) discovered, in 1665–66, the refractive properties of the prism. Discovery of the nature of white light as a combination of lights of various colors. This was the beginning of spectroscopy. In 1704, he published a *Treatise on Optics.*

Wollaston, William Hyde (1766–1828), in 1802, observed absorption lines in the solar spectrum; first spectrum of the Sun. In the following year, the optician Joseph Fraunhofer, in Münich, improved the spectroscope.

Niepce, Joseph Nicéphore (1765–1833) and

Daguerre, Louis Jacques Mandé (1787–1851) both contributed beginning in 1827, to the discovery and the development of photography. This technological discovery, which "exploded" between 1840 and 1850, completely changed the way of observation at the eyepiece (now therefore an obsolete appellation) and allowed the observer to reach the UV part of the spectrum. Many beautiful astronomical photographs were taken in this decade, in particular photographs of the Sun by Fizeau and Foucault in 1845.

Fraunhofer, Joseph (1787–1826), a lens maker, improved the technique of glass chemistry, but was also a theoretical scholar in optical studies. He was a European master in this field. He found (1809–1813) many lines in the solar spectrum, only a few of which had been previously seen by Wollaston in 1802. He used a diffraction grating (1823), with 260 parallel lines, like a prism. It enabled him to obtain a spectrum of a light source. He noted the coincidence of some solar lines with emission lines of terrestrial light sources. He also noted the difference in the spectra of the Sun and other stars, notably Sirius. Many astronomers such as Struve in Dorpat and Bessel in Königsberg used instruments built by Fraunhofer.

Bunsen, Robert Wilhelm Eberhardt (1811–1899) explored the possibility of chemical analysis of salts through the color of flames. From 1860, he worked with Kirchhoff.

Kirchhoff, Gustav Robert (1824–1887) of Heidelberg showed that at the same position in the spectrum as the Fraunhofer lines, gaseous flames can produce either absorption lines (if observed through) or bright lines. Roughly speaking, it is now understood that the occurrence of dark Fraunhofer lines is due both to the opacity of the solar matter, greater at the lines, and to the decrease outwards of the solar temperature.

Photoelectricity, experimentally explored during the 19th century by several physicists, beginning with Philipp Lenard (born in Pressburg, now Bratislava, 1862–1947) around 1899, paved the way for Einstein's theory of the photon (see pp. 387). It extended the study of the spectrum to its near infrared part, and permitted an increased sensitivity. The modern CCD allows us to obtain very good images, at a very high gain.

6.1.2.c
Brightness and Polarization of Light Measurements

Bouguer, Pierre (1698–1758) proposed a determination of the brightness of a star outside the atmosphere, suggesting the use of *Bouguer's straight line*. He may be considered the "father" of photometry.

Herschel, William (1738–1822) improved the determination of the brightness of stars by the use of photometric sequences (1796–1799). In doing so, he discovered many variable stars. One of Herschel's discoveries was the variability of α Herculis (1796). See above p. 318 and below, p. 328.

Arago, Dominique *François* Jean (1786–1853) was an excellent optician. He discovered the chromatic polarization of light, invented the polariscope, and later in 1811 the polarimeter. In 1815, he invented an instrument (the cyanometer) intended to measure the degree of blue of the sky and in 1833 he invented a photometer.

Airy, George (1801–1892) England. Many studies about polarization and polarizing devices.

Pouillet, Claude Servais Mathias (Doubs, 1790–1868), at about the same time as John Herschel (see following entry), built a similar instrument and undertook to measure solar heat and atmospheric absorption (1837).

Herschel, John Frederick William (son of William, 1792–1871) invented a solar photometer using a calorimeter (called an "actinometer"). With it, he determined the *solar constant*, i.e. the energy received from

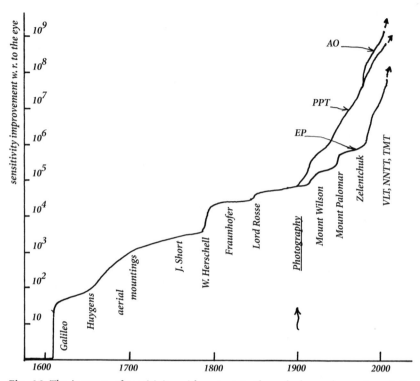

Fig. 6.2 *The increase of sensitivity with respect to the naked-eye observations (after Martin Harwit, adapted and extended).* This figure takes into account the increased size of the instrument, but also the properties of the receiver. (This includes integrating possibilities of photographic emulsions, and more recently of photoelectric and electronic devices, the progress of the latter illustrated by an arrow labelled PPT. The development of active and adaptive optics is also taken into account and labelled AO.) The visual observation of the focal image with an eyepiece is noted EP.

the Sun on one square centimeter of surface, at the distance of the Earth. More than anyone else, he developed a scale of brightnesses, photometric sequences, permitting very fine determination of brightness.

Langley, Samuel Pierpont (1834–1906), in the U.S., developed (in 1879–1881) improved techniques, and constructed the bolometer (1879–81). He founded the Smith Astrophysical Observatory in 1890. He could measure a difference ΔT of 0.00001 °C, in a 1 second exposure, with an error of less than 1%. Determination of the solar constant, taking into account the non-visible parts of the solar spectrum.

Photoelectric methods, applied to the determination of stellar brightness, now result in an accuracy of 1/1000 of a magnitude.

Figure 6.2 shows the dramatic improvement of the accuracy in brightness determinations.

6.1.3
Basic Physical Data Specific to Astronomy: Velocity of Light c, Gravitational Constant G, etc.

Roemer, Olaus (or Römer, Ole Christensen, 1644–1710) at the Paris Observatory, used the Cassini tables of eclipses of the satellites of Jupiter, notably Io, and their observations, to explain inequalities suggested by Cassini in his 1675 Jupiter satellite observations. Roemer showed (1676) that the velocity of light c is finite. He estimated the velocity of light (for him, light took 22 minutes to cross the diameter of the Earth's orbit, instead of about 16 minutes).

Mason, Charles (1728–1786) made observations of the passage of Venus and measurements of the Earth's density.

Cavendish, Henry (1731–1810), stimulated by John Michell, in Cambridge, England, performed a famous experiment in 1798 showing the existence of an attraction between two heavy lead balls. He determined the attraction constant G (which should be called the Cavendish constant, not the Newton constant), and the density of the whole Earth, 5.5.

Maskelyne, Nevil (1732–1811) and Hutton, Charles (1737–1823), both from Scotland, derived (1778) from the orientation of a plumb line besides a mountain, a density of the Earth with respect to sea water of about 4.5.

Fresnel, Augustin (1788–1827) arrived at the hypothesis of the wave nature of light, and with the help of Arago, developed an understanding of its consequences and checks. One of the greatest physicists of his time.

Doppler, Christian (1803–1853) showed in 1842 that if a source of sound waves is moving away from the observer, the frequency of sound is decreased, whereas it is increased in the opposite case. He suggested that the color of stars could be modified by a similar effect, resulting from light waves.

Pogson, Norman Robert (1809–1891), in 1856, suggested that the energy radiated by a star of any magnitude m bears a fixed ratio to that of a star of the next magnitude m + 1 (the so-called Pogson's law)[28]. Pogson's magnitude scale has been a standard scale for a long time.

Fizeau, Hippolyte (1819–1896), in Paris, developed an independent method to determine the velocity of light *c*, in 1849, using a dented wheel, over the distance Suresnes-Montmartre (see Chap. 7, p. 387 ff.). Independently of Doppler, who announced his discovery in 1842, Fizeau showed the same effect is true for light-waves. He, in fact, rediscovered the Doppler effect, and explained in a correct way the Doppler effect on light. It is said sometimes that Doppler is responsible for both discoveries, but Fizeau is certainly the author of illuminating papers on this "Doppler-Fizeau effect."

Foucault, Jean Bernard *Léon* (1819–1868) demonstrated, by the use of the pendulum, the motion of Earth with respect to an absolute system of reference. This experiment (performed in the Panthéon, in Paris, in 1851) made a strong impression on public opinion. He also measured the velocity of light using a rotating mirror (1850).

Newcomb, Simon (1835–1909) in the U.S., arrived at a measurement of *c* (1880–1882) using an improved version of Foucault's method. For a long time, Newcomb's value was the standard value adopted by astronomers.

Cornu, Marie Alfred (1841–1902), France, measured the velocity of light using Fizeau's methods in 1874 and 1876.

Michelson, Albert Abraham (1852–1931), U.S., arrived at a measurement of *c*, using a rotating mirror prior to Newcomb (1878).

6.1.4
Aperture of the Spectrum

Visual observers were limited to white light. Although Newton had discovered white light is a combination of light of all colors of the rainbow, no one studied the details of the spectrum until rather late. But even before that stage:

Herschel, (Frederick) *William* (1738–1822) already mentioned, a German who lived in England from 1757 on, initially as a composer, discovered the infrared part of the spectrum as *thermal radiation*. This

---

[28] This ratio has to be equal to 2.512, because the decimal logarithm of this number is 0.4, thus $2.512^{5/2}$ is equal to exactly 10, the exponent $5/2$ coming from the physical relation between magnitude and luminosity.

was, more clearly put, radiation that was not visible, but able to increase the temperature on a thermometer (showing only that the thermometer was opaque to the radiation, not that radiation conveyed any thermal energy more than rays of ordinary light!).

The *UV part of the spectrum* was discovered through photography, in the middle of the 19th century.

In 1932, *Karl G. Jansky* (1905–1950) discovered the radiation in the metric wave lengths of the Milky Way, predicted by Henri Deslandres, but not observed by Charles Nordmann, in Nice, in 1903. In 1944, *Grote Reber*, an Australian engineer, mapped the radiation from the Milky Way in long radio waves. In 1942, *James Stanley Hey* and *George Carl Southworth* independently discovered the radiation of the Sun in centimetric and decimetric wavelengths. Beginning in 1946, many radio galaxies were discovered. Radioastronomy quickly developed in a window of the spectrum (Fig. 6.3) ranging from submillimetric to kilometric wavelengths. Instruments are necessarily of a huge size. A single dish can be 300 m in diameter (at Arecibo, Puerto Rico). Associations of dishes in large interferometers (VLA, VLBI) can extend over thousands of kilometers.

The opening of space to artificial satellites, after the launching of very high-reaching rockets (Fig. 6.4) beginning in 1950, permitted the exploration of the whole spectrum, in particular the high energy part of it, gamma-rays, X-rays, and of course the whole UV field.

6.1.5
The Astrophysics of Particles
(solar particles, cosmic rays, neutrino astronomy)

This field began to develop at the end of the 19th century. Aurorae polares were indicators of particles flowing onto the Earth. With the discovery of cosmic rays in the 20th century, particles were identified. Their origin is in the Sun or in our Galaxy. Electrons and protons, guided by the magnetic fields of the Earth, produce aurorae, geomagnetic disturbances and storms. They may have climatic influences. Stars emit neutrinos. We now know how to observe solar neutrinos of various energies, as well as neutrinos from a supernova located in the Magellanic clouds.

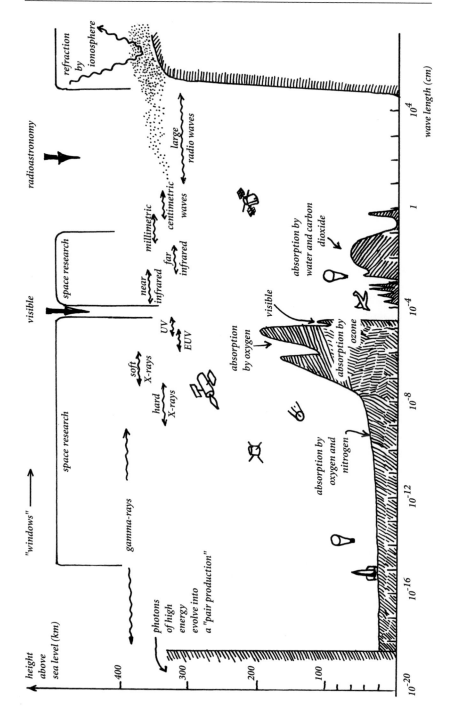

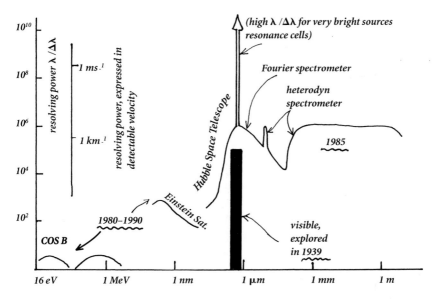

Fig. 6.4 *Progress of spectral exploration over time (adapted from M. Harwit and P. Léna).* The graph indicates actual spectral resolution obtained in various domains, with the date of acquisition. One can see the possibility of elaborate techniques, such as resonance spectrometry, at least for the study of very bright sources (such as the Sun). It also becomes clear that in the X-ray and gamma-ray domains, the resolving power is starting to become promising. There is little doubt that this part of the graph will indicate significant improvements in the coming decades.

Fig. 6.3 *Atmospheric and astronomical limitations of the spectral observations (adapted from Martin Harwit).* The extention from the very narrow zone of the "visible" has been very slow. Now, because of radio astronomy, infrared techniques from the ground and from balloons, and UV rays from space, the whole spectrum is open. On the abscissae, the wavelength in centimeters is represented on a logarithmic scale. For each wavelength, the altitude at which no observation (or very little) is possible is given in gray. In ordinates, we see the altitudes from which all observations are essentially feasible. Note the two black arrows pointing towards the "windows" of the spectrum, the "visible" window and the "radioastronomical" window.

6.1.6
Mathematical Techniques and Theoretical Tools for Astronomy

Sacrobosco, Johannes de (also John of Holywood, also John or Johannes Holyfax, Holywalde, Sacroboscus, de Sacro Bosco, ?–ca 1256). Geometry of the sphere, spherical motions.

Regiomontanus (1436–1476) reintroduced trigonometric quantities, used previously by the Arabs.

Halley, Edmund, published the writings of Apollonius on conics, and geometry in general, used later by Newton.

Newton, Isaac (1643–1727). Development in series (of a binomial), differential calculus in 1665.

Leibniz, Gottfried Wilhelm (1646–1716) was the creator of modern analysis, differential and integral calculus.

Bernoulli, Jakob (Jacques I, 1654–1705), Swiss mathematician. Series, probabilities, mechanics, etc... The first of a dynasty of brilliant mathematicians.

Bernoulli, Johann (Jean I, 1667–1748) was a teacher of Euler.

MacLaurin, Colin (1698–1746) was a student of Newton. Analytical theories of series, theory of tides.

Machin, John (?–1751). Mathematical description of the nutation discovered by Bradley.

Chevalier de Liouville (1671–1732), in 1720, John Bernoulli (1667–1748) in 1730, and Pierre Louis Moreau de Maupertuis (1698–1759), in 1732, developed Newtonian mechanics.

Bernoulli, Daniel (1700–1782), son of Jacques I, the first "Newtonian" mathematician outside England, was a teacher of Euler, as was his father. Theory of tides.

Euler, Leonhard (1707, in Basel –1783, in St. Petersburg), a Swiss national, brilliant mathematician, who lived primarily in Berlin and in St. Petersburg, developed methods used in Mayer's Moon theory, theory of tides, three body theory. Progressive improvements of perturbation theory, first proposed by Clairaut. Orbits of comets and planets. Method of determination of solar parallax.

Clairaut, Alexis Claude (1713–1765) was a very versatile mathematician. Newtonian theory of the figure of the Earth. Theory and tables of the Moon. Motion of comets. Method of perturbations by Jupiter and Saturn applied to the return of Halley's comet in 1759 (actually discovered by an amateur, Georg Palitzsch).

D'Alembert, Jean Le Rond (1717–1783) published the *Recherches sur la précession des équinoxes et sur la nutation de l'axe de la terre*, in 1749, and the *Recherches sur différents points importants du système du monde*, from 1754 to 1756. Elaboration of dynamics. Three body problem. Lunar and planetary theory. Theory of precession and nutation. Improvements in perturbation theories. Together with Denis Diderot, the main author of the *Encyclopédie*.

Lepaute, Nicole-Reine Etable de la Brière (1723–1788) was the wife of Jean-André Lepaute, a reknowned clock-maker (1720–1788). She observed with Lalande. One of the first women known as an astronomer.

Montucla, Etienne (1725–1799) wrote an epoch-making *History of mathematics* (including astronomy). This book, in four volumes, was finished after Montucla's death by Lalande.

Lambert, Jean Henri (Mulhouse 1728–1777) found perturbations in Saturn's motion. Photometry, building of celestial and terrestrial maps. Better known for his *Lettres Cosmologiques*, published in 1761 (see pp. 363).

Lalande, Joseph Jérôme le François de (1732–1807) and Nicole-Reine Lepaute extensively applied Clairaut's methods to Halley's comet, to the perturbations of Mars by Jupiter, of Venus by the Earth.

Lagrange, Joseph Louis (1736–1813) developed a general method of treating problems in dynamics (using *Lagrangians*). Chairman of the commission that established the *système métrique* in 1790.

Titius, J.D. suggested 1772 an "empirical law" ruling the distances of planets to the Sun as a function of their rank (1 for Mercury, etc..)

Bode, Johann (1747–1826) computed the orbits of possible asteroids (before their discovery by Piazzi) based on the empirical "law" (known as the law of Titius-Bode) of planetary distances to the Sun, as suggested by Titius in 1772.

Laplace, Pierre Simon (1749–1827) was remarkable as a mathematician. He developed "celestial mechanics", a designation he coined himself. Better known to astronomers for his evolutionary theory of the planetary system (the nebular hypothesis, proposed in *Le système du Monde* 1796), independent of and somewhat more detailed than a similar theory by the philosopher Immanuel Kant (1724–1777) published in 1755.

Bürg, John Tobias (1766–1834, in Vienna). Successful use of Laplace's method for building lunar tables.

Damoiseau, Marie Charles Théodore (1768–1846). Lunar theory using Laplace's methods.

Burkhardt, John Charles (1773–1825, a German in Paris) was a successful student of Laplace. Constructed tables of the Moon which superseded previous ones.

Whewell, William (1794–1866). Theory of tides.

Hansen, Peter Andreas (1795–1874). Celestial mechanics, perturbation theory applied to Jupiter and Saturn. Tables of the motion of the Moon. Theory of astronomical instruments.

Airy, George (1801–1892). Theory of tides. Theory and measures of diffraction.

Lubbock, John William (1803–1855). Theory of tides.

Delaunay, Charles (1816–1872, Paris). Mathematical theory of the Moon, implying frictional forces.

Gylden, Hugo (1841–1896). Pure research connecting planetary theory with pure mathematics.

Tisserand, François Felix (1845–1896). His *Treatise on Celestial Mechanics* (1889) is still very widely in use.

Poincaré, Henri (1854–1912) was a pure mathematician, involved in many other fields, and partly oriented towards cosmogonical-cosmological research.

6.2
⋆ Important Astronomical Discoveries After Galileo

In order to help the reader, we have noted the discoveries according to their sub-field, by a symbol at the left, as follows:

Earth measure: ♂; Sun: ☉; planets and comets, etc.: @; Moon: ☾; stars: ⋆; Milky Way: W; galaxies: S; universal structure: Ω
Important key objects or key discoveries are in italics.

The list is chronological by author. Here and there, authors of different generations have contributed simultaneously to a group of important efforts. This will be signalled by a sign (XX) at the beginning of an intermediary paragraph. Discoveries included in Harwit's list (Fig. 6.5a,b and Table 6.1) are indicated, in double parentheses, and in italics, by their *Harwit number* at the place where they are mentioned.

The first name to note in modern astronomy is, of course, that of its main founder, *Galileo*. We have given ample room to his observational discoveries (see above pp. 233ff., 315): Sunspots, Jupiter's satellites, lunar mountains, Milky Way stellar structure. We will not go into further detail here.

Table 6.1 (Abridged after M. Harwit) Forty-three Astronomical Discoveries. We present here, in an abridged form, and modified at a few points, Martin Harwit's table, which is the basis of Figs. 6.5 a and b.

Note that the great astronomers of the past were well-known for making discoveries of new objects, and identifying as objects of a new sort, some previously known sources of light which had been considered stars. In contrast, many of the modern astronomers are almost unknown to the ordinary reader, even to the specialist in another field. This reflects two trends. One is the tendency towards a higher and higher degree of specialization. The other is the emphasis (note here the methods of attribution of Nobel Prizes) on the "discovery," as against the conceptual understanding and astrophysical synthesis, perhaps, in our opinion, even more important.

Note also that Harwit has not included any discoveries concerning the Sun or the Moon, for no apparent reason. Note also some inconsistencies between table 6.1 and Fig. 6.5a: The numbers do not correspond between the two presentations of the same history.

1.	Interplanetary Matter, Meteors	1798	Brandes, Benzenberg
	meteorites	1803	Biot
	zodiacal dust	1934	Grotrian
	radar echoes	1946	Hey
2.	Planets/discoveries	BC	shepherds, sailors
	disk is resolved	1610	Galileo
	aberration (Earth's motion)	1728	Bradley
	radio	1955	Burke, Franklin
	exploration	1967, etc.	NASA, USA; Soviet Union
3.	Asteroids	1801	Piazzi
4.	Moons (of planets)	1610	Galileo
5.	Rings (of planets)	1655	Huygens
6.	Comets (as "astral bodies")	1577	Tycho Brahe
	(as periodical)	1705	Halley
7.		1717	Halley
	Main sequence stars (positions)		
8.	Subgiants, red giants (parallaxes)	1830–40	Bessel, Struve, Henderson
	(spectra)	1890	Pickering
		1910	Hertzsprung, Russell
9.	Pulsating stars	1596	D. Fabricius
	(photometry, spectra)	1912–1914	H. Leavitt, Shapley
10.	Multiple stars	1672	Montanari
11.	White dwarfs	1834	Bessel
	(Sirius B)	1862	Clark (Alvan)
12.	Galactic clusters	1754	Lacaille
13.	Globular clusters	1781	Herschel
14.	Planetary nebulae	1790	Herschel
15.	Ionized hydrogen regions	1865	Huggins, Miller

16.	Cold gas clouds	1903	Hartmann
17.	Interstellar dust	1914	Slipher
18.	Supernovae	1064	Chinese astronomers
	(as astral)	1572	Tycho Brahe
19.	Eruptive variables[1]		
20.	Nebular variables	1861	Hind
21.	Infrared stars	1949	Neugebauer, Leighton
22.	Flare stars	1949	Luyten
23.	Magnetic stars	1947	Babcock
24.	Cosmic masers	1965	Weaver, Weinreb, Barrett
25.	Pulsars	1968	Hewish, Bell
26.	X-ray stars	1962	Rossi, Giacconi
27.	Supernova remnants	1937–39	Mayall, Duncan
	(as such)	1942	Oort, Duyvendak
28.	Interstellar magnetic fields	1957	Dombrovsky, Mayer McCullough, Sloanaker
29–30.	Galactic (structure)[2]	1845	Lord Rosse
	(distance)	1925	Hubble
	(rotation)	1927	Oort
31.	Clusters of galaxies	1785	Herschel
32.	Radiogalaxies	1932, 1946	Jansky, Hey
33.	Unidentified radio-sources	1974	Kristian, Sandage
34.	Cosmic apparent expansion[3]	1912, 1929	Slipher, Hubble
35.	Quasars	1960, 1963	Palmer & Matthews, Hazard, Schmidt
36.	Superluminal radio sources	1971	Shapiro
37.	X-ray galaxies & clusters	1966	Friedmann, Chubb, Byram
38.	Infrared galaxies	1966–70	Johnson, Low
39.	Gamma-ray bursts	1973	Klebesadel, Strong, Olson
40.	Microwave background	1965	Penzias, Wilson
41.	X-ray background	1962	Rossi, Giacconi
42.	Gamma-ray background	1968	Clark, Galmire, Kraushaar
43.	Sources of cosmic rays	1912	Hess

[1] I disagree with Harwit about the insertion of this point as distinct from the discovery of supernovae.

[2] Here again, I group 29 and 30 together, and I disagree with Harwit on some details.

[3] The word "apparent" which I have added Harwit's is present throughout Hubble publications. It opens the way for cosmologies in which the cause of the observed redshift is not an expansion.

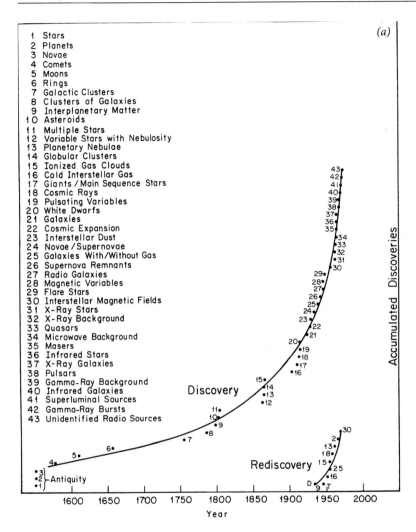

Fig. 6.5a *Important astronomical discoveries in history (after M. Harwit).* Since the beginning of astronomical inquiry, but particularly during the course of our century, a few dozen discoveries are noteworthy for being of prime importance. Of course, the choice is subjective, hence rather difficult, and may be strongly biased by individual interests. Harwit's opinion, however, reflects some general consensus in the world of astronomy. For (b) see next page.

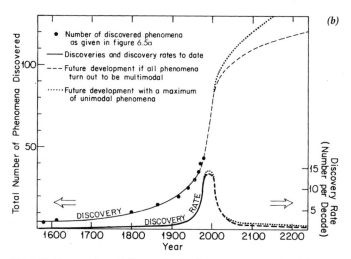

Fig. 6.5b *Progression of discoveries in the coming centuries (after M. Harwit).* Again, this is a very personal interpretation, by Harwit, of the observed trends. According to his point of view, the curve of discoveries could very well now be at its climax, and could regularly decrease in the coming centuries. One must be very cautious, however, in adopting that type of conclusion.

∗   *Fabricius, David* (1564–1617), ((9)) discovered (1596) the *variability of the brightness of Mira Ceti,* but not its periodicity.

☉ @   *Scheiner, Christoph, s. j.* (1575–1650) thought at first that the sunspots were planetoids. They were shown by Galileo to be on the solar surface. Observation of conjunction of the Sun and Venus. Confirmation of the heliocentric motion of Mercury and Venus. Great continuity in solar observations. Demonstration of the *rotation of the Sun around its axis,* and of differential rotation. Measurement of the solar rate of rotation.

(XX) Sunspots were actually observed, independently, by Galileo (Venice), Scheiner (Ingolstadt), Harriot (Oxford), Johann Fabricius (Wittenberg), D. Passignani (Rome). They were then identified as solar, not of terrestrial origin. Subsequently, they became a favorite subject of studies. (Wilson, in the 18th Century, found them to be "holes".) At the end of the 18th century, the activity cycle was discovered (Schwabe, Wolf), and later correlated with geophysical phenomena (Sabine, Wolf, Gautier).

☿ Snell, Willebrord (Snellius, 1591–1626). Measurement by a triangulation method (proposed by Gemma Frisius in 1533), of one degree of the meridian, between Alkmaar and Bergen-op-Zoom (130 km), Holland (1615, published in 1617): First measurement since Erathosthenes and the Arabs. Best known for his study of light refraction, later studied also, more accurately, by René Descartes.

☿ Norwood, Richard (1590?–1675), in England, near London (1636).

✶ Boulliau, Ismaël (1605–1694) determined, in 1667, the period of Mira Ceti.

@ ☾ ✶ *Hevelius* (1611–1687) constructed from his own drawings a beautiful *atlas of the Moon, seen at different libration states*. Named craters and other features. Drawings of Saturn. List of comets of the past. Catalogue of 1564 stars. The accuracy he reached was of 50″ (against 1′40″ for Tycho, 40″ for Flamsteed).

✶ Holwarda, Philocides (1618–1651). Recognition, after Fabricius, of the *variability of Mira Ceti* (first *variable star* discovered as such, after the supernovae discoveries, notably by Tycho Brahe in 1572 and Kepler in 1604).

(XX) An important effort was undertaken (Picard, La Hire, etc.) to measure the *solar apparent diameter*. These studies are now the object of a renewed interest, as the variation of the solar radius is often regarded, at the end of the 20th century, as a possibly real and significant phenomenon.

☉ ☾ @ ☿ *Picard, Jean* (1620–1682) used the movable-wire micrometer to measure the diameters of the Sun, Moon, and planets. Measurements of the *solar diameter* (at its passage through the meridian circle) in 1666–1682. Measurement of an *arc of meridian* near Paris in 1668–1670.

☿ @ ☉ *Richer, Jean* (?–1696). Measurement (at Cayenne, Guyana, 1672–73) of gravity acceleration (implying the *flattening of the Earth*). With J.-D. Cassini, J. Picard (Paris) measurement of Mars parallax in 1673, which led to the determination of the *solar parallax* by Cassini (9″.5).

@ ☿ *Cassini, Jean Dominique* (1625–1712) made tables of Jupiter's satellites (1668 and 1693). *Divisions of Saturn rings* – Cassini's division is well-known (1675). Discovery of four satellites of Saturn (1671–84); Mars parallax (1672–73), hence solar parallax, together with Richer at

Cayenne and Picard at Paris; Map of the Moon, 1679; *Zodiacal light,* identified as a celestial phenomenon, 1683. Measurement with La Hire of the *meridian arc* (1683) from Paris down to Bourges, and later (1700) with his son Cassini II (Jacques), and Maraldi I, down to Perpignan.

@ **W**   Huygens, Christian (1629–1695) discovered the *ring nature of the Saturn environment* (1656). Discoveries of *Titan,* first known satellite of Saturn (1655), of the Mars period, and of the *Orion nebula.*

@   Wren, Christopher (1632–1723) in 1677, after Hooke (see below). Suggestion of the attraction of planets by the Sun.

✶   Montanari, Geminiano (1633–1687), *((10))* discovered *variability of Algol* (i.e. β Persei) in 1667–70. Observed Mira Ceti, discovered in 1596 by David Fabricius (the father of Johann).

@   *Hooke, Robert* (1635–1703). Suggestion of the attraction of planets by the Sun in 1674.

⊙   *La Hire, Philippe de* (1640–1718). Measurements of the solar diameter (1683–1718). Picard's and La Hire's data were published later by P.C. Le Monnier in 1741.

@ Ω   *Newton, Isaac* (1642–1727). In 1687, he published the *Principia* (*Philosophiae Naturalis Principia Mathematica*), in Latin. In it, he detailed the discovery of the *law of universal attraction,* its mathematical derivation from Kepler's laws, and its main consequences (see pp. 267–278, and 295–307). Newton is without a doubt the astronomer who most influenced astronomical thinking for the next three centuries.

@ ✶   *Halley, Edmund* (1656–1742), *((6, 7))*. Rediscovery, after Wren and Hooke, of the *inverse square law of attraction.* As a friend of Newton, he encouraged him to finalize his discoveries in their perfect form. *Observation of the Comet of 1682, and prediction of its return, around 1757–58, after a study of possible past passages of what he assumed to be the "same" comet.* In 1715, observation of the solar chromosphere during a total eclipse. Halley also predicted the *passage of Venus in front of the Sun in 1761–69,* paving the way for future progress, and in particular suggesting the use of this phenomenon to determine the solar parallax (an idea first suggested by Gregory). He tried to do so in 1677 using the passage of Mercury. He suggested many ideas in his work, in particular the existence of stellar proper motions, and was a great stimulus to astronomy.

(XX) The building of stellar catalogues, first single stars, then binaries, then stars of decreasing brightness, etc. has been a major activity of most observatories since the beginning of the 18th century (one could even say, since the *Almagest!*). Starting as a systematic enterprise with Flamsteed, it is even now ongoing. *Flamsteed's catalogue of stars* was first published by Halley in 1712 without Flamsteed's authorization, reedited in three volumes by *John Flamsteed*, his assistants Abraham Sharp (1651–1742) and Joseph Crosthwait, and published after Flamsteed's death in 1725 and 1729. In 1712, 1935 stars were listed with an average error of 10" compared to Tycho's errors of about 1'; in 1725–29, 3000 stars. Slightly later, *Lacaille*, (1713–1762) created another catalogue.

@ Ω   Cotes (1682–1716). Second edition of *Principia*, in 1713.

@ ☉   *Delisle, Joseph-Nicolas* (1688–1768). After many conversations with Halley in London (1724), he was one of the promoters of the *systematic observation around the world of the passage of Venus in 1761*. Invited to Russia by Peter the Great (1721), he stayed there from 1727 to 1747, and produced many achievements, including the request for systematic observation of solar eclipses and of passages of Mercury.

@ ✶   Molyneux, Samuel (1689–1728) discovered (1723–1727) stellar aberration, together with Bradley.

@ Ω   Pemberton, Henry (1694–1771), third edition of *Principia*, 1726.

(XX) A considerable collective effort was undertaken, during the 17th and 18th centuries, *to check Newton's theory*, or to disprove it. Hence, two types of observations were made, centered around a measurement of the Earth and of its shape (Cassini, Picard, LaHire, LaCaille, Richer, etc ...). The Paris Observatory and the Académie des Sciences of Paris were primarily involved in these determinations.

@ ♂   *Bradley, James* (1693–1762) was the third Astronomer Royal (after Flamsteed and Halley). Discovery of *stellar aberration*, with Molyneux, in 1727. This was the first indirect proof of the validity of the Copernican system with respect to the Ptolemaic one. Discovery of *stellar nutation* (1727–1747). Bradley and, independently, Wargentin, Pehr Vilhelm (1717–1783), discovered the period of 437 days in the motion of the four satellites of Jupiter (period of the return of any given configuration).

♂   Bouguer, Pierre (1698–1758). See (along with next three entries), after Cassini de Thury, the paragraph on the measurement of the length of some arcs of the meridian circle for the contributions of Bouguer, Maupertuis, La Condamine and Godin.

Ⓞ̇   Maupertuis, Pierre Louis Moreau de (1698–1759).

Ⓞ̇   La Condamine, Charles Marie de (1701–1774).

Ⓞ̇   Godin, Louis (1704–1760).

@   Lomonossov, Mikhail Vasilievitch (1711–1765) was better known for his non-astronomical contributions to science. He used the 1761 passage of Venus to determine the effects of the refraction by the atmosphere of Venus on the solar light, thus discovering the very existence of such an *atmosphere around Venus*.

@ Ω   Thomas Wright of Durham (1711–1786). First suggestion of the rocky nature of Saturn's rings. A prophetic view of the Herschel island-universes (1750) (see p. 359).

Ⓞ̇ ☾ ✶ S   *Lacaille, Nicolas Louis de* (1713–1762), *((12))* measured the French *arc of meridian* in 1738. In 1750–1754, Lacaille went to the Cape of Good Hope, observed and described the southern constellations and nebulae and constructed catalogues of southern stars. Measurement of lunar parallax, together with Lalande who was in Berlin. Resolved the galactic (or open) *stellar clusters* into stars. Cometary studies.

Ⓞ̇   Cassini de Thury, César-François (or Cassini III, 1714–1784) was a student of G.F. Maraldi (his great-uncle). Measured the Earth in France, together with Cassini II, his father, in order to serve as a basis for the map of France. At his death, French cartography was essentially finished (only Britanny remained to be mapped).

(XX) The validity of Newton's theory, well accepted in England, was not so in France, mostly because of Cassini's influence. Therefore, *the French Academy of Sciences organized (for the first time in history, perhaps) a systematic expedition of astronomers to both Peru and Lapland*, to measure two arcs of meridian and determine whether the Earth is flattened (as predicted by Newton's theory) or not (as still claimed by Cassini until very late). Voltaire ˙(François Marie Arouet, le Jeune, dit-) (1694–1778), the famous French philosopher and writer, and *la Marquise du Châtelet (Gabrielle Émilie Le Tonnelier de Breteuil*, 1706–1749, an efficient scientist, translator of the *Principia* into French) played a major role in fostering the serious consideration of Newton's ideas in France. But the expedition was still considered necessary. Pierre Bouguer (1698–1758), *Charles Marie de la Condamine* (1701–1774), and Louis Godin (1701–1760), were the main participants in the expedition organized in Peru *(1735–1745)* by the (French) Academy of Sciences; *Pierre Louis Moreau de Maupertuis* (1698–1759) and Clairaut, then a young man, were the main participants in the expedition sent by the same Academy to Lapland *(1736–37)*. The result was quite convincing: *the flattening of the Earth was definitely proven and Newton's law became generally accepted.*

⊙ *Wilson, Alexander* (of Glasgow, 1714–1786) suggested (1774) that the Sun is a dark sphere surrounded by a bright envelope; sunspots are indeed holes in this envelope. Although outdated, a part of this idea is confirmed now by the measurement of lower pressure at sunspots, and by the *Wilson effect*, which describes *the spot as a depression in the Sun.*

☾ @ ⊙ Le Monnier (1715–1799) published (1746) new tables of the Sun and Moon. He systematically observed the Moon in order to determine the irregularities of the lunar motions.

☾ ⊙ *Mayer, Tobias* (1723–1762). Detailed *map of the Moon* (published in 1775). Complete description of its *libration*. Complete lunar theory. Solar and lunar tables, published by Charles Mason (1730–1787).

✶ Michell, John (1724–1793). Assumption of the existence of many binary stars (first actual discovery in 1803). Possible methodology to find stellar parallaxes.

☿ @ ⊙ Mason, Charles (1728–1786) made observations of the passage of Venus; measurement of the *Earth's density*; tables of the Moon and of the Sun.

S W *Messier, Charles* (1730–1817) discovered and listed *many nebular objects,* some being star clusters. His catalogue contains 103 such objects.

☿ Ω *Cavendish, Henry* (1731–1810) definitely confirmed the Newtonian theory of universal attraction.

☿ ✶ Maskelyne, Nevil (1732–1811) and Charles Hutton (1737–1823) made an important contribution to the determination of the Earth's density. Maskelyne also performed stellar studies, constructed catalogues, and obtained well-determined stellar proper motions.

@ ✶ ☿ *Lalande, Joseph Jérôme le François de* (1732–1807). His long and active life led him to measure the Moon's parallax (with Lacaille), to compute the passage of Halley's comet, with Clairaut and Nicole-Reine Lepaute, to coordinate the studies of the passage of Venus, to build a large stellar catalogue, and to exert a great influence on the astronomy of his time.

(XX) *The passage of Venus (1761 and 1769)* was observed by several groups of astronomers, at the request of governments, and academies, upon the initial suggestion of Halley and actively promoted by Delisle. *It was probably the first internationally organized enterprise in astronomy.* The aim was primarily the determination of the solar distance to Earth (through the *solar parallax*). There were 120 observing stations in 1761 and 150 in 1769. The analysis of data was done by Lalande and redone by Johann Franz Encke (1791–1865) in 1824. He found a solar parallax of 8".571, using the combined data of the Venus parallax, determined during the passages, and of Kepler's third law. In 1870, corrections were made to the longitudes used in the computation. This led to a value of 8".83. In 1890, Simon Newcomb found from these data (with some added corrections) a parallax of 8".79

@   Bailly, Jean-Sylvain (1736–1793) studied the inequalities of the motions of Jupiter's satellites. He wrote a famous *History of astronomy.*

@ ⋆ S W ⊙ Ω   Herschel, William (1738–1822), ((10,13,14)) was the major astronomer of his century. Many planetary observations. He discovered *Uranus,* a discovery confirmed by, among others, Anders Johann Lexell (1740–1784), who showed it was indeed a planet, not a comet, as first believed by Herschel. With his larger telescope, and the help of his devoted sister Caroline Herschel (see later), he discovered two satellites of Saturn, and two of Uranus. (He is thought to have discovered four more satellites of Uranus, but this is doubtful.) First *stellar catalogue* in 1782.

He also engaged in solar studies, and developed a theory of a cold solid Sun surrounded by the bright photosphere, in which the spots are holes (see Wilson).

One of Herschel's most important contributions to astronomy was his description of the sidereal system of which the Sun is a member, i.e. the Milky Way, using a method of star gauging he devised. Determination of the solar apex in 1783, from proper motions determined by Lalande and Maskelyne. Discovery and cataloguing of double stars (269 were listed in 1782, and many more in further publications).

Study of Mira Ceti, of Algol. Discovery of many new variable stars.

Herschel's hypothesis on the nature of nebulae and star clusters, more than 1500 of which were discovered by him, is famous. He thought some nebulae were stellar families (clusters). Some others are seen as *planetary nebulae* around a central star. For most of these, however,

Herschel's theory was that he had observed indeed some 1500 new universes, *"island-universes,"* comparable to our Milky Way. It is remarkable to note that this point was not firmly established until the work of Curtis, in the 20th century, after the famous battle with Shapley.

@   *Piazzi, Giuseppe* (1746–1826) stimulated by the network of observers organized by von Zach, and the calculations by Bode, discovered in 1800 the *first known asteroid, Ceres.*

♂   *Delambre, Jean-Baptiste* (1749–1822). Commissioned by the National Convention, he measured the arc of meridian of Paris, between Paris and the north of France.

@ Ω   *Herschel, Caroline* (1750–1848) worked extensively with her brother William in the systematic search for comets, and the discovery of eight of them. Discovered several new nebulae, notably "true" nebulae, or planetary nebulae.

✶ W   *Prévost, Pierre* (1751–1839) researched the motions of the solar system as a whole. Solar apex determination from Tobias Mayer's data, a few months after William Herschel, in 1783.

✶   *Goodricke, John* (1764–1786). In 1783, he measured the period of the variable star Algol (discovered as variable in 1669).

@ Ω   *Olbers, Heinrich* (1758–1840). Discovery of Pallas in 1802. Elaboration of the famous Olbers' paradox (see p. 408).

(XX) The initiatives of von Zach, and the discoveries of Piazzi and Olbers were the beginning of the *hunting for small planets, or asteroids* (between Mars and Jupiter). In 1804 Juno was discovered; in 1807 Vesta. In 1845, Karl Ludwig Encke (1793–1866) discovered a fifth one. Photographic methods allowed for massive discoveries (Max Wolf 1863–1932). In 1847, eight small planets were known, in 1897, 432 minor planets were known, 92 discovered by Charlois at Nice Observatory, 120 by Johann Palisa (1848–1925) of Vienna. Their distribution was studied and *Daniel Kirkwood* (1814–1895) discovered gaps in the distribution (number vs distance to the Sun), due to the resonant perturbation by Jupiter. The asteroid Eros is used to determine the solar parallax. Edward Emerson Barnard (1857–1923) at Lick, the discoverer of the 5th satellite of Jupiter, measured in 1894 and 1895 the diameter of the disks of the largest minor planets. In 1903, 512 asteroids were known, 2048 in 1978, and now more than 10000.

ð *Méchain, Pierre* (1774–1804). Commissioned by the Convention, he measured, out from Paris to Carcassonne, the meridian arc of Paris (see also Delambre, Biot, Arago).

ð Biot, Jean-Baptiste (1774–1962) was basically a physicist, specializing in radiation, polarization of light, and propagation of sound. He was commissioned to measure, together with Arago, the arc of meridian of Paris from the Pyrenées to the Balearic islands.

(XX) A revival of interest in the measurement of the longest possible arc of meridian was stimulated, primarily in France, by the definition of the meter as a universal unit of length by the French Convention (1792). This stimulated new expeditions. Delambre and Méchain started in 1792–1799 and measured the arc between Dunkerque and Carcassonne (Paris meridian). The measurement was later completed by Arago and Biot to the Balearic islands. It was the largest arc of meridian then measured.

ð Svansberg, Jöns (1771–1851) corrected in 1801–1803 measurements of the meridian arc length made by Maupertuis in Lapland.

@ *Poisson, Siméon Denis* (1781–1840). Basically a mathematician and a statistician, he was among the first to develop the *planetary theories and the lunar theory* on the basis of Laplace's methodology.

✶ *Bessel, Friedrich-Wilhelm* (1784–1846), ((8, 11)). Observation of the *parallax of 61 Cygni*, discovery of noticeable proper motions. His main works concern the theory of functions.

ð ✶ ☉ *Arago, François* (1786–1853) was a very influential French astronomer (see pp. 319, 325). As a young man, he took part in the measurements of the Earth by finalizing the measures of the Paris meridian started in 1792 by Méchain and Delambre, together with Biot.

☉ ✶ *Joseph Fraunhofer* (1787–1826), an optician from Munich, discovered approximately 600 dark lines in the *solar spectrum* (the so-called Fraunhofer lines). He mapped the location in the spectrum, from violet to red, of 324 of them. Some of them are still known (lines H, K, and b) by the letter with which Fraunhofer identified them. In 1823, Fraunhofer saw stellar spectra.

(XX) The importance of *spectrography* to the history of modern astrophysics cannot be under-estimated. It is the development of spectrography in the 19th century that results in the division of astronomy into two main branches: astrometry and astrophysics.

☉ ⚥   Sabine, Edward (1788–1883) discovered the relationship between sunspot numbers and geomagnetic disturbances (see also Gautier, Wolf).

☉   *Schwabe,* Samuel *Heinrich* (1789–1875), in Dassau, estimated, in 1843, the length of the *solar cycle (ca 10 years)* from his own sunspot observations. The result was noted widely only in 1851.

@ ☉   Encke, Johann Franz (1791–1895) determined the *solar parallax* (from passage of Venus in front of the Sun in 1761 and 1769 and from the constant of aberration as determined by Bradley). Other determinations from other values of the constant of aberration were made later by Nyren, Struve, Loewy, and Hall.

⋆   *Herschel, John Frederick William* (1792–1871) made a systematic study of double and multiple stars. Some of this work was done together with James South (1785–1867). Published a catalogue of 380 double stars in 1824. Study of the Southern sky from the Cape of Good Hope (1834–1838).

☉ ⚥   Gautier, Jean-Alfred (1793–1881). Discovery of relations between sunspot numbers and geomagnetic disturbances (see also Sabine, Wolf).

⋆ S   *Struve,* Friedrich Georg *Wilhelm* (or Vasily Yakovlevitch, 1793–1864) was the founder of the famous Struve dynasty (after the Cassini and Herschel dynasties), which lasted 150 years, and grouped together six astronomers, in Russia, Germany, and the last, Otto Struve (1897–1963), in the U.S. The founder of the Pulkowo Observatory, for which Struve ordered the largest telescope of that time, a 15-inch refractor. Measurement of *double stars*; catalogue (1822) of 795 double stars; new catalogues in 1827, 1852. Study of the stellar distribution in the sky; opacity of the interstellar medium. Discovered that the Sun is not at the center of the Milky Way. Observations of the parallax of α Lyrae (Vega) in 1835.

(XX) *The discovery of stellar parallaxes* between 1830 and 1840 by Struve, Bessel, and Henderson, independently, is a key point in the history. It was the first direct proof (Bradley's aberration was indirect) of the fact that the Sun is indeed the center of the planetary world, that indeed the Earth turns around the Sun, and that indeed, Copernicus' system does, unlike Ptolemy's, conform to the physical reality.

@ ☾   Hansen, Peter Andreas (1795–1874). Lunar and planetary theory.

✶   Felix Savary (1797–1841) observed in 1827 some double stars; showed that they follow Kepler's laws.

✶   Henderson, Thomas (1798–1844). Observation of the parallax of α Centauri.

@   William Lassell (1799–1880) discovered at Liverpool (1846) the first satellite of Neptune.

S Ω   *Lord Rosse* (William Parsons, 1800–1867). Observations of nebulae (1848–1878). He discovered the spiral form of many nebulae, such as the well known Canum Venaticorum *spiral galaxy*. This discovery started a great many speculations about the formation of worlds, following the Kant-Laplace hypothesis. We shall come back to the nature of spiral nebulae at a later stage in this book.

✶   Pritchard, Charles (Oxford, 1808–1893) built an Observatory. Many measures of stellar positions, using photography, several parallaxes of stars. Catalogue of all naked-eye stars (from $l = -10°$ S to $+90°$ N), in 1866.

@   *LeVerrier, Urbain Jean Joseph* (1811–1877) renewed the attack on the planetary theory, in particular the stability of the four-bodies system (Sun, Jupiter, Saturn, Uranus). He incorrectly predicted an infra-mercurian planet, and, quite rightly, a post-Uranian one, – *Neptun, and its orbit* (1846). The same prediction, but unpublished, was made by John Couch Adams (1843, 1845). Note also LeVerrier's method of determining the solar parallax from gravitational methods.

@   *Galle, Johann Gottfried* (1812–1910) discovered *Neptun* in a telescope, in 1846, at the request of LeVerrier. Another validation of Newton's triumph!

⊙   *Wolf, Johann Rudolf* (1816–1893) established in 1852 the *periodicity of the solar cycle,* using 17th and 18th century observations. Discovered the relationship between sunspot numbers and geomagnetic disturbances. See also Sabine, Schwabe, Gautier.

(XX) *The lunar and planetary theory, along the lines drawn by Newton and his followers, the mathematician Clairaut, Euler, Lagrange, and Laplace* was developed, little by little, by several astronomers. Among them were the mathematicians *Siméon Denis Poisson* (1781–1840), Giovanni Antonio Amadeo Plana (1781–1869), in 1832, Philippe Gustave Doulcet de Pontécoulant (1795–1874) in 1846, and John William Lubbock (1803–1865), in 1830–34, mostly based upon the Laplace methodology. Peter Andreas Hansen (1795–1874) published the most accurate lunar tables (1838, 1857, and 1862–64). *John Couch Adams* in Cambridge, and G.W. Hill, in Washington, published new methods of dealing with the lunar theory. *Charles Delaunay* (1812–1872) derived a new mathematical theory partially based on friction. William Ferrel tried to calculate it. Newton's universal gravitation theory arrives at its brilliant and perhaps final achievement in the discovery of Neptune, by the separate efforts of LeVerrier, Adams, and Galle.

⊙ @ ✶ Ω   *Secchi, Angelo* (1818–1878) was the author of many discoveries concerning the Sun, such as *solar granulation,* and solar flares. He proposed a classification of stars (five classes) based on their color, and of nebulae (planetary, elliptical, irregular), etc. He obtained spectra of comets, planets, and bright stars. He also proposed the existence of canals on the Martian surface, in 1859, some seeming to be divided in two (more than a thousand observations). These Martian observations were later proven to be artifacts.

☾ @   *Adams, John Couch* (1819–1892) developed new methods of dealing with the lunar theory. He predicted the existence of Neptune, but he was discouraged from publishing by Airy, the director at Greenwich, permitting the glory of the discovery to go to LeVerrier.

✶   Struve, Otto Wilhelm, or Otto Vasilievitch (1819–1905), son of Wilhelm, born in Dorpat (now Tartu), worked in Pulkowo. Micrometric observations of stars, and binaries. In 1850, he made a determination of the precession constant, universally used till 1895, when Newcomb determined a better value.

ꙮ   *Lord Kelvin,* actually *Thomson, Sir William,* baron Kelvin of Larys (1824–1907) was a physicist known for many important discoveries, notably in the fields of thermodynamics, and mechanics. World observations of *tides,* 1863.

⊙   *Janssen,* Pierre *Jules* César (1824–1907), at the eclipse of 1868, obtained the *spectrum of solar prominences,* as a spectrum of a very few bright lines. One day later, observation of prominences *outside the eclipse.* At Mont-Blanc, in 1888, he showed the dark lines of oxygen to be of telluric origin. He constructed the Observatory of Mont-Blanc in 1891. In 1897, conducted an expedition at Mont-Blanc to determine the solar constant. Photographs of solar granulation.

✶ W ⊙   *Huggins, William* (1824–1910), ((*12,13,14,15*)), like Angelo Secchi, found dark lines in the spectrum of all stars that were bright enough. Secchi, in 1863, *classified stellar spectra* according to the intensity and distribution of dark lines in the spectrum. Huggins, in 1864, *identi-fied most of the dark lines* with those of known elements, hydrogen, iron, calcium, sodium. Some stars have strange spectra, with bright lines, sometimes variable in their presence or intensity. In 1864, Huggins ob-tained the *first spectrum of a nebula*, and found in it three intense bright lines (those of the so-called *"nebulium"*), which confirmed the idea (ex-pressed by William Herschel) of a shining fluid.The element "nebulium" is actually known now to be ionized oxygen and ionized nitrogen. Hug-gins discovered a nova in Corona Borealis (1866). In 1868, Huggins mea-sured the velocity of some stars with respect to the Sun, as did Hermann Carl Vogel (1841–1907), in Potsdam. Huggins, in 1869, observed in a gi-ven line (Hα of hydrogen), and by opening the slit of the spectrograph broadly, the shape of a given prominence. In the 1880s, William Huggins identified the *spectrum of comets* with carbon compounds.

@   *Bond, George Phillips* (1825–1865) discovered, at Harvard Obser-vatory, the eighth satellite of Saturn (Hyperion), in 1848 and, in 1850, the *"crêpe ring," of Saturn.* The same discovery was made 10 days later by William Rutter Dawes (1799–1868).

(XX) The completion of the solar system was a constant preoccupa-tion. After Uranus and Neptune, searches were made for satellites of planets, and for rings. Now in the space age, we know a rather large number of satellites for each of the major planets. The four larger planets have rings, the asteroids number 10000 and more, and comets are pas-sing by, ten a year or so. Planetary surfaces (Mars primarily, and Jupiter) began to be investigated in detail and as a function of time.

⊙   *Carrington, Richard Christopher* (1826–1875) discovered the *dif-ferential rotation of the Sun* from observations conducted between 1853 and 1861; first description of a flare, in 1859.

@   Donati, Giovanni Battista (1826–1873) applied the spectroscope to the study of comets for the first time in 1864.

@   Hall, Asaph (1829–1907). Discovery of Phobos and Deimos, satel-lites of Mars.

@   Maxwell, James Clerk (1831–1879), one of the greatest physicists of all time, and only incidentally an astronomer, discovered *the nature of the rings of Saturn,* a large number of small solid bodies revolving inde-pendently around the planet. Of course, Maxwell is much more well-known for his discovery of the unified theory of electromagnetism, which made of him one of the fathers of modern physics.

⊙   Young, Charles Augustus (1834–1908) discovered the *reversing layer of the solar atmosphere* in 1870.

@   Schiaparelli, Giovanni Virginio (1835–1910) discovered some asteroids and comets. Study of Mars at opposition. Discovery of the rotation axis and of the polar caps of Mars (1877). His map showed the "seas and continents."

* ⊙   *Lockyer, Sir Joseph Norman* (1835–1920) proposed an evolutionary theory linking all types of stars to each other. Total eclipse of the Sun in India; discovery of the *emission lines of prominences*, independently of Janssen. J. N. Lockyer, in 1868, using the Doppler-Fizeau effect, interpreted the distortion of spectral lines in the solar spectrum, in the spectrum of prominences. Velocities up to 300 miles per second are found.

(XX) The behavior of the Sun had been studied so far using primarily solar surface observations. A new method, using coronal observations during a *total eclipse of the Sun,* now revealed the true nature of the corona. It was solar, and not lunar or terrestrial, as previously thought. Sir Norman Lockyer, and Janssen, independently and simultaneously (1868), observed the *prominences* during a total solar eclipse in India, and one day later, *prominence spectral lines outside eclipses,* with a slitless spectrograph aimed at the region just outside the solar disk.

*   Burnham, Sherburne Wesley (in Chicago, 1838–1921), discovered many binaries and measured them. In 1900, produced a catalogue of 1290 double stars.

⊙   Duner, Nils Christopher, of Uppsala (1839–1914) found, in 1887–89, the rate of rotation of the solar surface at latitudes larger than those of the observations made by Carrington.

*   Vogel, Hermann Karl (1841–1907) found the doubling of lines in the spectrum of *Algol,* showing that the star, known as variable ("photometric variable"), is indeed an *eclipsing binary,* meaning that one star is, at regular intervals, hidden by the other one.

ŏ @   *Darwin, George Howard* (1845–1912). World observations of tides. Also cosmogonical theory of the origin of the solar system, based upon a detailed dynamic analysis.

✶   *Pickering, Edward Charles* (1846–1918), ((8)). The head of the important group of observers and computers of the Harvard Observatory, his photometric data cover one fourth of a century (the *Revised Harvard Photometry* contained 45 000 stars, up to 1908). Pickering made systematic use of *photographic photometry* and was operational in the launching (1887) of the project of the "Carte du Ciel." His spectroscopic data, published (1918–1924) as the *Henry Draper Catalogue* (where stars are given numbers preceded by the two initials HD) are the basis of the *Harvard spectral classification*. In 1889, the doubling of lines discovered by Vogel in some binary star spectra (Algol= β Persei; and Spica= α Virginis) led him to the discovery of other eclipsing binaries.

(XX) The *chemical composition of the Sun and stars* became a major preoccupation of astrophysicists, beginning with Fraunhofer's discoveries and the need to interpret the spectral lines of solar and stellar spectra. Of course, many lines were unknown in the spectrum of Earth light sources. So hypothetical elements were suggested as being responsible for these lines. This was the case at first for helium. *Helium* was discovered (and named) by Sir Norman Lockyer in 1868, in the spectrum of the prominences and of the chromosphere. It was discovered on Earth in 1895 by William Ramsay. *Nebulium* was found also in the line spectrum of bright nebulae. These spectral lines were later shown to be due some to ionized oxygen and some to ionized nitrogen. A new element, *coronium*, was discovered by William Harkness (1837–1903) and Charles Augustus Young (1834–1908), in 1869, in the spectrum of the corona. A new element? Its lines will be shown (much later, in 1942, by Bengt Edlén) to be due to highly ionized elements, iron, calcium, and others.

ꙮ   *Chandler* (1846–1913) discovered, in 1891 and subsequently, the "Chandlerian" *motion of the pole of the Earth*, with a period 427 days, amplitude up to 30 feet.
✶   Cannon, Annie Jump (1863–1941) measured spectra, under the direction of Pickering, and created a 12 spectral class system, would later give way to the famous and much-used Harvard classification.

(XX) The development of new techniques mentioned above, pp. 325 ff., had already allowed more and more precise measurements of the solar brightness, and of the *solar constant*, by John Herschel, and by Claude Pouillet in 1837 (see above p. 325). Better data were obtained in 1880–81 by Langley, still better data obtained on Mont-Blanc in 1897 by J. Janssen.

W ✶ Hartmann, Johannes Franz (1865–1936) ((*16*)). Discovery (1903) at Potsdam, of *interstellar ionized calcium lines*, in the spectrum of photometric double stars (eclipsing binaries), first in δ Orionis.

(XX) *Astronomy began to be organized on a world scale.* The *systematic study of the night sky* bloomed in the International organization of the *"Carte du Ciel"* (1887); *solar studies* in the *"Commission Internationale du Soleil,"* time determination in the *"Bureau International de l'Heure"* (or BIH) created in 1912 as a result of the Paris conference organized by the Bureau des Longitudes. Later (1920), these organizations were enlarged, transformed, coordinated and partially fused into the *"International Astronomical Union,"* which has published, since then, its regular reports at intervals of three years in general, and has grouped scientific questions into about 40 different specialized commissions. Progress is, of course, very rapid, and covers a huge number of important new discoveries, and new fields. In Fig. 6.5a, Martin Harwit has listed the discoveries he considers important (the Harwit numbers from this figure are repeated between double parentheses in the preceding list of discoveries and achievements; but most of them were made more recently. We have stopped to people born before 1865!). In Fig. 6.5b, he describes the projected trend of this evolution, which might very well be reaching a stage of levelling off at the end of the 20th century according to some of his views, a position with which I lend to personally disagree.

6.3
Conceptual Consequences of the Broadening of Horizons from Galileo to Einstein

During the Middle Ages, many attempts were made, beginning quite early on, to account qualitatively for the system of the World as a whole. We have mentioned, for example, the efforts of the Fathers of the Church. At the time of the great astronomers of the Renaissance, however, building a system of the World was still, as it was for Plato or Aristotle, essentially building a "realistic" quantitative model of the solar system. Going beyond that was very risky. Only a few daring philosophers tried to look at the stellar universe, or at the Universe as a whole, and these were qualitative attempts, often very far from reality. Indeed, they were comparable to the medieval attempts, in their naivety, their simplicity, and often indeed, in their prophetic orientation.

The stars were not much more than points in the sky, useful as markings, as mile-stones, so to speak. The sky had no depth. We must remember the views of Kepler in this context. The great astronomer, influenced by what later became known as Olbers' paradox, thought that the number of stars had to be finite (Fig. 4.5a, p. 199 and Fig. 6.6). In order to achieve that, to permit the sky to be dark at night, he put all the stars between two spheres, separated by a ridiculously small distance of a small fraction of the Earth's radius! Beyond it was emptiness. The space was still amply wide enough to accommodate deities!

It seems that one of the first astronomers to suggest that stars could be distributed in depth, and present at all depths, was Thomas Digges, in his re-publication (1576) of his father Leonard's *Prognostication everlastinge* (Fig. 4.36, p. 235). This is a long way from the microcosmos-macrocosmos earlier discussions, and the continuing views of the local universe which still preoccupied Newton's followers to a great extent. However, ideas had begun to change after Copernicus.

Fig. 6.6 *Finitude of the Copernican starry Universe.* According to Hans Holbein in his illustration of Erasmus' *Praise of Folly*

Copernican ideas produced a profound change in the minds of the philosophers. Copernicus retired the Earth forever from its very special status of being "the" center. Now that the center of the solar system was admittedly the Sun, or now that at least both descriptions properly accounted for all the observed phenomena in the same way, could we not now ask several other questions?

The Copernican point of view, at one and the same time, led the philosophers and theologians to consider the overall distribution of stars and the non-singularity of the planet Earth. Why should the Sun be "the" center? Is it not at the ridge of some larger structure, revolving around its center as well? Additionally, if there is nothing special about the Earth, would it not be possible to find living inhabitants on other planets? Many philosophers, up to the present, have unceasingly discussed these matters.

Still, the microcosmos-macrocosmos analogy often resurfaces and the mythical view of the Earth leads to various mystical views of the Universe. A study of our Galaxy, of the extragalactic Universe will lead in due course to the modern cosmologies, which we will study in the following chapters. These cosmologies are now concerned with the Universe as a whole (not only the World, our World, the main concern of scientists before the 19th century). In the last section of this chapter, we shall try, however, to describe some of the the fantasies that originated in the minds of philosophers of the 17th to the 19th centuries, often prophetic and generally extremely tentative and without much observational basis often also quite reasonable. There is indeed a strange continuity between these attempts and the modern cosmologies (generally quantitative), which we shall present at the end of this book, see pp. 479 to 526.

6.3.1
Giordano Bruno's Cosmology

Giordano Bruno was born in Nola, near Naples, in 1548. In 1576, he left the monastery and embarked on a life of travels and adventures across Europe, publishing in the next fifteen years three main books, in the form of dialogues. He then returned to Italy, was imprisoned for his heretical views, and condemned to death. He was burned in 1600, a date almost symbolic of the end of the "dark ages".

"All that exists is one and knowledge of this unity is the aim and term of all natural philosophies and contemplation: allowing the highest degree of contemplation, which transcends nature, and which for him without faith is impossible."[29]

A Platonic view of the Universe, strongly influenced by the poet-philosopher Marsilio Ficino, was the foundation of Bruno's views. He asserted not only the unity of the Universe, but he believed in the existence of a soul of the Universe. Whereas Ficino made a hierarchical distinction between *mens* (the soul of the Universe) and *anima* (the source of universal life) this soul is indeed God himself, for Bruno. The attitude of Bruno with respect to the nature of matter is clear. He is definitely an atomist, and a great admirer of Democritus. The Universe is infinite, an idea strongly influenced by those of Nicola da Cusa and the poet Marcellus Palingenius Stellatus (Pietro Manzoli, 1502–1543). For these poets, the Universe (i.e. the "World" of Copernicus) could be surrounded by a great void (Bradwardine, 1290–1348), by light (Palingenius), by other universes (Nicola da Cusa). It is the last point of view which was Bruno's. He definitely believed in the plurality of worlds, similar to our own solar system, a view expressly forbidden by the condemnations of 1277 issued by Bishop Étienne Tempier, which we have already mentioned (see p. 176–177). Bruno was not only a philosopher. He revised the Copernican system and found some mistakes in the Copernican description, such as the need for a "third motion" of the Earth. Bruno himself, however, made several errors, and more or less ignored many observational properties of heavenly bodies, in particular the distance of comets, as determined by Tycho. His real contribution lay in his detailed description of the plurality of worlds (Fig. 6.7). He demonstrated that the impression of a celestial vault, of a ceiling, is an illusion. Instead, space is boundless. The celestial bodies can move therein quite freely, provided there are sufficiently large distances between them to make all risk of contact avoidable.

Indeed, Bruno went far beyond this purely mechanical description. The Universe was, according to his views, a living Universe. The unity of that living Universe demanded that other worlds be populated by ani-

---

[29] G. Bruno, *De la Causa, principio* et Uno, 4th dialogue, translation into French by Emil Namer, Paris 1930. This translation into English is taken from P.H. Michel, *The Cosmology of G. Bruno*, Hermann, Paris, 1973, in English p. 58.

Fig. 6.7 *The plurality of the inhabited worlds.* (here according to Fontenelle, but very close to the concepts of Giordano Bruno).

1. Mercure. 2. Venus. 3. La Terre. 4. Mars. 5. Jupiter. 6. Saturne.

mated organisms, similar to animals and men on Earth. Actually, Bruno, perhaps influenced then by the Pythagorians, suggested that all Suns are "male" and all planets (including the Earth and Moon) are "female." Bruno seems to deny (contrary to Aristotle) the existence of a universal time. For him, the Universe is eternal, but the worlds can evolve. They can die, hence the need for different times. The Brunonian cosmology excluded any creation a nihilo, and excluded also the end of the Universe, or even of a world inside this Universe.

The idea of the plurality of inhabited worlds returned only less than a century after Bruno, with Cyrano de Bergerac or Fontenelle, for example. The Enlightment popularized these ideas in novelistic form in the work of Voltaire, Jonathan Swift, and others. These ideas were extensively developed in the middle of the 19th century by Camille Flammarion and later by Svante Arrhenius.

Clearly, Bruno was much in advance of his time, especially with respect to the dogmas of the Roman church. He was not a very learned astronomer. He read widely, but almost at random, and he ignored many important discoveries. His death by fire has been seen as a symbol of the refusal of the Church of the Counter-Reformation to entertain Coperni-can ideas, but still more of its rejection of the post-copernican views of an infinite and eternal universe, perhaps inhabited by other men.

Although he was not a pupil of Bruno's, we should note another *auto-da-fé* similarly showing the obstinate attitude of the Roman Church. Marko Antonije (1560, in Rab – 1626), known as Bishop Dominis (in Senj, then archbishop in Split, on what is now the Croatian Dalmatian coast), had seen all his books burned in public. He was exiled to his par-ish, forbidden to appear in Rome. He was, undoubtedly, a follower of Co-pernicus, but his books were mostly treatises on optics, considerations of lenses and telescopes, or concerned with the rainbow and tides. Still he was punished severely, for being openly a Copernican.

6.3.2
Swedenborg, Emanuel (1688–1772)

The case of the Swedish scientist Swedenborg is quite apart from the flow of philosophical ideas. Swedenborg had a very learned and strange mind. From 1736 on, it is possible that he was partly insane, having fan-tastic visions during which he entered (so he said) into communication with the souls of the dead. He published his verbose *Arcana Coelestia* in 1749–1756. It was translated into English in 1873.

Having begun from a point of view quite materialist, even positivist, he published many papers and books on astronomical questions, and is often said to have been far in advance of his time, although no historian of astronomy (to the knowledge of the author) dared to quote him. Swe-denborg's thought progressively evolved into a mystical description of the World, where the material Universe is nothing but a reflection of the spiritual world, or a sort of image of it.

His World is organized according to the microcosm-macrocosm dua-lity.[30] Correspondences dominate the system. God, with celestial powers, regulates spiritual affairs, whereas with spiritual powers, He regulates natural affairs, and with natural powers, finally, corporeal affairs. This is what constitutes the "order." The ultimate goal of a man must be to ac-

---

[30] We borrow most of this analysis from George P. Conger's *Theories of Macrocosms and Microcosms in the History of Philosophy*, 3rd edition, Russel & Russell, New York, 1967.

quire a corresponding order of moral perfection. He then becomes an image of the heavens. Swedenborg believed that the heavens were indeed the seat of souls, a society of angels occupying the planets. The mystical world of Swedenborg had little to say about the organization of the Universe.

6.3.3
Thomas Wright (1711–1786), Immanuel Kant (1724–1804), Jean Lambert (1728–1777).

There is little doubt that Kant was one of the most scientifically minded among the philosophers of the Enlightenment. His views, influenced by Wright, and which in turn influenced Lambert, are central to this discussion.

Let us begin with *Thomas Wright*. This man, originally an astronomer, then an adventurer, a sailor, specialized in navigation methods. An artist, essentially a layman, his work remained completely unknown to the community of astrophysical scholars, and came to light only more than a century after the publication of his book *An original theory or New Hypotheses of the Universe* (1750). This book, translated into German, was commented on in 1751 by Georg Christian Gring, in Hamburg.[31] Obviously, an essential preoccupation for Wright was to reconcile his views of the Universe with his concept of Deity. This book and the following one, his *Universal Architecture* (1755) are broadly illustrated with engravings which doubtless strongly influenced the views of his readers and followers (Fig. 6.8). He tried to unify in a single view what he knew about planets and their satellites, the solar system, and the Milky Way. For him, a center existed to the Universe. As the solar system was flat, so was the Milky Way flattened. He even suggested that the rings around Saturn were made of separate distinct rocks, a prophetic view in this matter indeed! The center of this universal hierarchy of systems is, unlike Saturn, or the Sun, essentially supernatural. The center is divine. The unity of religion (first, save the Holy Scriptures!) and science was, in his view, thereby achieved. But he went further, of course, in many elaborate discussions of the details. For him, in his last papers, the Universe consisted

---

[31] *Freye Urtheile und Nachrichten zum Aufnehmen der Wissenschaften und der Historie überhaupt*; achtes Jahr, I. Stück, Hamburg, 1751. Republished in English in the volume quoted hereafter.

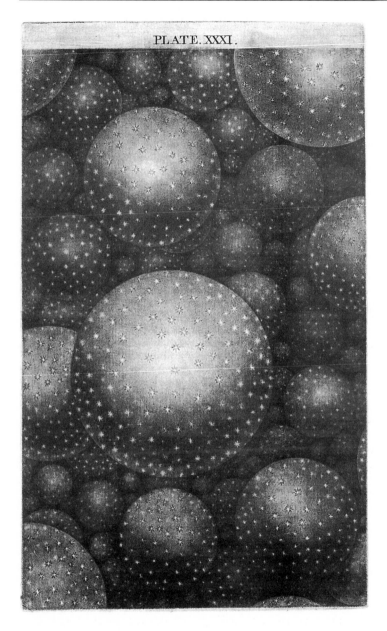

Fig. 6.8 *One amongst several Wright's fantasies.* The hierarchy of celestial structures appears in many forms.

of an infinite sequence of concentric shells (a hierarchy, so to speak) surrounding the (necessarily) divine center. Our sky would be one of these solid shells. Viewed from within, it is studded with volcanoes, which appear to us as stars. This hierarchical system is combined with a somewhat "dantesque" vision of the various parts of the Universe where good and bad souls are wandering, which would seem to contradict his earlier views.

In 1751, *Immanuel Kant* read the comments published in Hamburg and was at first attracted, even seduced, by the hierarchical views of Wright, which of course he tried to free from contradictions and mysticisms. He never fully realized, however, that the "centers" of Wright's system were basically supernatural, a view opposed to his own, as his God was only a Creator. His views on the Universe, were expressed in his famous (anonymously published) book *Universal Natural History and Theory of the Heavens* (1755). The adjective "universal," and the way Kant announced the book in a previous publication (under the likely title: "Cosmogony, or an Attempt to Deduce the Origin of the Universe, the Formation of the Heavenly Bodies, and the Causes of Their Motion, from the Universal Laws of the Motion of Matter, in Conformity with the Theory of Newton") conveyed initially the prospect of a very broadly, perhaps too broadly, ambitious treatise. Of course, we now know the nebular theory, later made more precise by Laplace, which explained, in a Universe otherwise not under consideration, the formation of the solar system from a primitive nebular flattened cloud; but Kant's *Universal Natural History* contained an entire cosmology.

Kant had read Epicurus, through his follower Lucretius. He was strongly influenced by the atomist view of the Universe. He wrote: "I assume, like those philosophers, that the first state of nature consisted in a universal diffusion of the primitive matter of all the bodies in space, or of the atoms of matter, as these philosophers have called them. Epicurus asserted a gravity or weight which forced these elementary particles to sink or fall; and this does not seem to differ much from Newton's attraction, which I accept ... Finally the vortices which arose from the disturbed motion, is also a theory of Leucippus and Democritus, and it will be found also in our scheme."[32] In other words, Descartes is reconciled with Newton, which shows how qualitative Kant's ideas were. Laplace la-

---

[32] Immanuel Kant, *Universal Natural History*, English translation 1900 by Hastie. Quoted by Munitz in the introduction to the 1969 edition, U. of Michigan, p. xi.

ter demonstrated that the theories of Newton were falsifiable, in the Pop-perian sense, while the vortices of Descartes' theory were not.

It was not the solar system that was Kant's main preoccupation, but rather the extension of its structure (and evolution) to larger structures of the Universe. The Galaxy itself (still called the Milky Way) turned around a center, stars in rotation, like planets around the Sun; and the Milky Ways, the other galaxies, themselves turned around a material center. Kant called this arrangement a "systematic constitution." "When, therefore, a number of heavenly bodies which are arranged round a common center, and move around it, have at the same time been so limited to a certain place that they have freedom to diverge from it on both sides only as little as possible; when the divergence takes place gradually and only in those bodies which are furthest removed from the center, and therefore have less participation in the relations indicated by the others: then I say that these bodies are found combined in a Systematic Constitution." This sort of self-referential structure of the Universe, from the smallest to the largest structures, can be called Kant's cosmological principle, a term used by Kant himself in his *Critique of Pure Reason*.

The Universe, like the solar system, must evolve and undergo basic changes. It is there that Kant locates Creation. Unlike Newton, who just put the solar system in the midst of an existing Universe, of eternal duration, Kant admits the reality of a Creator, hence an evolution of the Universe. He believed that the Universe came into existence as an act of Creation by a transcendent deity. Kant remained a theist. He went so far as to consider the astronomical description of the world as the only proof for the *Existence of God* (in a book of this name published in 1763).

The conclusion of Kant's book ("Of the creation in the whole extent of its infinitude in space as well as in time") is interesting in that it comes back to the heart of the ideas behind Kant's construction. "All nature, which involves a universal harmonious relation to the self-satisfaction of the Deity, cannot but fill the rational creature with an everlasting satis-faction, when it finds itself united with this Primary Source of all perfec-tion. Nature, seen from this center, will show on all sides utter security, complete adaptation. The changing sciences of the natural world will not be able to disturb the restful happiness of a spirit which has once been raised to such a height. And while it already tastes beforehand this blessed state with a sweet hopefulness, it may at the same time utter itself these songs of praise with which all eternity shall yet resound." These are

Kant's last words. But the book ends with the addition of a poem by Addison, a prayer rather: "... Eternity's too short,/To utter all Thy praise."

We are now indeed quite far from astronomy.

Johann Heinrich (or Jean Henri) *Lambert* (1728–1777) does not refer to Kant, although he was published in 1761 (translated into French, at the initiative of Lalande by Antoine Darquier, an astronomer from Toulouse, in 1801).[33] This is due to the fact that Kant's book was unavailable, the book's publisher having gone bankrupt. So Lambert's ideas were indeed independently expressed from those of Kant. Nevertheless, they are strikingly similar in essence.

Lambert's book, written in German (*Cosmologische Briefe über die Einrichtung der Weltbaues*, 1761, abstracted in French in 1779) was well-known to the scientists of the following period. Lambert was then often considered the "Alsatian Leibniz." His "Briefe" are indeed full of interesting ideas about many problems. Strongly influenced by Leibniz indeed, but also by Voltaire, Lambert was certainly much less mystically oriented than Kant, not to mention Wright. Still, and although Lambert was well aware of the developments in the astronomy of his time, derived from Newtonianism, his thought rests on metaphysical, or teleological ideas, as one might expect from a follower of Leibniz.

These ideas are simple: The Universe must be populated by the largest possible number of objects, or species, planets, or stars, or milky ways (an idea of "*maximum content*"). In particular, it must also contain inhabitants other than us; the multiplicity of inhabitable worlds is part of Lambert's thinking. The Universe must be "*harmonious*," hence no catastrophic events, thus excluding from it collisions and disorders. Finally, the idea of maximum implies that of *variety* and diversity among the positions and motions of the various components of the Universe.

Lambert adopts a "relativistic" point of view, after Galileo, in the same sense and reconciling, so to speak, the Ptolemaic description and the Copernican one. This multiplicity of points of view does not exclude the notion of the unity of the Universe and the necessity of the existence of an absolute center for it (as suggested by Leibniz), outside, of course, of any relativistic consideration. Combining Copernican mechanics and this metaphysical point of view led to a *hierarchical description of the*

---

[33] J.H. Lambert, *Lettres Cosmologiques sur l'Organisation de l'Universe*, facsimile reimpression in 1977 by Alain Brieux éd., Paris, with an Introduction by Jacques Merleau-Ponty.

Universe, similar in many ways to Kant's or Wright's. However, whereas Leibniz and Kant envisioned an infinite Universe, Lambert saw it as finite. In his 17th letter, he expressly speaks about "*the most external frontiers of the Creation.*" It is interesting to compare the Kant-Lambert opposition on this with the Kepler-Digges opposition already mentioned (Fig. 4.5a, p. 199 and 6.6, p. 354 above). In both cases, Kepler's and Lambert's attitude were dictated by a consideration of the dark sky paradox, later known as Olbers' paradox.

There is nothing in Lambert (unlike in Kant, or later Laplace) implying the evolution of the heavenly bodies.

Lambert, as before him Wright and Kant, believed in the existence of millions of comets (a very high estimate for that time!). Lambert stated very early, as a consequence of the Copernican point of view, that once the Earth is no longer the center of all motions, there is no reason why the Sun itself should be so. The Sun is obviously in motion, the "fixed" stars are not fixed, the Milky Way has an enormous depth, and is only one galaxy among many others. Unlike many other philosophers and scientists, Lambert dared to suggest distances to stars. He estimated the closest star (Sirius?) to be at a distance of 500 000 astronomical units from us. The distance of the "system" closest to us in the Milky Way could be 1500 times that distance. At the time of Lambert, these estimates appeared fantastic, but beginning in the period 1830–40 (Struve, Bessel, Henderson: the first stellar parallaxes), they appeared indeed to be rather small. Pascal said "… the Center is everywhere," but instead, Lambert tells us: certainly, there is "a" center, an absolute center, but astronomers will identify it only during the … 50th century !!

The essential part of Lambert's system still remained his hierarchical view of the Universe, "the" way to escape the gravitational and optical paradoxes described later as Olbers' and Seeliger's paradoxes, but known essentially since Kepler, as we have seen. This description, the characteristic unicity of the solution, the necessity of a single language (perhaps still to be discovered) to properly describe the Universe, permeated all his work, which emanated from a metaphysical (teleological, one would say) intuition.

Again, Lambert's ideas were prophetic in several ways, but sometimes quite misleading. But whatever they were, Lambert was quite aware that observations were, in the end, more important tests for any theory than teleological intuitions. His "Briefe" concluded as follows: "I shall not go further, and I shall end my task here, until your new observations will have told me whether or not we are on the road to Truth, or whether we must modify or reorient our ideas; you know how exacting I am in what concerns the rigor of the proofs, and I have had the satisfaction to show you how much astronomy gained already from it. It is to the observations of posterity to improve and to perfect it."

6.3.4
Edgar Allan Poe (1809–1849)

In the middle of the 19th century, when Newtonian physics began to encounter problems, and long before Einstein started to solve them by a synthesis of the the old Newtonian concepts with modern physics, ideas of a similar nature were circulating. They were essentially qualitative notions coming from poets and philosophers. The critical return to the lessons of the history of science exerted a great weight in many respects. Both the Aristotelian method, and the Baconian method of successive deductions and inductions, both based on a strict observance of the logical paths and a strict respect for the observations, were coming up against the wall of the "unknown." Attempts to scale the wall did not meet with any obvious success. The old methods did not offer much more in terms of mystical vision than Newton's point of view, in which, as in the visions of Kepler, intuition played a great part. This intuition was somewhat mysterious, and sometimes clearly dominant over the methodology of research proposed by the great proponents of "natural philosophy." Consequently, mystical visions, guided purely by intuition, with no real basis in either observations or logic, emerged in number. As typical of these we have seen the ideas, the dreams, of Swedenborg. Much closer to 19th century science was *Eureka*, the philosophical poem by Edgar Alan Poe, published in 1850, translated into French by Charles Baudelaire, in 1856, but not published until a few years later. Poe considered *Eureka* his major work. Three months before his death in in 1849, he wrote "I have not more the desire to stay alive, as I wrote *Eureka*."[34] But the book was badly received (in France), to the disappointment of Baudelaire, by the "scientists" of the time. *Eureka* has, however, been read more in its French version than in the English one.

For Edgar Poe, God is a way to express the basic unity of nature, and its basic simplicity. We are back, seemingly, not only to the theistic atti-

---

[34] July 7, 1849, letter to Mrs. Clemm.

tude of Kant, but even to the Occamist point of view! Poe conferred to the concept of Beauty as much value as to that of Truth. Indeed, Beauty was even considered as a proof of Truth, and Poe claimed that this vision of the Universe was absolutely, rigorously, demonstrated: "I offer this book of Truths, not especially because of its character of Truth, but because of the Beauty which flourishes in its Truth, and *which confirms its truthful character.*" (italics by the author).

It is, in fact, fascinating to read Poe's *Eureka*, informed as it is by a very good and rather complete knowledge of the development of the astronomy of the time, by a deep admiration for Kepler, Newton, Laplace, and others. He says about Kepler: "Yes! These vital laws, Kepler guessed them, we can even say he *imagined* them" (italics by E. A. Poe). It is remarkable to see in Poe many premonitions of discoveries which were scientifically formalized later, such as the hierarchical structure of the Universe (already described by Wright and Kant, and much later reintroduced in the description of the Universe by Charlier), as an explanation for Olbers' paradox; or again the need for unifying time and space in a single concept, premonitory of General Relativity. Instead of the Big Bang, which entered the literature a century later, Poe conceived of a Big Crunch. He was clearly opposed to the classical concept of an infinite and stable Universe (not yet expressed, as in Einstein's 1917 cosmology, but implicit in many Newtonian publications). Many of his other deductions are known to have been quite wrong, but on the whole his thinking was extremely well documented and provides a unified, if not true, view of the Universe.

This quasi-Pythagorian view is exemplified by the following typical quotation from *Eureka*: "This being understood, I claim that an absolutely irresistible intuition, although impossible to define, pushes me to conclude that what God has originally created, that this Matter which He has created, by the strength of His Will coming out of His Mind, or from Nothing, cannot have been anything else but Matter, in its purest state, in its more perfect state, state of ... of what? Of Simplicity."

It is interesting to note that *Eureka* is, in its form, its style, and its spirit, quite close to the modern popularizations of science, which have so much appeal to the public (if not to scientists), in part because of their poetic qualities, in part for the number of limpid descriptions of facts difficult to understand easily, and in part for their semi-mystical content; but they do not belong to the history of science.

6.3.5
Flammarion, Arrhenius, and the Plurality of Inhabited Worlds

Camille Flammarion (1842–1925) started his long life as an astronomer. The many disagreements and clashes he had with LeVerrier led to his expulsion from the Paris Observatory. He then turned into a very prolific writer of popularized astronomy and the organizer of amateur astronomy in France. Known throughout the world, his fame was due to the poetic style in which he described the marvelous astronomical landscapes. His *Astronomie Populaire*, first published in 1880, with a great many further editions in many languages, was still in print at the time of the centenary of the first edition. It has been a best-seller for more than a century.

Flammarion's first important book was devoted to the plurality of inhabited worlds. It was based on ideas such as those of Giordano Bruno, but augmented with a deep knowledge of modern astronomy. In particular, the homogeneity of composition of the Universe appeared then as, if not evident, at least very probable. If life appeared on Earth, why should it not have appeared on many other planets, in the glorious immensity of our Galaxy alone, and still more in other galaxies? The philosophical convictions of Flammarion led him to believe so. Still it was only a "belief."

Svante Arrhenius (1859–1927) was a physicist of great reputation in the field of electrolytic processes. He went further than Flammarion and others in that he was one of the first real scientists to propose a mechanism which could explain the beginnings of life on Earth. His famous book, *Evolution of Worlds,* was published in Stockholm in 1907. This book covered many fields. In particular, Arrhenius invoked a repulsive force due to radiation. He was the first to have suggested the notion of "radiation pressure" (a notion very common to us, at the end of the 20th century) as a way to explain cometary tails. He discussed the Kant-Laplace hypothesis on the origin of the solar system. The equilibrium between attractive (Newton) forces and repulsive forces explained, according to Arrhenius, all the particulars of the planetary motions.

The most original part of the book, however, concerns life in the Universe. Arrhenius being convinced of the reality of the nebular formation of stars, hence of their rather general nature, believed that there were a great many planets on which life could exist. Arrhenius was a follower of Charles Darwin. Therefore, he looked for any elementary form of life, which, ultimately, would give rise to some more elaborate form of life (mankind ...?). Kelvin had assumed that life could appear after a catastrophic fall of a meteorite, coming from the Galaxy, but Arrhenius refuted this opinion. He assumed instead that radiation pressure forced micro-organisms to travel from one part of the Galaxy to another, to the Earth in particular. These "germs of life" would be extremely cold during this travel, but Arrhenius showed that they could survive. Dust grains could be the efficient vehicle of their transportation.

The Arrhenius theory was indeed quite a sound theory for its time. It is not impossible that we will soon return to similar views. It barely touched on cosmology, except for its assumption of an infinite time, which allowed the transportation of life from almost anywhere to any other place; but what is the difference between an infinite time (forever), and merely a very long time, the time it takes light to travel from the more distant parts of the Universe (15 billion years, at present, i.e. at the end of the 20th century)?

6.3.6
The Cosmogonic Theories as Limitations to the Cosmological Stream; The Contribution of Poincaré

The huge change of scale of the observed Universe, from the solar system, only partially accessible to the pre-galilean observers, to the immensity of the extragalactic Universe, took some time to be fully realized. It is now an obvious fact that the Sun and the solar system are relative newborns in the life of the Universe, in the present day almost unevolved Universe, so to speak. Furthermore, the solar system is completely independent of the past, and in any case unknown, history of the Universe, completely independent of the cosmological concepts, of ideas about Creation, and the infinite extent of space and time.

We have briefly evoked the theories of Kant and Laplace concerning the formation of the solar system. This part of science is now called cosmogony, as opposed to cosmology, the science of the Universe as a whole. The two were confused until the beginning of this century. One can see in the *Leçons sur les Hypothèses Cosmogoniques,* by Henri Poincaré (1854–1912), the state of affairs, as it was in 1911. Even in the first page of its Preface, Poincaré states clearly that the problem of the origin of the World (i.e. of the solar system) is identical to that of the origin of the Universe, which is, of course, quite untrue! The language has evolved. In the following chapter, for historical reasons which we have mentioned previously, we shall use primarily the word «Universe».

Poincaré based the need for cosmogony (and cosmology) on the second principle of thermodynamics, as expressed by Carnot's principle. For him, the very idea of the evolution of the World is common to all scales; and it is true that at that time, whereas geologists gave the Earth an age of a billion (10^9) years or two, no one dared to speak about the age of any other body in the Universe, be it a star, a star cluster, a planet or a galaxy. Therefore, Poincaré spoke about the Kant-Laplace hypothesis of condensation of a nebular cloud. He discussed the hypothesis of Hervé Faye (1814–1902), an improvement on Laplace, and that of du Ligondès. For Sée, the planets were captured (later Jeans will suggest that they were ejected!). The role of tides in planetary evolution was entered as evidence by George H. Darwin (1845–1912).

Unsolved, at that time, was the problem of the origin of solar energy. The solutions then assumed (Kelvin, Helmholtz) were not satisfactory.

Poincaré finally examined some theories of stellar evolution, that of Sir Norman Lockyer, that of Schuster, that of Arrhenius etc., now outdated, but showing the progressive extension of the concept. In the last chapter, he looked into the (fantastic!) formation of spiral galaxies, according to Sée, and the theory of Emile Belot, which interpreted the origin of the solar system as an interaction between a whirling dense nebula (the star?) and the nebulous medium it crosses. Actually, we should not speak of these conceptions as theories. They were no more than "working hypotheses."

Although Poincaré gave quite a clear account of several attempts to interpret the evolution of the World, his book only added some question marks to the description of the Universe as it appeared to Copernicus, Kepler, Galileo, and Newton: Has the Universe really evolved, as it seems that the World has evolved? And has the evolution of the Universe slowed down in much the same way as it seems that the World no longer sensibly evolves? Does our World actually even continue to evolve at all? What arguments can be enlisted to prove or disprove a theory? Obviously very little support can be expected from observations. Cosmogonic hypotheses concerning only the World (i.e. our solar system), and even quantitative theories have indeed continued to develop more and more successfully (von Weizsäcker, O. Schmidt, Kuiper, etc.), but they are now definitely out of the realm of studies covered by this book.

6.3.7
Fournier d'Albe, Seeliger, Charlier, and the Hierarchical Structure
of the Universe

It is quite interesting, a century or more after Wright, Lambert and Kant,
to see a return to a conceptualization of the hierarchical structure of the
Universe, a return which we shall see (p. 463) emerging again in the 20th
century, as a very credible alternative cosmology based on observations
(de Vaucouleurs, etc.).

Etienne Fournier d'Albe, an Irish gentleman, published in 1907[35] a
very interesting book, divided in two parts: the "infra-world" and the
"supra-world," a nomenclature reminiscent of the macrocosm-micro-
cosm duality, except that here, these new worlds designate any two successive steps in the hierarchy. Beginning with the atomic and molecular
universe and from it to the "following universe" (Earth), the ratio of
sizes is, according Fournier d'Albe, of "10000 trillion to 1" (actually
10^{22}).[36] As he says, this number "is nothing less than the ratio of the
scales of the successive universes." The Universe (with a capital U) contains several, many, universes (with a small u), which contain stars, and
solar systems. Universes are enclosed in each other, a little bit like the
dolls of the Russian matriotchkas. There are universes of increasing order.

A careful examination of the observations of stellar motions led Fournier d'Albe to propose a hierarchical distribution of matter in which "the
mass comprised within a world sphere increases as its radius, and not as
its volume, or, in other words, that the density within a world-sphere
varies inversely as the surface of the sphere." By "world-sphere" Fournier
d'Albe means a sphere enclosing a "visible universe" of any order. He described what we would now call a *fractal* distribution of matter or mass,
with a *fractal index* actually equal to unity.

This fractal Universe has a few interesting properties, a priori assumed as principles: (1) A luminiferous ether pervades all space (see the
discussion of this question in Einstein's time, p. 407); (2) the number of

---

[35] *Two New Worlds,* Longmans, Green & Co., London, 1907.

[36] Note (to understand texts of the period) that a "trillion" is equal, in English, to
10^{18}, a "billion" to 10^{12} and not 10^9 as in American usage. A "million" is 10^6. The
number 10^9 is called a "milliard" in English (as in French), but a "billion" in
American English.

dark bodies is comparatively small; (3) the stars are irregularly distributed; (4) the luminous stars have an eternal existence, as does the Universe itself. This system would account completely for the apparent limitation of the Universe (a consequence of Olbers' paradox). The Universe is infinite in a three-dimensional space. The law of gravitation holds good throughout infinite time and space.

Although Fournier d'Albe referred to an "over-soul," and described man's situation in the Universe in a quasi-religious way, we could consider his work quite materialist on the whole, strongly inspired by the philosophical ideas, and scientific common knowledge of his time.

Karl Wilhelm L. Charlier (1862–1934) was also preoccupied with the notion that he had to explain Olbers' paradox. His hierarchical Universe, fractal like that of Fournier d'Albe, was indeed widely discussed in scientific circles, and accepted by some contemporary scientists. We allude to it hereafter, p. 411.

The case of *Seeliger*'s (*Hugo von*, 1849–1924) research is in a way more important. Seeliger, on the one hand, tried to solve the problem of the observed departures from Newton's law (see p. 375), not by any new concepts such as those of General Relativity, but by modifying Newton's law of attraction by a multiplicative exponential term, reaching zero asymptotically at an infinite distance. This actually helped him to solve what could be called the Seeliger's paradox, similar to Olbers', but applied to the gravitational field. Indeed the similarity between the two is only in the $1/r^2$ dependence of either the illumination of a terrestrial surface by a given light source, or the attraction exerted on a terrestrial mass by a given attractive other mass. Actually, the two problems are not really identical. In the case of the Olbers paradox, the light is a scalar, and the sum (at any considered point, on Earth typically) of lights coming from all directions, whatever the distribution of the sources, is positive. On the other hand, the gravitational field is vectorial. If the two halves of the Universe are equal to each other, the attraction by one half-Universe destroys the attraction by the other. Still the paradox exists, but for a second-order reason. In the Universe (whatever it is), there are fluctuations from here to there. The local effect of very distant fluctuations in the distribution of matter is far from negligible, if the Universe is a Euclidian Universe. If it is closed and bounded, the problem may be less severe, but it still exists. The Seeliger exponential term was a way to solve the dilemma.

Although in the line of the Kant, Wright, and Lambert hypotheses, or those much later of Fournier d'Albe, which were quite qualitative, not to say fantastic, Charlier and Seeliger's works were highly rigorous in terms of mathematical formulation, and do not really belong in this chapter, but rather in the following ones (see pp. 408–412).

6.4
Conclusion

It may very well be, as suggested by Harwit, that we are now reaching a climax of discoveries. The cost of large equipment, and the need to have some international means to finance them, limit the development of astronomy. The next step may be a telescope, or some interferometric devices, on the Moon. But when? At the opposite extreme, the data so far gathered are far from being well understood. The synthesis of our ideas concerning the Universe is still out of reach. If I characterize the 20th century as the one during which we have, roughly speaking, properly understood the evolution of the stars, the 21st might be the one which will allow us to understand the evolution of galaxies. The knowledge of the evolution of the Universe (the subject of this book!) will need these basic first steps to proceed safely, contrary to what several scientists think, convinced as they are that they have, now, the only solution, the right cosmology, the description of the absolute Truth. I am much more skeptical!

Chapter 7

Towards Modern Cosmology

7.1
Failures and Difficulties of the Newtonian Description

The return of Halley's comet, the explanation of the perturbations of Saturn, the discoveries of Uranus and Neptune, and the discovery of asteroids, all gave an enormous weight to the Newtonian construction of the Universe. The study of double stars further reinforced that triumph.

However, all was not for the best in the best of all possible worlds. LeVerrier himself had noted the discrepant behavior of the planet Mercury.

We know, according to Kepler's law, that the orbit of a planet is an ellipse. Perturbations by other planets affect this, leading to a description of the orbit as either like a spiral, or more simply like an ellipse the axis of which is regularly turning slowly, but in a measurable way, around the Sun (Fig. 7.1).

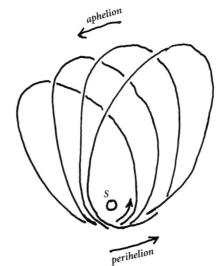

Fig. 7.1 *The advance of Mercury's perihelion.* In this schematic diagram, the eccentricity of Mercury's orbit is largely exaggerated, as is the angular displacement of the perihelion from an orbit to the following one. See text for exact values.

Table 7.1 *Abnormal advance of perihelion of some planets* (in seconds of arc per century)	Theorical	Observed
Mercury	43.03 ± 0.03	43.11 ± 0.45
Venus	8.63	8.4 ± 4.8
Earth	3.8	5.0 ± 1.2

It is easier to measure this effect on Mercury than on other planets, since Mercury has the largest eccentricity of all the planets, apart from Pluto, which was not yet discovered at that time. LeVerrier indeed calculated the effect, comparing (from about 1855 until his death in 1877) the observed motion of the perihelion of Mercury's orbit with Newtonian theory at an accuracy better than 1" of arc. The change in the longitude of Mercury's perihelion is 5"27 per year, according to LeVerrier's computations, or 527" per century. But he found that the observations showed a precession of 565" per century, i.e. 38" per century faster than the calculated precession. In 1882, Simon Newcomb studied the motion of Mercury again. His estimate of the discrepancy was 43" per century, confirming in essence the value found by LeVerrier. A similar phenomenon had been found for Venus. But this last effect disappeared almost entirely (see Table 7.1, according to data quoted by Max Born), when new determinations of the solar parallax using Encke's comet allowed the measurement of the Earth's mass to increase slightly. The relatively large effect observed on Mercury, however, remained non-reducible, and was estimated to be of 41"± 2".1, according to Newcomb's *Nautical American Almanac*.

What could be the origin of such discrepancies? LeVerrier thought first of the possibility of a planet (or some planets) between Mercury and the Sun; but the search for "Vulcain" (the planet was baptized even before being discovered!) was a complete failure, in spite of considerable effort including the assertion by Lescarbault that he had indeed observed the inframercurian planet. Another suggestion, made by Newcomb, was that the zodiacal light, this well-known extension of the solar corona, could provide enough matter of cometary-meteoritic nature, to produce the observed perturbations. This, however, would have led to contradictions with other features of the motion of Mercury and even Venus. Seeliger proposed in 1896 another distribution of zodiacal matter, which would not affect Venus, but this was not acceptable from the point of view of solar observations.

In 1895, Newcomb admitted that perhaps Newton's law should be modified, and was not strictly in $1/r^2$. A proposal was made to change the power 2 to $2+[(1/6)\times10^{-6}]$. Another one (by Seeliger) assumed an exponential term going to zero at a very large distance. Actually, it was logical to assume that any perturbation to Newton's law would be more observable near the Sun than far away from it, because of the strong decrease, with distance, of the gravitational force and of course of its eventual perturbations as well. However, this line of thought was not tenable because of some of its consequences, and Newcomb finally decided that the best way was to keep the law of gravitation as it was, i.e. purely Newtonian, and to wait for some local explanation of the Mercury mystery.

We have to wait for Einstein's General Theory of Relativity (1917), specifically to an earlier form of it (1914), to solve the problem, as demonstrated in 1926 by Chazy.

7.2
Criticisms of Newton's Theory: The Mach Discussion

In the *Principia*, Newton expressed clearly a few bases of his mechanics. In particular, he referred to the existence of absolute and relative motions. In a scholium which follows immediately after the definitions, even before book I of the *Principia*, Newton explicitly says:

"I do not define time, space, place, and motion as being well known to all. Only I must observe that the common people conceive these quantities under no other notions but from the relations they bear to sensible objects. And thence arise certain prejudices, for the removing of which it will be convenient to distinguish them into absolute and relative, true and apparent, mathematical and common." This single sentence implies that one can indeed speak of *absolute* space, or time, of *true* motion, of a *mathematical* orbit.

The Newtonian conceptions thus refer to "absolute" space and to "true" motion, concepts which, already in Newton's time, were opposed by Leibniz's ideas, and by Huygens' remarks with respect to the relativity of motion. Actually, this problem calls for a frame of reference for all motions, which would permit us to speak of absolute space, of true motion, of a mathematical orbit. This is not a new problem. Remember Osiander's preface to Copernicus' *de Revolutionibus*: Is it equivalent to say: "the Sun turns around the Earth," and "the Earth turns around the Sun?" It was, but it is no longer the case, when one considers the fixed stars as a system of reference. Remember also the Galilean "principle of relativity," according to which "the motion is like nothing," illustrating

Fig. 7.2 *The Galilean principle of relativity.* One can see why, in Copernican times, some people said (for example, Osiander, in his preface to Copernicus' *de Revolutionibus*) that to say: "The Sun turns

around the Earth" is equivalent to saying: "The Earth turns around the Sun," as long as one considers only two bodies!

the fact that, for example (Fig. 7.2), the motion of a man in a boat as seen from that boat is different from the motion of the same man, in the same boat, relative to the shore, if the boat is moving.

One could say, more scientifically, and as expressed by Galileo himself, that the laws of mechanics are not changed by a change in coordinates of the Galilean type in a rectilinear motion, implying a simple, linear addition (or subtraction) of velocities (see Chap. 7.7). The motions only appear different.

So the choice of a system of reference appears to be an essential choice for the description of mechanical phenomena, of motions. For centuries, it was, quite naturally, the Earth, then it was the Sun, then the "fixed stars." The stars, however, are also subject to motions with respect to larger entities (the Galaxy, first), which naturally gives rise to two questions: Can we define any "absolute frame of reference?" Can we define it, even in a closed box, without looking at the Universe, only at its manifestations in the box?

In fact, Newton gave an example by which one can indeed distinguish an "absolute" motion from a "relative" one. Let us quote him, in a rather simplified way. "We may," says Newton, "distinguish rest and motion, absolute and relative, one from the other, by their properties, causes, and effects. It is a property of rest that bodies really at rest do rest in respect with one another ... It is a property of motion that the parts which retain given positions to their wholes, do partake of the motions of those wholes ... A property more akin to the preceding is this that if a place is moved, whatever is placed therein moves along with it; and therefore a body which is moved from a place in motion, partakes also of motion of its place." (Note that this is nothing else but the Galilean principle of relativity expressed by Newton.)

Then Newton, as an illustration, referred to an experiment he performed himself, that of a bucket put in a closed room (Fig. 7.3).

First twist the rope by which the bucket is hanging (1). Fill it with water (1); and then (2) let the rope untwist; while the cord is untwisting the bucket turns in the opposite direction to that of the initial twisting. At first the surface of the water will remain flat, horizontal; but gradually the bucket communicates its motion to the water, and the surface of the water forms (3) a concave figure, with a hollow region in the center, the water ascending along the sides of the bucket. A centrifugal force is acting. If we stop the bucket, the water remains concave, for a while. Then, this effect slowly decreases, by viscous forces communicated to the water by the bucket. The centrifugal forces exerted are the manifestation of the fact that the whole experiment is linked to a reference space, – the room, not the bucket. Newton's conclusion is that centrifugal force cannot be explained by the relative motion of water and room. In (2) and (4), a rel-

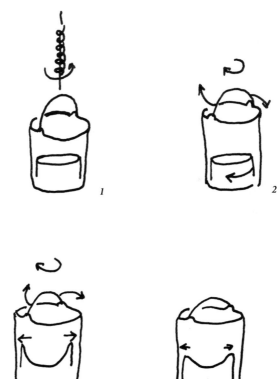

Fig. 7.3 *Newton's bucket.* This experiment demonstrates the need for an absolute system of reference. See text for explanation of symbols 1, 2, 3, 4.

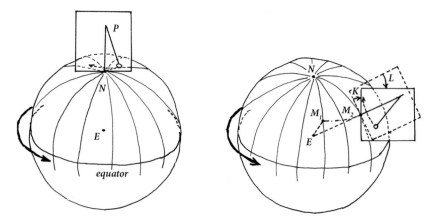

Fig. 7.4 *Foucault's pendulum.* In essence, the Foucault experiment, actually performed in a closed room, demonstrates the rotation of the Earth with respect to some absolute system of reference. The plane of oscillation of the pendulum stays the same with respect to an absolute system of reference, whereas the Earth turns. Hence, the inclination of this plane with respect to the local meridian plane changes from M_1 to M_2 (on the right) and this change (shown by arrows K and L) measures the time. One can actually compute that: $T = 24 / \sin \phi$ hours where T is the period of the pendulum, and ϕ is the latitude of the experiment. At the pole (case P, on the left), one gets, obviously, exactly 24 hours. There, it is clear that the Earth is turning under the fixed pendulum plane. At the equator, the period becomes infinite.

ative and identical motion exists; but its is only in (4) that the water alone does indeed take a parabolic shape. At this point (according to Newton), the centrifugal force of the bucket is zero. Of course, with respect to the bucket, considered as a reference space, the water would not behave that way, and would remain plane and horizontal. The room, however, could suffer some distortion; but Newton did not take this effect into consideration. In any case, it is too small an effect to be measurable. Therefore, he claims to have proven the existence of absolute motion.

This brings us to a discussion of Foucault's pendulum (Fig. 7.4), more interesting in a way than Newton's bucket.

According to Newton's point of view, a pendulum, oscillating in a given plane, must maintain this plane of oscillation permanently, in absolute space (provided one excludes all perturbating deflection forces, like magnetic forces for example, if the pendulum is metallic). Therefore, on Earth, which rotates, the plane of oscillation remains fixed with respect to the Sun and the "fixed stars," which means it turns with respect to the laboratory. This experiment can be done any place on Earth. It was performed by Foucault to demonstrate (to the general public) the rotation of the Earth (as if this were not obvious, from the observations of stars, then known to be very distant suns!). This experiment proves the "absolute motion" of Earth, as the motion of water in the bucket proves the "absolute motion" of the bucket.

Mach (in his *Mechanik*, published in Germany in 1883) argued against the conclusions of these experiments and claimed that one cannot say anything, at least in such a simplified way, about "absolute space" or "absolute motion." Mach's ideas later exerted a strong influence on Einstein's ideas. Even now, this is a very useful book to read, and quite representative of the very positivist attitude which developed at that time. To quote the first words of the preface written by Mach to the French translation of the book, in 1903: "The present book is not a textbook intended to teach the theorems of mechanics. One will find rather in it a work of critical examination inhabited by an anti-metaphysical spirit."

Newton, says Mach, went much further than his intended adherence to the observed facts. In Mach's mind, the explanation for Newton's behavior lay obviously in his mystical tendencies. For Mach, there is no such thing as absolute motion and after all, he argued, Newton overlooked the fact that the masses of the Earth and Sun or fixed stars would have to be taken into account. We could ask further, along the same lines: What if the rotating bucket had been at the end of a rope oscillating like the Foucault pendulum? Actually, for Mach, neither the bucket nor the room is a bona fide system of reference, even "at rest." Inertial (centrifugal forces) are linked to the *whole of the massive Universe*. They cannot be linked with anything else. The centrifugal force, demonstrated in the bucket experiment, is indeed a dynamic gravitational effect of rotating masses. Only relative motions are implied in this experiment.

The law of inertia must therefore be modified. In its Galilean and Newtonian formulations, it is, notes Mach, based on the fact that most motions on Earth are of such duration and of such a small deviation that it is quite unnecessary to take into account either the rotation or the progressive variation of velocity of the Earth, to account for them. But "to say that a body keeps its velocity and direction in space is an abbreviate way to refer to the whole Universe." Concerning the motion of bodies, to assert that one knows something else but their behavior with respect to the heavens is – and I quote Mach literally – "an act of dishonesty."

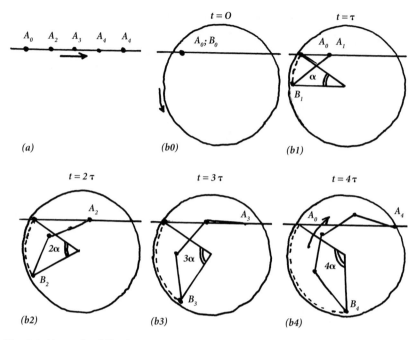

Fig. 7.5 *About the difficulty of defining a law of inertia, according to Max Born.* A body runs uniformly from A_0 to A_4, in 4 time intervals t, by successive jumps. The stationary observer in (a) sees a uniform motion along a straight line. Now think of a moving observer, at the center of a rotating disk, for example at time t_0 (b0). After time t (b1), the body is at A_1, but its starting point is now B_1. After time interval $2t$ (b2), the body is at A_2, but the point from which the motion has started is now B_2, etc. For the rotating observer, the motion is as indicated on (b4). Choosing smaller time intervals would display a curved line, not a polygon, which is introduced here as a first approximation of it.

Max Born argued along the same lines as Mach. He is certainly another physicist who later influenced Einstein (and who was strongly influenced by him).

A rather good illustration of the difficulty of defining an inertial motion is given by Born (Fig. 7.5).

Born considered the motion of a body, a small rolling sphere, say, on a flat table. Leaving aside frictional forces, the sphere will move uniformly on the plane, taking successive positions 1, 2, 3, 4, ... (Fig. 7.5a), if no force is acting on it. However, there is no such a thing as an immobile table in the Universe. The Earth moves and the Sun moves, etc. Just think of the table in question as a turning table (Fig. 5b). Then, clearly, if it turns uniformly, irrespective of what causes the turning, the motion of the sphere will not appear to the observer (located on the table) to be linear, but rather curved. The law of inertia holds only when the reference space is well defined and when the observer is located within it.

Einstein himself argued along similar lines as Mach when he noted that an electric charge, at rest with respect to a given system of reference, does not exert force. When it is in motion, it exerts force on objects that are linked to the "fixed" geometrical system of reference.

From Newton's attempts (and illusions?) and from Mach's criticism, there emerged the idea that the only absolute system of reference that one can speak of is the aggregate of the masses distributed in the whole Universe. This is *Mach's principle*, perhaps one of the most powerful (but perhaps not really exploited, and very rarely mentioned) ideas in modern cosmology. In essence, Mach's proposal means that the average acceleration of a mass m with respect to all masses m_i of the whole Universe, located at distances r_i, is equal to zero, or in mathematical terms:

$$\frac{d^2}{dt^2} \left(\frac{\sum_i m_i r_i}{\sum_i m_i} \right) = 0$$

The influence of a nearby (r_i small) small (m_i small) mass of matter is obviously not significant, and does not alter this equation in a sensitive way.

This principle also leads to the recognition that the inertial force is linked to the local action of the massive Universe and perhaps with a new cosmological vision. This is quite difficult to conceptualize, as masses do move continuously, with respect to other masses and as gravitation, as we shall see, is not likely to propagate instantaneously.

There was not much further development of the new mechanics, based on Mach's principle. It was only a beginning and perhaps it still is. The new theory necessitated by the Mach discussion would have to explain motions in gravitational fields as well, whatever the relative description with respect to a given system of reference. Newton gave a mechanical description of the Copernican system in a gravitational field, the Sun being at absolute rest; but one should also be able to account for the Ptolemaic description as a gravitational phenomenon, assuming the Earth to be at rest, as noted by Reichenbach (1927) and by the neo-positivist School of Vienna.

This is, more or less, what Einstein did, going from Mach's concept of inertia to a geometrical theory of gravitation that encompassed all the preceding descriptions.

7.3
The Ether

Remember the reflections of Plato and Aristotle about the "aether," as it was conceived by the Greek philosophers. It was not matter, not made of one of the four basic elements (earth, fire, water, and air); but it filled the vacuum, as "nature abhors a vacuum". It was also supposed to be the nature of celestial bodies. However, with the discoveries of Tycho and Galileo, it became more and more clear that celestial bodies were made more or less of "ordinary" matter. The 19th century spectroscopists, analyzing the spectra of the Sun, stars, and planets, and finding in them the presence of sodium, calcium, iron, hydrogen etc., fully confirmed this now obvious fact, the basic unity of the nature of the celestial bodies. Nonetheless, the problem of aether (or ether, as we shall write hereafter, not to confuse two rather different groups of concepts) was still there. We have seen that Descartes' theory assumed some kind of continuous medium of vortices which would transmit from point to point, by contiguity so to speak, the motions of the celestial bodies, and in a way, guide them (see pp. 259–265). In opposition to Descartes, Newton's theory assumed an "action at a distance," without the expressed need for an intervening medium. Newton was actually quite conscious of this. In fact, he conceived of a "mechanical" ether, the "ether of gravitation," which transports the gravitational interaction. It lends a touch of causality to the theory, but formally, Newton did not use it and avoided speaking about it rather carefully. In any case, he still had strong doubts about the nature of gravitation.

Since, however, stars radiate light, the light has to travel in space. Can this space then be empty? Is this conceivable? This depends very much, of course, upon one's views concerning the nature of light. In any case, the vehicle for transporting light needed to be defined, as it did for gravitation. This vehicle is the "ether of light."

7.3.1
The Nature of Light

Newton had the idea that light was of a corpuscular nature. It is made up of very small moving objects, the motion of which, like that of the planets, is governed by the forces of gravitation and inertia. However, as light propagates in a straight line (unlike sound waves, in a way which could not be easily explained if light were made of waves), the gravitational forces are negligible, in view of the vanishing mass of the light corpuscles. They are able, of course, like the planets, to move in a "vacuum" (in the "ether of gravitation"). The fact that the attractive force of gravitation needs no support, no intervening medium, to be exerted on distant bodies like planets was not a property associated with any propagation, but was rather a property of space itself, propagation being instantaneous. However, this was rather strange and it should have been evident, even in Newton's time, that the propagation of gravitation is not necessarily instantaneous. This would hold true as well for the propagation of light (whatever the nature of it). Both would have to be transported by some medium penetrating all space, a medium acting through some local motions, from place to place, or perhaps by elastic deformation…the ether of gravitation, precisely!

It is true that Newton discovered many properties of light (his treatise on *Optics*, another remarkable book, was published in 1704). In 1666, he discovered the nature of white light (sunlight) to be a combination of lights of different colors. He discovered, if not refraction (known qualitatively for centuries, its laws having been written quantitatively by Descartes and by Snellius), at least its differential nature, the refractive index being different in material media (water, air) for different colors. Newton even discovered some properties, such as the deviation of light by massive objects, the diffraction of light (earlier discovered by Francesco Maria Grimaldi (1618–1663)) and the colored "Newton's rings." These rings were in reality discovered first by Robert Boyle (1627–1691), and by Robert Hooke (1635–1703), the friend and correspondent of Newton. All these properties are linked (we now know) to the wave nature of light.

Nevertheless, Newton held to his theory of corpuscular light, perhaps partially to contradict his enemies such as Descartes and Huygens, but primarily because of the obvious propagation of light along a straight line in a vacuum.

7.3.2
The Wave Theory of Light

Among the first to advance the idea that light is made of waves were
Hooke, on the basis of diffraction phenomena, and especially Huygens,
who spoke of light waves moving in a space-filling ether, of which the
"particles," by pushing one another, propagate the light waves. Huygens
made the fundamental discovery of polarization, while studying the double refraction properties of calc-spar, discovered earlier (1669) by Erasmus Bartholinus (1625–1698). Newton's explanation of these phenomena
was, in a way, prophetic. He assumed that light rays had "sides," and this
sort of transversality seemed to him a definitive objection to the wave
theory. At that time, only sound waves were known, the propagation of
which is longitudinal.

The discovery, in the beginning of 19th century, that the properties of
light were mostly those of waves, was indeed overwhelming. Several physicists, in the laboratory, succeeded in conducting splendid experiments
displaying most of the wave properties of light with considerable accuracy. Thomas Young (1773–1829) suggested in 1801 a mechanism for understanding interferences; Augustin Fresnel in particular succeeded in
giving a complete description of the light properties of refraction, color,
diffraction, and interferences. François Arago (1786–1853) confirmed
the Fresnel theory by finding a bright point in the center of the shadow
of a small circular disk. Malus, Fizeau, Foucault in France, and Young,
Hamilton, and Brewster, in England, contributed extensively to the firm
establishment of the truth of the wave nature of light. Actually, light was
found to be a transverse oscillation, unlike sound waves. If light were the
vibration of some elastic and wholly inertial "ether of light," it should be
comparable to a solid, in some way, as one really observes polarized
light. Transverse oscillations are not possible in a fluid. Hence, there
emerged the idea of a quasi-rigid ether for light, "at rest" everywhere
else, hence linked with the long-sought after "absolute frame of reference," whatever it might be.

A crucial experiment (Fig. 7.6) was performed by Fizeau and Bréguet,
in 1851, and repeated more or less by Hoek.

The corpuscular theory implied (for obscure reasons) a greater velocity in the denser mediums. The wave theory implied the contrary.
Therefore, it was necessary to measure the velocity of light, both in air
and in water. Up until then, the velocity of light had not been measured

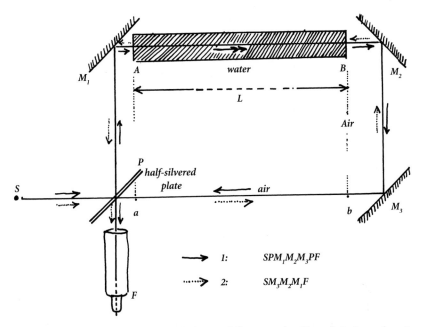

Fig. 7.6 *The Hoek experiment.* The light can follow a path of length L through water (and the water can flow there with some velocity V, indicated by a double arrow), or a path through air, of the same length. The two light rays interfere, at the focus of the interferometer, in F. The velocity of light appears smaller in the water (whatever the actual velocity of the water), as predicted by wave theory. Note this is a "first order" effect.

except in the solar system, using the satellites of Jupiter. Römer, in 1675, discovered the finite speed of light and measured this velocity by observing occultations of its satellites by Jupiter. Bradley, with his discovery of the aberration of light, provided a strong (but inaccurate) quantitative argument in favor of the value found by Römer. However, in the laboratory, this proved to be more difficult. One must use interferometric devices such as those shown in Fig. 7.6. The results clearly supported the prediction of the wave theory.

A step forward of great importance was the promotion of the concept of *field* by Faraday. The idea of action at a distance was a difficult one to conceive without ether, be it the ether of gravitation or the ether of light. Faraday introduced the idea of a field, which is the manifestation, instantaneous and at a distance, of some force.

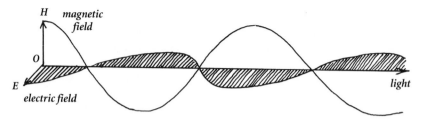

Fig. 7.7 *The Maxwell theory of light waves.* The light waves originate from an oscillator, located at O. The oscillator creates oscillations of the magnetic and electric fields, perpendicular to each other. They both propagate with the same velocity, the velocity of light *c*, in a vacuum. The light is an electromagnetic wave.

The wave theory of light was achieved (Fig. 7.7) in the Maxwell theory of the propagation of transverse electromagnetic waves. The local *electromagnetic field* is oscillating. The theory of light is indeed the theory of the propagation of this field.

Light is thus identified with an electromagnetic oscillating field. This pushed people still further towards a belief in the need for some ether capable of propagating the transverse oscillations of the electromagnetic field. The theory of ether, however, was still inconsistent. One difficulty arose from the fact that in Maxwell's theory, *c*, the velocity of light in a vacuum, was identified with a ratio of electromagnetic units to electrostatic units, and was therefore independent of the system of reference. In other words, Maxwell's theory was in apparent contradiction with the principles of Galilean relativity. Could a mechanical model of ether account for Maxwell's laws? In the 1880s, a radical notion of ether was developed by H. Hertz, who suggested a basic dualism. On one side, he envisioned a purely mechanical conception of nature, on the other side, the existence of electromagnetic fields as fundamental quantities. Ether was for Hertz the necessary substratum of mechanical phenomena, motions, etc., but also of electromagnetic fields, as such fields exist in a vacuum, and this indeed permits an adequate description of the propagation of light. This ethereal substratum is entirely coexistent with matter, quite similar to it. In matter, it is dragged along by the motion of matter itself, an hypothesis which, unfortunately, is contradicted by the experiments we shall describe later. Hertz's conception was indeed untenable. H. A. Lorentz has shown that we must abandon the hertzian dualism and deprive ether of its electromagnetic properties. All actions of an electromagnetic nature were reducible to the Maxwell equations. Lorentz admitted only that ether was at rest, in the Newtonian sense.

We shall not here go into all the arguments for the wave theory and for the corpuscular theory. In the 19th century, as we have said, the evidence was overwhelming in favor of the wave theory, explaining, as it did, interferences, diffraction, polarization, etc. It was only later that it appeared that the two concepts were not mutually exclusive. The particles of light (or *photons*) are indeed associated with trains of waves. This is a result of the photoelectric effect, as analyzed by Einstein, and quantum mechanics and wave mechanics, as elaborated by Louis de Broglie, Erwin Schrödinger and others. Notably, the prediction of a greater velocity for particles in a denser medium was shown to be completely wrong, as we have seen. Packets of waves, the difference between group velocity and wave velocity, were therefore defined to solve these apparent discrepancies by techniques that were used first with heavier particles, such as electrons or protons. The investigation of the wave properties of light, using interferometric devices, was actually quite operational in the development of a pre-relativistic physics of the Universe.

7.3.3
The Problem of Ether. Is it Necessary?

With this in mind, what can we now say about ether? The very use of the notion of field made it a priori somewhat obsolete, but ideas do not permeate the scientific world very quickly and ether, as conceived in the mid-19th century, remained a substratum the properties of which were not entirely clear. An important point when considering the velocity of light is that its velocity in a given laboratory, in a given medium, and in astronomical space, might differ, the laboratory being in motion with respect to the ether of the astronomical space, i.e. the ether of gravitation.

It was necessary, therefore, to imagine many new devices to measure the velocity of light on Earth with sufficient accuracy. Fizeau used a toothed wheel method as described in Fig. 7.8, Foucault another method with turning mirrors (Fig. 7.9). They derived the following values: 313 300 km/sec and (with a better accuracy) 298 000 km/sec. Otto Struve, using the motions in the solar system, and notably the aberration constant, determined a value of 308 000 km/s. At that time, these discrepant values were nevertheless used to re-open the discussion of Römer's observations and to determine the value of the solar parallax.

Fig. 7.8 *Fizeau's toothed wheel.* The light source S projects an image, after a reflection on a half-reflecting plate SM, at T, in the interval between two teeth of a toothed-wheel containing n teeth and turning at a rate of N revolutions per second. The light is thus interrupted Nn times per second. An image of S is made at F at a large distance L from the point T. The light is then reflected towards the observer E. If the travel time TFT is an integer number of the time $t = 2L/c$ between two passages of light, the observer can see the light. If it is equal to an integer + 1/2 times t, then the observer will not see anything. Knowing n, L, and t, one can adjust N in order to pass from "visibility" to "nonvisibility." In the actual Fizeau experiment, $L=8633m$; $n=720$. The minimum time between two successive passages of the light is $t =2L/c=N_0 n$, which corresponds to $N_0=22.6$ revolutions per second. All multiples of that value will allow observation of the passage of light. Then, one can derive c from the number N_0 deduced from the values $N_1=mN_0$ and $N_2=(m+1)N_0$.

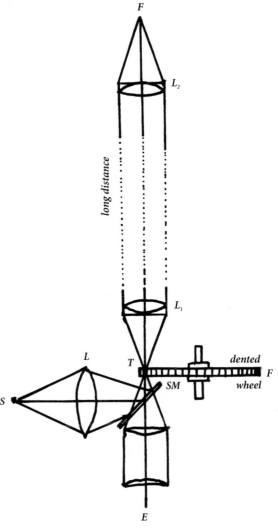

Fig. 7.9 *The Foucault*
turning mirror experiment
(principle). The source S is
focused on the point S_1,
through one lens, one
half-reflecting plate P, and
– at a large distance *L* –
one fixed mirror A (the
turning motion of which
is symbolized by an ar-
row) and fixed mirror M.
During the time $\Delta t = 2L/$
c, the light goes from A to
M and backwards. Light,
coming back, finds that
the mirror has turned
during the time Δt by

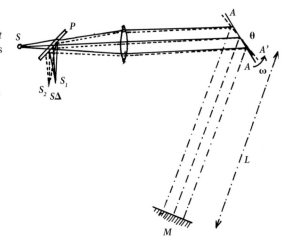

the small angle θ. Then it is focused on S_2, instead of S_1. The distance $S_1S_2 = \Delta S$
is a measure of θ. If the angular velocity of the mirror M is ω (rad/second), then
$\theta = \omega \Delta t$.
Knowing ω and q, Δt is measured. Knowing L, c is thus easily deduced from the
data.

Now our problem is different: is light dragged away by the motion of
the medium in which it propagates? Another Fizeau experiment in water
(Fig. 7.10) showed that light is indeed not dragged away by the water,
hence that the ether (linked with the light?) was "at rest" whereas the
water was moving.

The phenomenon of aberration of light, discovered much earlier by
Bradley, can be interpreted in roughly the same way (see p. 285ff.). The
planet in motion does not carry the ether with it. The ether is at rest.

However, a question remains unanswered. Is the velocity determined
by astronomical methods the same as that measured by laboratory techniques?

Michelson (1879) was probably the first one to draw attention to the
importance of measuring very accurately, in the laboratory, the velocity
of light with respect to the ether, the "wind of ether," the ether being a
substratum linked at the very least with the solar system. Indeed, if the
ether is at rest, planets move with respect to it, as does the Earth, notably
through its own rotation around its axis, and in a still more clear way,
through its revolution around the Sun. If light has a velocity c with respect to the ether (Maxwell's theory), should we detect a difference in

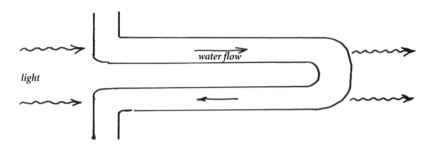

Fig. 7.10 *The Fizeau experiment in water flow.* This experiment, shown here only in a schematic way (compare with Fig. 7.6), shows that the interference fringes do not move. It appears as if the two paths are equivalent. The water does not "drag" the light – at least in a significant way. The experiment primarily allows for the determination of the velocity of light in water. It depends upon the refractive index n and it is a first order effect: $V_+ = V_- = c/n \pm V(1-1\,/\,n^2)$.

the apparent velocity of light when the motion of the Earth is different, the source of light also moving, therefore, with respect to the ether, creating an "ether wind?" Let us assume that the Earth is not at rest with respect to the ether, c being the velocity of light as measured on Earth, not by by Römer, but in the laboratory as in Fizeau's and other interferometric experiments. The velocity of light on Earth should then be $V-c$ at point A, and $V+c$ at point B (Fig. 7.11). The measured value should indeed not be constant but oscillating with a period of one year between the extremes: 299762 km/s and 299822 km/s (estimated with modern val-

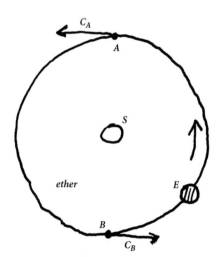

Fig. 7.11 *The wind of ether, as it should appear.* See text.

Fig. 7.12a–c *The Michel-son-Morley experiment.*
(a) The instrument. The light can follow two paths, SPMF or SNPF, involving the half-reflecting plate P, and the mirrors M and N, assuming (for convenience) that the 4 successive parts of each path are equal to *L*. Interference fringes can be obtained between the two paths. If there is some difference in the lengths of the two paths, the fringes will be displaced.
Note that if the instrument is moving, the mirrors and plate move also with the velocity *V*, from P_0 to P_1, to P_2, to P_3, to P_4 , from M_1 and N_1 to M_2 and N_2, to M_3 and N_3, to M4 and N_4, and so forth (and for S and F as well), as the whole instrument has done between S and F. This induces a correction to the time taken by the light to cross the instrument from S to F, a difference which is a function of the velocity *V* and its orientation. If *V=0* (no motion), the total length of the path, in all cases, is *4L*. For **(b)** and **(c)** see next pages.

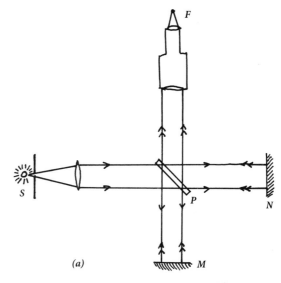

ues of *c* and *V*). The Fizeau experiment should have demonstrated this, but it was not accurate enough. Note that this is different from the Fizeau experiment with water, different but an extension of it. In the latter, the main phenomenon is a result of the difference of refractive index between air and water; the fact that moving water does not affect the result comes perhaps from the slow motion of water. The motion of the water is slow; the accuracy of velocity determinations by Fizeau is not very high. Perhaps, with a very accurate interferometer, one could use the much larger velocity of the Earth in its revolution. (The difference between the velocity of the Earth at night and Earth at day is only about 0.5 km/s, but on its orbit, the velocity of Earth with respect to "rest" is about 30 km/s.) However, the very precise Michelson experiments (1881) and the more elaborate Michelson-Morley experiments (1887) (Figs. 7.12a–c) were completely negative. *The velocity of light is constant and well determined.* However a small residual has been observed by Morley and Miller, and is either spurious or unexplained.

Fig. 7.12b The figures (b$_t$, top) and (b$_b$, bottom) show the origin of the difference in the lengths of each of the possible paths. In (b$_t$) the path may be either $Lc/(c-V)$ or $Lc/(c+V)$. The path from P$_1$ to M$_2$ and back to P$_3$ is therefore: $Lc\,(1/(c-V) + 1/(c+V))$, $\approx 2L(1 + V^2/c^2)$. In (b$_b$), the length of the path from P$_1$ to M$_2$ and back to P$_3$ is easy to compute. The time between positions P$_1$ and P$_3$ is $2L/c$. Therefore, from a simple construction, one gets: $2L(1+V^2/c^2)^{1/2}$ $\approx 2L(1+V^2/2c^2)$. The difference between the two path-times involved (on b1, or top of of b, and on b2, or bottom of b) is therefore: $\Delta L = LV^2/c^2$.

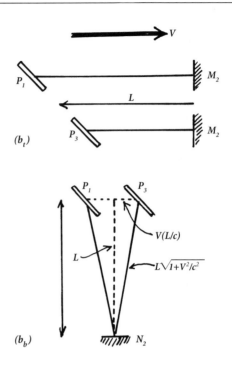

These measurements were remade many times, on different types of interferometers (Sagnac, 1913, Fig. 7.13), with higher and higher accuracy, later using lasers, etc. No value found for c ever differed from another, within the accuracy of the methods. It is now determined with many significant figures. $c = 2.99792458 \times 10^5$ km/s is even now taken as a *fundamental constant of physics*, on which, by definition, there is no error bar. It has facilitated the determination of other less well-known constants in physics.

From this, all we can say about the nature of ether is that *the ether wind does not exist!* The velocity of light is not influenced by the motion of the Earth, even to the extent of involving quantities of the second order (see Chap. 7.7). This does not prove that the ether does not exist and we shall come back to it much later. It does, however, raise an important question: Why is it that the velocity of light cannot be added to or subtracted from the velocity of its source, or of the laboratory? Why is it that this quantity, whatever way one measures it, has a constant value?

Fig. 7.12c, c' In (c), the line
SM is parallel to the as-
sumed instrument velocity
V with respect to the
ether. Actually, the instrument may be rotated. A
rotation of 90° can bring
the instrument in position
(c') where the line S'N' is
perpendicular to the direction of velocity *V*. If one
turns the instrument, as
shown from (c) to (c'), the
two paths are different,
as easily seen in parts
(b) and (c), by 2Δ*L* (see
Fig. 7.12b). One should be
able to observe that difference, through a move of
the fringes, that can be
shown to be measurable.
It is actually not observed.

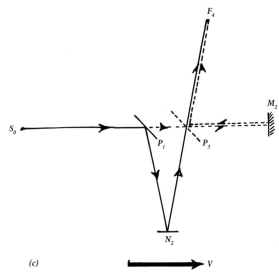

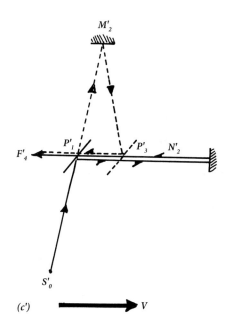

Fig. 7.13 *The Sagnac experiment.* On a rotating table, the orientation of the instrument velocity (with respect to the ether), changes continuously. The principle is comparable to that of the Michelson interferometer (Fig. 7.10), with two half-reflecting plates P_1 and P_2, and an additional mirror L, thus allowing more objectivity, in the sense that at rest, the two optical paths SP_2LMNP_1F (single arrows) and SP_2NMLP_1F (double arrows) are strictly equal to each other.

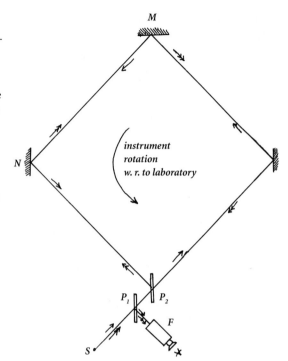

7.3.4
The Lorentz-Fitzgerald Contraction

A first reply to these questions was given by Fitzgerald (in 1892), and elaborated on by Lorentz. They assumed that the constancy of the velocity of light was an indicator of a real, physical, contraction, in the direction of the motion of objects moving fast enough, macro-objects, or electrons! This contraction is not observable by an observer linked to the motion (of the Earth, in this particular case). Of course, this also implies that, seen by the observer, it is the system of reference, not the observer, that contracts. We could give, as a pleasant illustration of this phenomenon, the humorous cartoons drawn by George Gamow in his famous popularized version of the problem, *Mr. Tompkins in Wonderland* (see Fig. 7.18 p. 405). We must again at this point discuss the meaning of the ideas of relative and absolute motion, and the principle of relativity. The invariance law of mechanics through the Galilean transformation of coordinates was broadly accepted. Is the contraction hypothesis sufficient to justify the "principle of relativity?" Lorentz established that this was not, in fact, the case, but rather that a new measure of time would be necessary in a system which moves uniformly with respect to the reference system. Thus, a change of coordinates, from a system at rest to a moving system of references, implies changes in time coordinates, as well as space coordinates. Lorentz expressed the formulae for the change of reference systems from one to another, moving in a relative motion of translation with respect to each other. Larmor and Poincaré reached similar conclusions at about the same time.

In the new theory, implying the Lorentz transformation of coordinates, the principle of relativity is still valid, and in conformity with the results of experiments, including all electromagnetic events, and light propagation. As Max Born said: "An observer perceives the same phenomena in his system no matter whether it is at rest in the ether or moving uniformly and rectilinearly. He has no means to distinguish the one from the other."

The consequence of this statement is that there is no need at all to speak about an "ether of light" (or an "ether of electromagnetic phenomena"); no more indeed than there was any need, after Newton, to maintain the notion of an "ether of gravitation."

Thus, we shall either have to forget once and for all the concept of the ether, or we shall have to revive it later in a different way, as a physical description of space. Such attempts returned to the use of the word ether to describe empty space equipped with gravitational and electromagnetic fields. Einstein used the word with that meaning, and taking a long step forward, Dirac described an ether, its detailed nature, and its effects. Opinion is still divided about Dirac's ether.

7.3.5
Einstein's Special Theory of Relativity (1905)

Einstein was rather dubious about the physical reality of a contraction. He was, nevertheless, forced to admit that the velocity of light is independent of the state of motion of the observer. It always has the value c, as given above. He recognized that there was a contradiction between this constant value and the principle of composition of velocities. Where then is the fallacy?

The main criticism was directed towards the concept of *simultaneity*. This concept had been generally accepted. It is implicit when one speaks of the "instantaneous" transmission of gravitation. The conclusion of Lorentz, however, based on his transformation formulae, of the need for two different times, in different systems, imposes the idea of a "local time." Even if the causality holds (an important basis, a postulatum so to speak, of all physics, perhaps the postulatum *sine qua non*!), two phenomena simultaneous for one of the observers, in one of the systems, may not be so for an observer in another system. Of course, it would not be possible to observe the phenomenon before its cause.

This is too new a notion not to explain in somewhat greater detail. Although it is not really a cosmological idea, it had a great many cosmological consequences.

Indeed, until Einstein (apart from the little noticed Poincaré discussion on simultaneity, as early as 1900), concepts such as "moment in time," "simultaneity," "earlier and later" etc., had an absolute meaning. They were considered valid in the whole Universe, where only one time is conceivable, that of our clocks, regulated by the rotation of the Earth and its revolution.

But when Einstein thought about it, from the point of view of the quantitative physicist, of the observer, he realized that the statement that an event in A and an event in B are "simultaneous" is a meaningless statement. The reason is that the physicist has no way to check this assertion. It would be necessary to have clocks at every point which we could ascertain would go at the same rate (or, in other words, beat synchronously). Let us follow Max Born's presentation of this problem.

Let us conceive of two clocks at A and B (Fig. 7.14), at a distance L apart, both at rest in a given system of reference. There are two methods of regulating the clocks (if they are pendulum clocks, for example, by changing the pendulum's length) so that they turn or work at the same rate. We could put both of them *at the same point*, regulate them there,

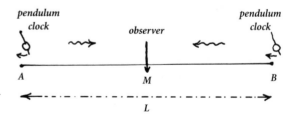

Fig. 7.14 *Defining simultaneity* (see text).

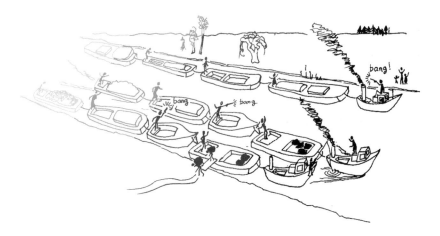

Fig. 7.15 *Barges moving and barges at rest, in the fog.* Is there an absolute simultaneity? (see text).

so that they are then truly synchronized, and then bring them back to points A and B, respectively. Alternatively, we could use *time signals* to compare the clocks at a distance, the signals being emitted at each beat by both clocks, and observed at the same point, perhaps M in the middle of the segment AB.

Both methods are actually used in practice at sea, but then the motion of the boat is very slow (compared to that of signals), and all effects are not visible, just as in the first Fizeau experiments, although they do not measure the same thing. The signal method is more reliable than the transportation method. The clock transported from M to A or B may very well vary, which would cause the smallest error in the beating to increase continuously. No real clock is free from error, but even if clocks were ideal, how could we check the equality of their beats if not by signals? We are indeed obliged to use the signals to define time, especially in systems that move with respect to each other. How to build an ideal clock is indeed another problem altogether!

Let us be still more specific. Consider (Fig. 7.15) a long series of barges B, C, D ... pulled by a steam tug A over a river.

There is no wind, but the fog is so dense that each boat is invisible to the others. If the clocks on the barges and on the tug are to be compared, sound signals may be used. The tug sends out a shot at 12:00:00. When the sound is audible on the barges, the bargemen set their clocks at

12:00:00. Of course, in doing so, they commit a small error, since the sound requires some time, a short time (still shorter if it were light signals or radio signals) to arrive from A to successively B, then C, then D. If one knows the velocity of sound s is equal to 340 m/s, the error can be eliminated. If B is at 170 m behind A, the sound will take $t = L/s = 1/2$ seconds to travel from A to B, hence the clock at B should be set at 12:00:005, at the moment the sound reaches it. But is this correct? Not completely! If the tug and barges are moving during the time interval during which the signal travels from A to B, it will require less time. We must know indeed the relative velocity of the ships with respect to the air, with respect to the medium which is the support, the substratum of the sound. This results from the velocities of the wind and the velocity of the boat train.

If sound cannot be relied upon, perhaps light signals might be used as a way to synchronize the clocks. Light is much faster than sound. Now, the case of a train of boats may not be very interesting, but if we replace them with a heavenly body moving in the sea of ether, then our thinking remains quite valid. Therefore, determining the correct comparison of various times in various moving systems requires that we know the velocity of light in the ether and the velocity of these systems with respect to the ether. However, as we have described above, the result of all measures mentioned earlier is that it is impossible to detect any motion with respect to the ether. It would follow from this that the simultaneity of two events cannot be ascertained.

However, aren't we moving in a vicious circle? After all, if we compare times by means of light signals, we must know the exact value of the velocity of light c, but the measurement of the velocity of light requires the determination of a duration in time.

Thinking about this, Einstein suggested a definition of *relative simultaneity*, in essence a pragmatic definition. Let us stick to the example of the tug and the barges. When they are at rest (with respect to the shore and with no wind), we place a boat M exactly halfway between boats B and C and send off a shot from M. It must be heard simultaneously at B and C. Now, if the series of boats (S = A, B, M, C, D, ...) is in motion, we can apply the same method. If the bargemen do not realize that they move with respect to the air, they will be convinced that clocks B and C are synchronized. (The three boats keep the same relative distances.) Now consider another series S' (A', B', M', C', D' ...) of boats whose barges are exactly at the same mutual distances as the barges of series S.

Compare their clocks. Let us assume that series S overtakes series S', presumed to be at rest with respect to the shore. At a certain moment, A will coincide with A', B with B', etc. The bargemen, on both series, will compare their clocks and they will find that the clocks of S and of S' do not agree. Even if A and A' were (accidentally!) beating simultaneously, that could not be the case for B and B'.

This is obviously due to the following fact: when the boats are in motion, the signal from M will take more time to arrive at the preceding ship B and less to arrive at the following ship C, because B is moving away from the the signal wave, while C moves towards it, whereas this does not occur in the boats B' and C', assumed to be at rest with respect to the shore. The difference of velocity between the two trains regulates this difference between C and B. This has nothing whatever to do with the velocity of the barges with respect to the water, to the air, (or to the ether), but *only with the velocity of one train of barges with respect to the other.*

So there is no such thing as absolute simultaneity. Developments of that idea led Einstein to reformulate the hypotheses of dynamics as follows, using the transformation of coordinates he discovered, seemingly independently of Lorentz, which was called the Lorentz transformation, at the suggestion of Poincaré.

(a) *Principle of relativity.* There are an infinite number of systems of reference (or inertial systems) moving uniformly and rectilinearly with respect to each other, in which all physical laws assume the simplest form, the one originally derived from the concepts of motion with respect to absolute space, or to ether at rest.

(b) *The velocity of light c is a constant of nature.* In all inertial systems, the velocity of light has the same value when measured with rods and clocks of the same kind.

The Lorentz transformations can be easily derived from these two principles (whereas Lorentz found them from the contraction hypothesis; see Chap. 7.7 for more details on these important relations).

They imply apparent dilation or contraction of time, and of space, in the direction of the motion. The time read from a clock in the system of reference in which the clock is at rest is the "proper time" of that system. It is the same as Lorentz's "local time." But whereas local time is more or less an auxiliary mathematical quantity, in contrast to true absolute time,

Einstein established that there is no way of determining absolute time or of distinguishing it from the infinite number of equivalent proper times of the various systems of reference that are in motion. Absolute time, for Einstein, has no physical reality.

However, Mach's point of view should not be forgotten. Matter in the Universe defines one "absolute" space system of reference. In it, there is only one time, the "universal time" (which may differ from Earth's time). This universal time is the time of cosmologists when speaking about the evolution of the universe, and is an aspect of the "cosmological principle," when applied in the framework of General Relativity.

From Einstein's kinematics, we can derive a law of velocity composition (see Chap. 7.7). This law is rather simple, but not as simple as the one used since Galileo.

In system S, the velocity of a mobile body M is W. In system S', moving with respect to system S with velocity V, the velocity of the moving body M is now W', and the composition law is:

$$W' = \frac{W+V}{1+VW/c^2} \quad \text{or} \quad W' = \frac{c+V}{1+V/c} = c, \text{ if } W \text{ is equated to the velocity of light } c.$$

The first expression shows that no velocity can be larger than that of light, even if V and W are very large, unless there is one velocity V or W which can be shown to be larger than that of light. No such velocity has ever been found. In other words, *the velocity of light is a limit velocity*. The second relation expresses the *invariance of the velocity of light*, in full agreement with all experiments performed in order to determine the ether wind (Michelson, etc.), and described here above, pp. 387 ff.

Einstein's special theory of relativity is, so far, a purely kinematic one, in essence comparable to Kepler's law. It is necessary then to add some dynamic principle. It was Newton who systematized Kepler. Einstein had only Einstein! The first step was to perform dynamics within the concept of special relativity.

Of course, one may ask whether Einsteinian kinematics is completely correct. Why is c the limit for all the velocities? Why, instead, could we not have used a larger velocity such as that of gravitation, which is (perhaps!) larger than that of light (although Poincaré, in 1902, suspected that gravitation was indeed moving at the velocity of light), which is larger than that of sound? Using sound signals and maintaining that the velocity of the sound was constant (which we know, by experience, to be wrong!), would have led to absurd (i.e. not experimentally verified) results. Gravitation was actually not found to be travelling faster than light. As a matter of fact, in Einstein's conception, gravitation is instead a property of space geometry. The modern approach to the Lorentz transformations replies to this paradox. In that context, c appears as an integration constant, which can be identified with the velocity in a vacuum of any particle of mass zero.

Unless we discover some new fact, the velocity of light is indeed a limit value. The velocity of light is constant, according to the current level of observations.

An important consequence of Einstein's constructions is that there is no "absolute" unit of time or length. A material rod is not a spatial thing, but a *space-time* configuration. A material clock is not a time-measuring machine, but a *space-time* device. One should no longer think in terms of three space coordinates and one time coordinate, but of a four coordinate space-time. Clearly, it is difficult to draw figures to explain phenomena in that sort of 4-dimensional logic. Therefore, our figures will be drawn using only one or two space coordinates, x (or: x and y) and one time coordinate, not t of course, but its length "equivalent" ct, length covered by a light ray in time t (Fig. 7.16).

Let us (and we shall still follow Born here) consider a measuring rod MN. If we bring that rod into S, then into S', two different systems of reference, that rod represents the same length in each, provided it is affected in neither system by external forces. A fixed rod at rest in S and of length 1 m will have the length 1 m when at rest in system S', provided the physical conditions (gravitation, position, temperature, electric, magnetic forces, etc.) are the same in S' as in S. The same is postulated for the clocks. This is indeed a tacit assumption of Einstein's theory. We could call it the "principle of the physical identity of the units of measure." Armed with this principle, we can return to our previous experiment and discussion. We see that the velocity of sound is not a good unit to measure time, but the velocity of light is, because of its invariance, derived from the Michelson experiments.

Back to our material measuring rod MN. Again, it is not a spatial thing but a space-time device. Every point of the rod exists at this moment, at the next, still at the next, etc., at every moment in time. The proper picture of the rod, in the 2-dimensional (x, t) image of the 4-dimensional space (x, y, z, ct), is a strip of the plane (Fig. 7.17).

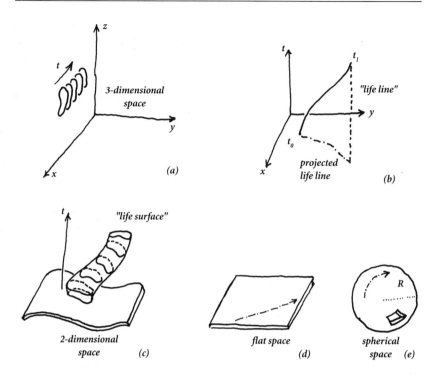

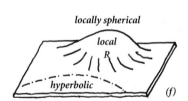

Fig. 7.16a–f *4-dimensional space-time: its 3-dimensional representation.* In a 3-dimensional space (x, y, z) an object (here a closed curve) moves with time t, and occupies successive positions, as in (**a**). If one replaces the 3- dimensional space, for convenience and illustration, by a 2-dimensional space (x, y), a moving point describes its "life-line" (**b**), of which the "projected life-time" is the equivalent of the surface described by the closed curve of Fig. 7.16a. A closed curve describes a sort of tube, composed of the life-lines of each of the points of the curve, in (**c**). Of course, the 2-dimensional space may be flat (Euclidian), as in (**d**), or it may be spherical, with one unique radius of curvature R as in (**e**), and it can also have only local curvatures as in (**f**). An inertial motion of a body is a straight line in (**d**), a circle on the spherical space, in (**e**), and a hyperbolic trajectory far from the curved area, in (**f**). In 4-dimensional space-time, 3- dimensional space may also be curved in many different ways, or it may be flat and Euclidian, but it is difficult to represent graphically on the page of a book!

The same rod is represented by two different strips in S and S'. There is no a priori rule as to how these 2-dimensional configurations of the x, ct plane are to be drawn, in such a way that they may correctly represent the physical behavior of one and the same rod, but at different velocities. Actually, each rod is represented by a strip parallel to the t and t' axes. In this representation, both strips have the same widths if measured parallel to a fixed x-axis, in the classical concept of measure. In Einstein's concept, they have the same width only if this is measured for each rod in the x-direction of the system in which the rod is at rest. The "contraction" does not affect the strip in itself, only the section cut out of the x-axis. The strip has a physical reality, not its cross-section! Thus the apparent contraction is only a consequence of our way of looking at things. It is not a change in physical reality. The contraction is purely an appearance, a perspective effect in a way!

The same remark applies to time. An ideal clock always has a well-defined rate of beating in a system of reference in which it is at rest. It indicates the "proper time" of the system of reference. From another system, the same interval of time, in the proper time, seems longer. This is the "slowing down" of the clocks. It is, in other words, meaningless to ask what is the "real" duration of an event (except in cosmology, where, being "Machian," we consider the whole of the Universe as a system "at rest" as the basis for the definition of some absolute time).

Einstein's concept led to several paradoxes, some even amusing, illustrated for example in George Gamow's previously quoted book, *Mr. Tompkins in Wonderland*, where the author exemplifies and draws what

Fig. 7.17 *World lines of two measuring rods moving relative to each other.* Each rod is represented by a strip of world lines (Sect. 8.1.2) parallel to their respective t and t' axes. The hatching lines represent the rods in their rest frame at different times (after Max Born). The t coordinate is replaced by a ct coordinate, which has the dimension of a space coordinate.

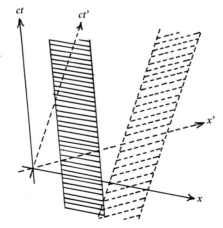

would happen if the velocity of light were only 30 km/sec (Fig. 7.18). The strangeness of the concepts appears there in full light, but Einstein's demonstrations are quite coherent and convincing.

In a four-coordinate space, where we cannot really separate time from space, an important notion can be derived. The Lorentz transformations of coordinates, as used also by Einstein, are indeed constructed so that the velocity of light is the same in two systems of coordinates each moving with respect to the other with a velocity V. Let us consider the "interval element" between two points of the space-time (Fig. 7.19). Its square is ds^2. It is easy to see, using the Einstein-Lorentz transformation (see Chap. 7.7), that $ds^2 = ds'^2$. The (squared) velocity of light c^2 is equal to the ratio of the (squared) path $dx^2 + dy^2 + dz^2$, to the time (squared) used by the light to cross this path (squared) dt^2, and equal, as well, to the ratio $dx'^2 + dy'^2 + dz'^2$ to dt'^2.

We shall see that the interval element, or, as one says for short, the "ds^2," is an essential quantity in the cosmological consequences of the General Theory of Relativity, defining what we call its "metrics."

The dynamics of Special Relativity are also quite new. The very notion of mass itself has to be changed. There is a mass m_0 in a system S, "at rest." But the mass of the same object measured in another system S', in

Fig. 7.19 *The space-time element.* Between two points in 4-dimensional space-time (here the z-axis has not been represented), one can define an interval element ds, of which the square ds^2, is the basis for the various model-dependent geometrical descriptions of space-time, its "metrics." Note that the t coordinate is replaced here by an imaginary coordinate ict, so that its square is $-c^2 t^2$, for a convenient illustration of the meaning of the velocity of light, c.

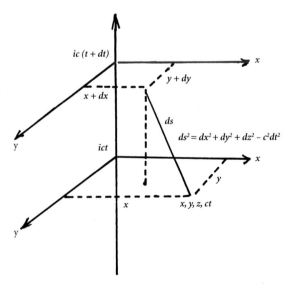

Unbelievably flattened

The city blocks became still shorter

Fig. 7.18 *The Lorentz-Fitzgerald contraction.* In a "universe" where the velocity of light is assumed to be $c=10$ km/sec, and where it is a limit velocity, as in the Special Theory of Relativity, the sidewalk pedestrians see a flattened cyclist, whereas the cyclist sees the sidewalk pedestrians very much flattened, like the blocks of houses in the town (reproduced from G. Gamow, *Mr. Tompkins in Wonderland*).

relative motion with respect to S, is different. We shall not go into the demonstration of this phenomenon, which is based on a very straightforward principle, that of the the the conservation of momentum. In ordinary mechanics, mass and momentum are both conserved. In Einsteinian mechanics, momentum is considered more physical than mass, and is conserved, but mass is not. On this basis, one finds

$$m = \frac{m_0}{\left(1 - V^2/c^2\right)^{1/2}}$$

A body moving at the velocity of light, has no mass! This is the case for the *photon of light*. One could of course argue with this point of view – and we shall, in the next chapter, Sect. 8.6.2

We have not mentioned the law of classical mechanics concerning the *conservation of energy*. How is it transformed in Einsteinian special relativity? Again, we cannot go into the demonstration of Einstein's results. In essence, the results demonstrate that the addition of an energy ΔE (be it of a kinetic or a thermal nature) to a body changes its mass by $\Delta E/c^2$. Hence the famous relation:

$$E = m\,c^2.$$

This expresses at the same time the laws of conservation of energy and conservation of inertial mass. This is the law of "inertia of energy," perhaps (Max Born dixit) "the most important result of the theory of relativity." (A parenthesis here: Many people in Hiroshima might be able to confirm Born's statement, were they still alive. At the same time, this discovery, the context of it and its date, discharge Einstein of all responsibility in this tragic event. Einstein is not the "inventor of the atomic bomb", as I heard in one science Museum, somewhere – and I will not say where! Pure science, Einstein's science, is a response to curiosity, to a need for rigor, for coherence. Applied science is guided by its potential effects.)

The essence of the new mechanics consists of the *inseparability of space and time*. The world is a 4-dimensional manifold. A point in this space is "an event," happening in one place àt one moment. This, however, did not suffice for Einstein. His theory, so far, did not really concern gravitation. The generalization extended to gravitation is the *principle of equivalence*. Moreover, all the demonstrations above concern relative motions that are purely rectilinear and uniform. This was a limitation of the Special Theory of Relativity of 1905. The generalization to all motions is the *principle of invariance*. These two principles are the basis of General Relativity, proposed by Einstein in 1915–1917.

We shall return to them in our discussion of General Relativity.

7.3.6
The Ether of Special Relativity

Let us come back to the subject of ether, the discussion of which led to the construction of Special Relativity. It has often been said that Einstein killed any ether theory. This is not really true. Einstein indeed first found that ether was *not necessary*. He assumed that electromagnetic fields are independent quantities, not linked to any substratum, no more than they are linked to ponderable matter. However, Einstein himself showed that what was needed was not a suppression of ether, but rather a conception of ether that did not attribute to it a well-defined state of motion. We shall come back to this question when speaking of General Relativity.

Actually, what is true is that Einstein, strongly attached to a complete account for the observations, showed himself from the beginning to be an "Aristotelian," as we have defined that term in the beginning of this book. He was determined to save the phenomena first, but people are not always consistent. If Einstein was so reluctant to accept the ether concept, it was in the sense that he did not like "dualistic" ideas and was aiming at some kind of unification of the laws of physics, quite like the modern advocates of the *grand unification theory*. In that sense, he was certainly not insensible, like the modern physicists, to the Pythagorian temptation, often hidden behind Occam's razor.

7.4.
Action at a Distance

A problem still present in scientific discussions at the beginning of this century is the following, which we have alluded to several times.

Why should light have a finite velocity and gravitation be instantaneous, a fact implicit, for example, in the calculation of the perturbations in the theory of comets?

Here again, General Relativity shed a new light and modern theories of perturbations take General Relativity into account. Modern observations of gravitational waves, a predicted complement to this theory, are not yet very successful, although no one doubts seriously that they will eventually brilliantly confirm this theory.

7.5.
Olbers' Paradox

Modern cosmologists have made extensive use of several points of view which were absent from the Newtonian discussion. One of the more important of them is the so-called Olbers' paradox, completed by the discussion, by Charlier and his successors, of the hierarchical nature of the observed Universe.

Olbers' paradox was actually formulated by Loÿs de Chézeaux, much earlier, in the 18th century. Kepler, still before, felt the need for it. It was his reason for keeping the stars in a thin spherical layer. Halley also, in a more modern way, considered this question. Olbers (1823), however, went much further than his predecessors in the statement and explanation of this paradox. There are actually several demonstrations of the paradox. We shall, hereafter, use only one of them.

Consider the surface of a star, at a distance r from the observer. A given instrument receives the pencil of light contained in the aperture $d\omega$, the aperture of the cone, the so-called solid angle defined by the base of this cone (Fig. 7.20 a–c). If dS is the surface from which we receive the light within the elementary cone $d\omega$, one defines $d\omega$ as $d\omega = dS/r^2$, in such a way that the total solid angle of the sky, of which the real "surface" at that distance is $4\pi r^2$, is actually $\Omega = 4\pi$. The whole sky then occupies a solid angle of 4π steradians.

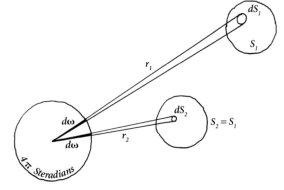

Fig. 7.20a–c *Olbers' paradox.*
(a) Brightness of a radiating surface as a function of distance (see text). For (b) and (c) see next page.

Fig. 7.20b If the stars are uniformly distributed in a Euclidian open space, the visual line will always eventually meet the surface of a star. The sky should, as demonstrated in Fig. 7.20a, appear to be of a uniform brightness comparable to that of a star.

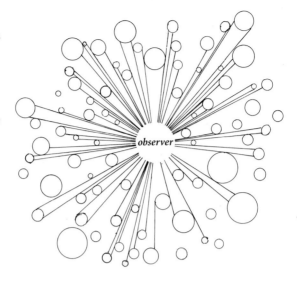

Fig. 7.20c Because the forest is there, we cannot see what is behind it, as in Fig. 7.20b.

Let us bring that star, initially at a large distance r_1 to a closer position at a distance r_2. The elementary solid angle is instrumental, then the same in both cases, leading therefore to $d\omega_1 = d\omega_2 = d\omega$. However, the "apparent" radiating surface of the star is different : $S_2 = S_1(r_1^2/r_2^2)$. What can we say about the *measured brightness* dB of the observed radiation, *within the pencil of light* $d\omega$, coming from either dS_1 or dS_2? In case 1, it is, say, dI_1. What is it in case 2? The *intensity* of the *total* stellar surface, *as emitted*, is not changed. What we observe in $d\omega$, however, is changed, according to the obvious relation:

$$dS_2/S_2 = (dS_1/S_1)\ (r_2^2/r_1^2).$$

The fractional *measured brightness, as observed*, dB, is really proportional to dS/S, and equal to $B_{total}\ (dS/S)$. Therefore, $dB_2 = (dS_2/S_2)\ B_{total\ 2}$, whereas $dB_1 = (dS_1/S_1)\ B_{total\ 1}$.

In case 2, the stellar total brightness B (as observed) is different than when the star is in 1. As was known since the time of Kepler or even before (since Scot Erigene actually), $r_1^2\ B_1 = r_2^2\ B_2$, therefore

$$\frac{dB_2}{dB_1} = \frac{(dS_2/S_2)}{(dS_1/S_1)} \times \frac{B_{total\ 2}}{B_{total\ 1}} = \frac{r_2^2}{r_1^2} \times \frac{r_1^2}{r_2^2} = 1.$$

In other words, the apparent brightness of a piece of stellar surface does not depend upon its distance. The value of the instrumental solid angle $d\omega$ does not enter into the conclusion.

To go further, we need additional hypotheses. They do not conform to our modern ideas of the Universe, but in a Newtonian-oriented period, they were straightforward. These hypotheses are as follows:

(a) The Universe is infinite in all directions (Digges, Newton, etc.).
(b) The Universe is uniformly filled with stars and stellar systems.
(c) All stars are similar to our Sun (Galileo) in surface intensity.

Then the first consequence is that in any visual direction of observation, one will (Fig. 7.20b) encounter, near or far, some stellar surface. This phenomenon is similar to the following fact: In a large forest (Fig. 7.20c), the observer sees only the trunks of the trees, ad infinitum, and does not expect to see through them. Our sight is limited in distance, whatever the direction of observation.

Then all pencils of light will contain the same amount of energy, and *the entire sky, even at night, will be approximately as bright as the solar surface.*

Clearly, this is not true (hence the "paradox"). The real sky is rather dark at night, so we must conclude that some of our hypotheses above are wrong, perhaps all of them! E. Fournier d'Albe and Charlier assumed that the hypothesis of uniformity was not valid. They built a hierarchical Universe, of a *fractal* nature, which means that each volume in the Universe is similar to a smaller volume of a "higher" degree. Essentially, one could say that they assumed the density of stars (as of luminous matter) to be insufficiently well defined at the scale of the Universe. The larger the volume in which density is measured, the lesser the density. This is easily achieved by fractal constructions, as demonstrated by many authors, in particular B. Mandelbrot. De Vaucouleurs has shown (Fig. 7.21, see also Fig. 8.17, p. 464) that the actual Universe, has, within the observable limits, a fractal structure of which the dimension is of the order 1.3 (the fractal dimension of a uniformly filled space is equal to 3). This means that the mass M contained in a volume of radius R is proportional to $R^{1.3}$.

The big bang Universe theories are not concerned with the uniformity hypothesis, but with the finitude hypothesis. They imply the existence of a real horizon, a boundary to our observations. A spectroscopic horizon is also implied, as far-distant galaxies cannot even be observed from the ground in the visible spectral range, as the bright part of the spectrum is indeed shifted (because of redshift) far from the observed visible light.

There have been other solutions to Olbers' paradox and there have been criticisms of the one we give above. In any case, these solutions (apart from the hierarchical Universe) contribute little to the solution of the "gravitational form of Olbers' paradox" (which we should call the "Seeliger paradox"), to which we shall return when discussing the modern theories of the "big bang."[37]

The Charlier discussion raises another problem of cosmological importance, linked to Olbers' paradox, but different. How is it that matter

---

[37] The "big bang" is an expression which, I must say, I do not much like, because of its appeal to other ideas, of a metaphysical, or even a journalistic character. Following Peebles, we could call it the "primeval fireball hypothesis," although that would imply the idea of a "ball." Perhaps the "general explosion" would be a better term.

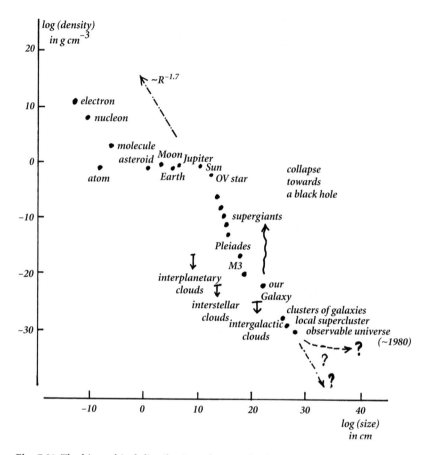

Fig. 7.21 *The hierarchical distribution of matter in the Universe* (drawn according to de Vaucouleurs). In the Universe, the average density ϱ decreases with the size R of the medium in which the density is measured, as $R^{-1.7}$. How is this trend evolving? So far, the exploration of our Universe stays below the line of collapse to a black hole, but does this imply a limit to the size of the Universe?

is so distributed, and very far from evenly? The hypothesis ϱ = const, used by practically all cosmologists of the Einsteinian period, leads to the distribution (no longer fractal!), $M = ϱ R^3$, M being the mass contained in a sphere of radius R. This has to be understood, one way or the other. We shall see that modern cosmology has some difficulty in replying to this question.

7.6
The "Significant" Facts of Cosmology and the Subjectivity of Their Choice

We have perhaps given this chapter a fallacious title, as cosmology has hardly been considered. Indeed, what occurred in the 19th century was a progressive awareness of the extent of the Universe, in space, in time, and in its stellar structure. The cosmology remained essentially "classical," as we will use that term later in this book. The Universe is considered infinite in space and unchangeable in time. It is isotropic. It is homogeneous overall (neglecting the small-scale accidents, such as stars or the Galaxy). It is governed by gravitation and crossed by electromagnetic waves.

Questions concerning the significance of time and space arose from the eversought for and never-observed ether wind, an observational fact. They led to Einstein's Special Relativity. Other observed facts led to some criticisms of classical cosmology, such as Olbers' paradox, another observational fact.

All concepts about the Universe are based on some observed phenomena and some preconceived ideas. One of the difficulties is that no cosmology really uses all the observed facts, because there are too many. Moreover, some observed facts may have no cosmological implications, in the sense that they do not affect our views concerning the Universe at large.

Some observations are clearly without cosmological significance. For example, the chromospheric eruptions, or the volcanic nature of Olympus Mons, or again the origin and evolution of neutron stars and white dwarfs. They may have some cosmological implications, but there is a great distance indeed between them and cosmology. Still one should be careful about that! For example, the deflection of light rays by the Sun, or the behavior of the redshifts of quasars may have, for some authors, a very high cosmological significance. Therefore, one cannot a priori study cosmology while ignoring most of observational astronomy, but one can do so after some careful consideration of the implications of the facts.

There are, however, facts that clearly concern the Universe as a whole. For example (see Chap. 8), the redshifts of distant galaxies, or the so-called *cosmological background radiation,* or again the abundance of light elements in the stars. Also to be considered are the distribution of matter in hierarchical structure, the Olbers-Seeliger paradox, the discovery of fundamental particles, the measure of the mass of the neutrino, and perhaps the phenomena showing that redshifts are not due necessarily to motion through Doppler-like effects.

For some cosmologists, certain of these phenomena are "more" significant than others. For example, the inhomogeneity of the Universe is hardly considered important by the big bang theoreticians, who regard the redshift as the really significant fact. Some consider, on the other hand, that the redshift is not that important, even if the Doppler-like effect remains the only plausible explanation of it, but that the inhomogeneous structure of the Universe is quite essential, perhaps the most essential of all cosmological facts.

It is clear, therefore, that cosmology is still an evolving science, which essentially began in the beginning of the 20th century, and which is still in the making. One should read the next chapter with that in mind. There is great subjectivity in the choice of arguments, hence there is great diversity in cosmologies. They may be internally consistent, but this may be a function of ignoring some obvious facts, considered by some scientists to be not cosmologically significant.

7.7
Appendix: Modern Cosmology
The Velocity of Light, Composition and Measurement

7.7.1
The Galilean Composition of Velocities

A vibrating signal (sound or light) has three characteristics: its *frequency* ν, its *velocity c*, and *the direction of its propagation*. The wave-length λ is defined by the relation $\lambda = c/\nu$. Moreover, a "train" or a "packet" of waves contains only a limited number of waves, *n*. How are these quantities modified in the case of an observer in relative motion with respect to the emitter of waves?

An observer O_1, in a coordinate system S_1, at rest in the ether, observes a train of waves emitted by a light emitter A (Fig. 7.22 bottom) along the *x*-axis in a system S_0, also at rest.

The train of waves moves with velocity *c* in the system S_1. It reaches x_0 at the time t_0. It leaves x_1 at the time t_1. S_1 and S_0 are presumed to be at

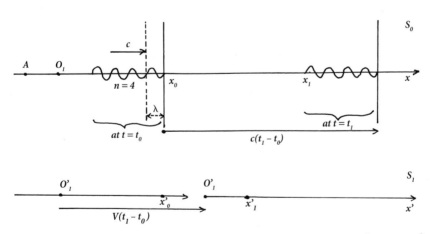

Fig. 7.22 *A packet of waves seen by an observer in motion with respect to the source of the waves (see text).*⋆

⋆ All the figures in this appendix are inspired directly by Max Born.

Fig. 7.23 *The Galilean transformation of coordinates.* $x=x' + Vt$, $y=y'$, $z=z'$, $t=t'$ (*V is here assumed to be parallel to the x-axis, in the positive direction). If a material point in the moving system* $S'(x'y'z't')$ *is in motion with velocity W with respect to the system S (xyzt) at rest, the law of velocity combination is U = V+W. The velocities U, V, W are algebraic quantities and can have both signs.*

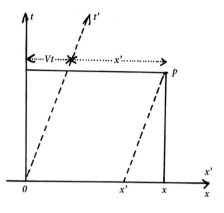

rest with respect to each other. From the four measurable data x_0, x_1, t_0, t_1, we can easily compute the number of waves n:

$$c\,(t_1 - t_0) = x_1 - x_0 + n\lambda \qquad \text{or} \qquad n = v\left(t_1 - t_0 - \frac{x_1 - x_0}{c}\right)$$

The time $T = 1/v$ (the *period*) is the time necessary for one wave to pass any point on the x-direction.

Assume now S_0 to be "at rest," but let S_1 move with respect to S_0 with a velocity V (Fig. 7.22). The observer O_1 at S_1 will observe n waves in the

Fig. 7.24 *The Doppler-Fizeau effect. The motion of the wave-emitter is defined with a velocity V along the axis BA. The frequency, as observed at A and B, is as computed in the text. At A, the velocity V of the emitter and that of the motion of the waves W (larger than V here) are*

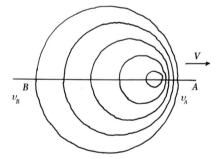

substracted. Hence, the frequency is increased, the wavelength decreased (*blueshift* in the case of light waves). At B, the velocity of the emitter and that of the waves must be added. Hence, the frequency is decreased, the wave-length increased (*redshift* in the case of light waves).

train of waves, with a frequency v and a velocity c. Another observer O'_1, moving in the x-direction with a velocity V, measures the same number of waves, of course. However, he perceives another frequency v' and another velocity c'. At the time t_0, the waves reach x'_0, at t_1 they leave x'_1. Therefore:

$$n = v' \left(t_1 - t_0 \frac{x'_1 - x'_0}{c'} \right)$$

By equating the two values of n ($n = 4$ in the figure, inspired by M. Born), assuming the origins $x=0$ and $x'=0$ coincide at time $t=0$ and assuming the two observers have chosen to observe at the same time $t_1 = t_0$, then one easily finds

$$v/c = v'/c'.$$

The observation may be made at a fixed space point, in the moving system $x_1' = x_0'$. Then the *Galilean transformation* (Fig. 7.23) can be applied (*of course, it is here that an error is introduced with respect to Special Relativity*), resulting in

$$v = (1 - V/c)\, v',$$

V being an algebraic quantity, positive or negative. This equation is the expression of the classic Doppler effect (Fig. 7.24), provided the orientation of motion with respect to the chosen axes is well defined. Note the intervention of a factor we shall often see, the ratio $V/c = \beta$. Here, because of the Galilean transformation: $c' = c - V$.

7.7.2
The Convection of Light By Matter (or Ether) from Römer to Michelson

When measuring the velocity of light, if we admit a priori the validity of the Galilean transformation, the relations above are valid. However, the difficulty of the measurement influenced physicists to use differential methods, hence to look for the second terms or the later terms of the development of the observed frequency in terms of the difference between it and the frequency of the moving source.

  Let us see, indeed, what occurs using different methods.

(i) Let us look at Römer's method of determining the velocity of light, using the observations of Jupiter's satellites.

We admit that the velocity of light is c in astronomical space, and $c' = c - V$, if V is the velocity of the observer towards the source of light (or vice-versa). Look at the astronomical situation (Fig. 7.11, p. 390). When Jupiter is at A, the eclipses of the satellites are displaced by $t_1 = L/(c + V)$, where L is the diameter of Earth's orbit. When Jupiter is at B, the delay is $t_2 = L/(c-V)$. What is measured is indeed the difference:

$$\Delta t = t_2 - t_1 = L\left(\frac{1}{c-V} - \frac{1}{c+V}\right) = 2\frac{LV}{c^2 - V^2} = 2\frac{LV}{c^2(1-\beta^2)}$$

Let us neglect β^2 with respect to unity, and introduce the time $t_0 = L/c$. If the solar system were entirely at rest in the ether, we could write:

$$\Delta t = t_2 - t_1 = 2\frac{LV}{c^2} = 2\,t_0\,\beta \quad (\textit{of first order in the small quantity } \beta).$$

This allowed Römer to determine c, knowing (from Kepler's laws, etc.) L and V relatively well. If we wished to obtain a better determination of c (terrestrial experiments?), it would be necessary to correct V, the difference being the motion of the planetary system with respect to the ether.

(ii) Terrestrial methods of determining the velocity of light.
In these experiments (Foucault and Fizeau), the light crosses a path from A to B, then to A again, which is the same in both directions. The motion of the Earth in the ether is not of concern, and cannot be detected this way, since the only velocity that can be measured is an "average velocity" over the totality of the path. However, the velocity V of the Earth with respect to the ether, is implied.

We can actually compute this, in somewhat more detail.

Along the path AB (measured by L) the time required for the light to travel is $L / (c-V)$. Along the path BA, the time required is $L / (c + V)$. The total time is

$$t = L\left(\frac{1}{c+V} + \frac{1}{c-V}\right) = \frac{2Lc}{c^2 - V^2} = \frac{2L}{c(1-\beta^2)}.$$

So the mean velocity $2L / t$ is $c(1-\beta^2)$, which differs from c by a *quantity of the second order in β, not detectable.*

(iii) The velocity of light in a refracting medium.

Let us now look at the experiments of Hoek, and Fizeau, and the Fresnel interpretation. We shall consider the matter from a slightly different point of view.

Let us return to the Hoek experiment (Fig 7.6, p. 385). Let L be the length of the water tank. Let the velocity of light *in the water* at rest be c_1, such that $c/c_1=n$ is the refraction index of the water. But if the light is dragged at least in part by the water, the velocity of light in water will be $c_1 + \phi$. The light along the water path A requires the time $L /(c_1 + \phi - V)$, influenced by the *convection coefficient* ϕ. The light along the air path, in the other direction, will take $L / (c + V)$. On path B, crossing the water in the opposite direction, the time will be: $L / (c_2 - \phi + V)$, and that crossing the same length in air will be: $L /(c - V)$. The experiment showing no difference in the time of travel whatever the orientation of the instrument, requires the equality of times of both the A and B paths, hence:

$$\frac{L}{c_1+\phi} + \frac{L}{c+V} = \frac{L}{c_1-\phi+V} + \frac{L}{c-V}.$$

Neglecting the terms of second and higher order, we find the Fresnel formula:

$$\phi = c\left(1 - \frac{1}{n^2}\right).$$

To link this with the "equation of state" of matter, remember the classical physical law according to which, in any elastic medium: $c^2=p/\varrho$ (p pressure, ϱ density). In ether, therefore, $p = c^2\varrho$. In matter, $p_1 = c_1^2\varrho_1$.

Let the velocity of light in the water at rest be c_1, such that $c/c_1=n$ is the refraction index of the water. But if the light is dragged at least in part by water, the velocity of light in water will be $c_1 + \phi$. The Fizeau experiment (Fig. 7.10, p. 390) shows that this is not the case.

(iv) The Michelson-Morley experiment.

Could we, by measurements such as the ones mentioned in (i), actually measure the quantity β in spite of its very small value?

It is very difficult! ... hence the ingenuity of Michelson and Morley (Fig. 7.12, p. 391–393). They imagined a system in which two rays of light are considered. One crosses a path A (backwards and forwards), as earlier. The other crosses a path B, of the same length L, but perpendicular to the Earth's orbit, hence to the Earth's velocity. During time t, the

Earth moves from M to M'. The time t is needed to cover the distance MN' crossed by the light in ether. Therefore, MN' = Vt, and

$$c^2 t^2 = L^2 + V^2 t^2, \quad \text{or} \quad t = \frac{L}{c} = \frac{1}{(1-\beta^2)^{1/2}} .$$

This time has to be multiplied by 2 from the path both ways, i.e. MN'M'.

The measured difference between time t_A and t_B or between the double of the computed values is

$$\Delta t = \frac{L}{c} \beta^2,$$

a quantity which depends *directly* upon the *second order of the quantity β*, i.e. β^2, and should allow us to measure it. The accuracy of the Michelson and Morley measurements were sufficient to detect it, but they did not. Their measurement resulted in the value $\beta = 0$. This implied essentially that *the Galilean transformations are not valid, in the case of ether motion, when one looks for the composition of the velocity of light with another velocity. There is no ether wind.* But some more recent measurments, by Morley and Miller seem to show that β, although very small, differs from zero. This is controversial, and not yet interpreted.

7.7.3
The Lorentz Transformations of Coordinates

(i) Transformation of coordinates.
If we work with a rectangular (Cartesian) system of 3 coordinates, we may consider the passage of coordinates of an object in system S into one in system S', when S' results either from a translation or a rotation of system S. Let us assume that S' is moving rectilinearly with respect to S

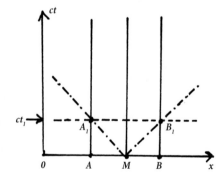

Fig. 7.25 *Lorentzian transformation of coordinates* (see text). S is at rest with respect to S'.

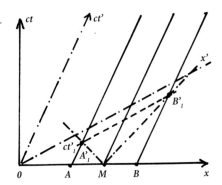

Fig. 7.26 *Lorentzian transformation of coordinates* (see text). S' is moving with respect to S.

with velocity V. We may well choose rectangular coordinates in both systems, so that the direction of motion becomes the x or the x' axis. Let us assume that, at time 0, the origins of the two systems coincide. In time t, the origin of system S' will have to be displaced by $a = Vt$ in the x direction. Then $x'= x-Vt$; $y'=y$ and $z'= z$. These are the *Galilean transformations* (Fig. 7.23).

Let us shift to the 4-dimensional systems of reference. Same thing! The system at rest will be represented by two rectangular coordinates t, x (assuming that the y and z axes are not involved, all motions being assumed to be parallel to the x-axis). The system S' is defined by two non-rectangular coordinates t', x'. The x'-axis coincides with the x-axis. The t'-axis, inclined with respect to the x'-axis, is the world line of the origin of the system of coordinates. This corresponds to the Galilean transformation.

If we have to abandon the Galilean transformations, we must find another system of transformations in which the velocity of light would be invariant, in which there would be no ether wind. We must abandon the idea that the x-axis and the x'-axis must necessarily coincide. In essence, we have already said that in different language when speaking about the impossibility of an absolute simultaneity. Let us still restrict ourselves to motions parallel to the x-axis (Fig. 7.25).

Points A, M, B (M in the middle of segment AB) are at rest on the x-axis (Fig. 7.25). In the x, t system S, their world-lines are represented as three parallels to the t-axis. In the system S, the light signal is moving both ways from M with velocity c. It is, therefore, represented by the straight lines $t = \pm x/c$. For convenience, we will give these straight lines slope unity, which means that the unit of the ordinates is not t, but ct.

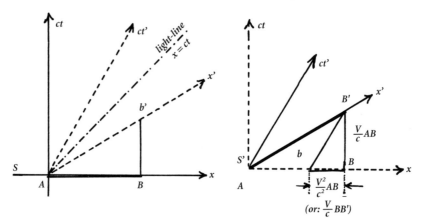

Fig. 7.27 *The measurement of a rod of length unity in the two systems* S *and* S', *in relative motion with respect to each other* (see text)

These straight lines are the "light world-lines." Note that the t values of the points of intersection A_1 and B_1 of the light lines with the world lines of points A and B are the times t_1 at which a signal emitted by the light-emitter M reaches the physical point-observers A and B.

If we now assume the three points A, M, B to move uniformly, like the system S', with the same velocity V in the x direction (Fig. 7.26), their world lines are parallel but inclined with respect to the x-axis. As in Fig. 7.23, they are parallel to the time axis, now ct'. The light signals are represented by the same light-lines, having proceeded from M. But the points A'_1 and B'_1 do not lie on a parallel to the x-axis. The orientation of this segment defines the x'-axis. They are not simultaneous in the system S (x, ct), but they are simultaneous in the system S' (x', ct'). In this case, both axes are inclined with respect to the original axes of the system S. This results uniquely from the definition of simultaneity, which corresponds to the physics of Special Relativity.

(ii) The Lorentz transformations.

If we now apply the principles of Special Relativity, let us consider two systems S and S', in a relative motion of translation with respect to each other, in the direction of the x-axis. Let the other axis be ct, leaving aside the y and z axes perpendicular to the motion. Assume the segment AB to be the representation (Fig. 7.27) of a rod of unit length, at rest in S. The world-lines of this rod with ends at A and B are the ct axis and the paral-

Fig. 7.28 *Same as Fig. 7.26, for the time-coordinate* (see text)

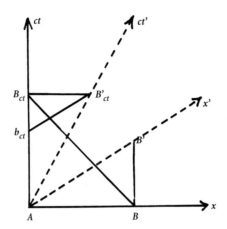

lel line through B. This line intersects the x'-axis at b'. The world-lines of the same rod at rest in S' will be the ct' axis and a parallel line through a point B' on the x' axis. The segment AB' represents the unit of length in the system S'. The world-line through B' cuts the x-axis at b. In essence, since we made A coincide with the origin of both coordinate systems S and S', we can use geometrical considerations rather easily.

The "meaning" of AB' is the following. An observer at rest in S', who wants to measure the length of the unit rod at rest at S, will find as a result of a "simultaneous" observation (from his point of view), points A and B' for the end-points of the rod. However, S is moving with respect to to S'. Therefore, we must have a simultaneous observation made by an observer at rest in S. Since, in the S system, the unit is AB, the result of the S' measurement is the Ab'/AB part of the unit at S. The observer at S

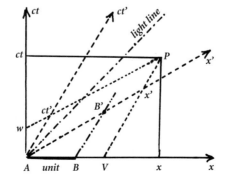

Fig. 7.29 *Lorentzian transformation of the coordinates of a point* P (see text)

would find Ab'/AB' instead. The same remark, of course, can be applied to the measurement of Ab at S: Ab/AB is now the factor relating the two measurements.

According to the principle of relativity, the two systems are equivalent. Therefore, the relative changes of measurements have to be equal:

$$\frac{Ab'}{AB'} = \frac{Ab}{AB} \qquad \text{or} \qquad AB \cdot Ab' = AB' \cdot Ab \qquad (\alpha)$$

This relation permits the construction of point B'.

The unit of time in the S' system $AB_{ct'}$ can be constructed in a similar way from $ABct$, the segment AB'_{ct} defining the unit of time in S. This is done in Fig. 7.28.

Clearly, from Fig. 7.27 and 7.28:

$$Ab'^2 = AB^2 (1 + V^2/c^2), \qquad (\beta)$$

$$Ab^2 = AB^2 (1 - V^2/c^2), \qquad (\beta)$$

We are now able to transform x and ct of any world-point P in the system S into the coordinates x' and ct' of P in the system S'. Fig. 7.29 explains how this can be done.

First, $x' = Ax'/AB'$, where Ax' is the length measured in a given unit U (a meter for example) and x' the abscissa of P in the system S. Also, $x = Ax/AB$, or $(x - Vt) = AV / AB$. The geometry of the figure leads to

$$\frac{x}{x-Vt} = \frac{Ax'}{AV} \frac{AB}{AB'} \qquad \text{and} \qquad \frac{Ax'}{AV} = \frac{AB'}{Ab} \cdot$$

Using the relation (γ) written above, one can write

$$\frac{x'}{x-Vt} = \frac{AB}{Ab} = \frac{1}{\left(1-V^2/c^2\right)^{1/2}}$$

Similarly, one can write the same relation for the time coordinates:

$$\frac{ct'}{ct-Vx/c} = \frac{AB}{Ab} = \frac{1}{\left(1-V^2/c^2\right)^{1/2}} \cdot$$

If we complete these two relations by the relations $y'=y$ and $z'=z$, we obtain four relations, which constitute the well-known *Lorentz transformations*, which permits the calculation of the coordinates of any world-point P in S' from those in S. Their usual form is

$$x' = \frac{x-Vt}{(1-\beta^2)^{1/2}}; \quad y' = y \; ; \; z = z'; \; t' = \frac{t-Vx/c^2}{(1-\beta^2)^{1/2}}.$$

There are many ways to derive these relations, many ways to invert them (reaching relations giving x, y, z, t as functions of x', y', z', t'). A consequence of these relations is the derivation of the Lorentzian law of composition of velocities. Without going into the demonstration, we can write this law as follows (compare with Fig. 7.23):

$$U = \frac{V+W}{1+VW/c^2} \quad \text{or, if } W = c, \text{ and } U=c': \quad c' = c,$$

a result which could be expected, according to the basic principle of *conservation of the velocity of light*.

Note that, when leaving aside $\beta = V/c$ in the Lorentz equations (which means giving to c an infinite value, thus admitting the instantaneous transmission at a distance of light – and gravitation), we found the formulae of the Galilean transformations.

7.7.4
The Cone of Light

In a system S (x, y, z, t), the locus of all light-lines is a cone of aperture 45°, if the units on the $x\,y\,z$ axes are the same and if the unit on the t-axis is c (or if this axis represents, on the same scale as x, y, z on the x, y, z axes, the quantity ct.

This is difficult to represent in a 4-dimensional space, but in a 3-dimensional space x, y, ct can be represented more easily (Fig. 7.30).

Let us note, therefore, that all events outside the cone are not linked causally to the event in O, at the origin of the coordinates. The upper part of the cone represents all the possible "futures" of O. It contains points that may receive a light signal from O. All points contained in the lower part of the cone can have sent signals to the event O. This is the "cone of the past." Hence, causality relations may occur between related events in the cone of the past and the event in O, or between the event in O and the points in the cone of the future. Points outside the cone cannot have any relation with O, hence no causality relations. Modern cosmologists often use this notion of a light cone in describing their theory of the Universe.

Fig. 7.30 *The light cone of point* O. Inside the cone, above the plane *t=0,* the future; below, the past; outside the cone, the unknowable

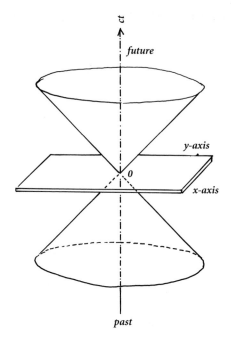

CHAPTER 8

Cosmologies of Today and Tomorrow

8.1
General Relativity

As we have seen, the cosmological theorizing at the beginning of the 20th century involved a discussion of Newtonian cosmology, and the concepts of ether and action at a distance. Einsteinian Special Relativity (1905), which provided a satisfactory reply to both questions, was, however, limited to uniform motions, and to linear motions of systems of reference with respect to one another.

There was at that point no solution to the problem of how to generalize the idea to systems of reference which could be affected by acceleration (hence by gravitation), and which would not be moving in a straight line (for example, in an astronomical context). Again, it was Einstein who responded to the challenge in the second step (1916) of his thinking, the theory of General Relativity (abbreviated hereafter GR).

As we have indicated earlier, the GR is based on two principles, the principle of invariance and the principle of equivalence.

8.1.1
The Principle of Invariance

This principle states that the general laws of physics can be expressed in a form that is independent of the choice of space-time coordinates. It extends the principle of the constancy of the velocity of light to all physics. This has many unexpected consequences; for example, we could not use Newton's law of attraction as an appropriate axiom of the GR, as its *expression* in an "invariant" language would be undoubtedly very complicated.

The principle of invariance imposes a suitable mathematical language, called *tensorial analysis*. We cannot go into the difficult details of the mathematics of how to obtain invariant expressions for the laws of physics. We can only say that they imply the use of tensorial calculus (see Fig. 8.8, p. 526), in which all quantities are expressed in the form of a 4-dimensional tensor. (A 4-dimensional tensor can be conceived of as more or less equivalent to a vector in Euclidian space, but a tensor is indeed a much more general concept, which does not necessarily reduce to a 4-dimensional vector.) Essentially, tensors permit the construction in space-time coordinates of invariant expressions which will not change with a change of coordinates, just as the length and orientation of a vector does not change with a change of coordinates in 3-dimensional Euclidian space.

More precisely, the "interval element" ds, or better its square, ds^2, is a very useful notion (equivalent to the square of the length of a vector). In order to give a positive sign to the square of the velocity of light, our ds^2 will become, instead of the expression used earlier in this book (p. 404–406, where for practical reasons it is given erroneously): $ds^2 = - dx^2 - dy^2 - dz^2 + c^2 dt^2$.

This is not, let us note now, the more general formula for a ds^2. It corresponds to Euclidian geometry and to the obvious use, in 4-dimensional space-time, of Pythagoras' well-known theorem. A more general formula is the general tensorial one, x^u, x^v... being the coordinates (u and v are, of course, not powers, but *indices* put above the line!):

$$ds^2 = \Sigma\, g_{uv}\, dx^u\, dx^v,\text{ noted by simplification } g_{uv}\, dx^u\, dx^v.$$

The simplified (quasi-Euclidian) expression which we had written before corresponds to $g_{11} = g_{22} = g_{33} = -1$; $g_{44} = c^2$; $g_{uv} (u \neq v) = 0$.

The form of the ds^2, its choice, is a basis for any cosmology, the geometry of which is defined by it. For that reason, we list in Sect. 8.8 a few definitions and properties related to tensors, and to the forms of ds^2.

8.1.2
The Equivalence Principle

Let us come back, once more (still following Max Born's clear exposition of these questions), to Newton. Newton, as we have said, conceived the idea of an absolute space and an absolute time. This was based on the existence of inertial resistances and centrifugal forces (as exemplified in Newton's description of the bucket, discussed at length by Mach, see p. 377). These inertial and centrifugal forces, in Newton's conception, do

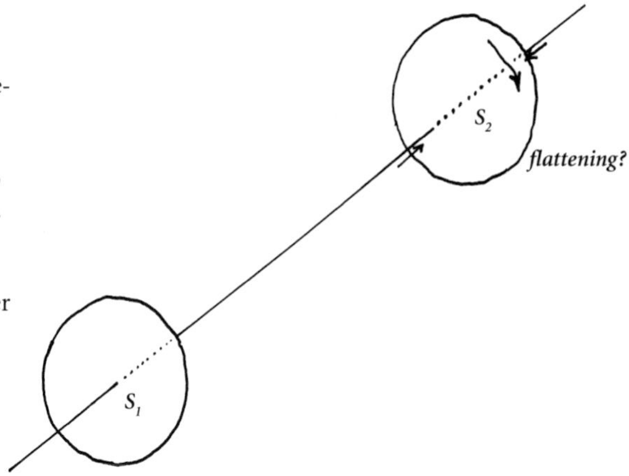

Fig. 8.1 *Two spheres in a vacuum.* Sphere 1 is assumed to be fixed with respect to an "absolute space." Sphere 2 is assumed to rotate, hence to flatten, but this is untrue. From both spheres, the other one appears to rotate. We cannot link either sphere with a particular absolute system of reference.

not depend on interactions between bodies, but on absolute accelerations. *Absolute space* is then a sort of fictitious cause of physical phenomena. The example of the water bucket, used by Mach, shows how unsatisfactory we now find some features of Newton's theory. Another example, used by Born, is perhaps still more straightforward (Fig. 8.1). Let us consider two bodies at rest, S_1 and S_2. They are of the same size, and spherical shape, and made of the same fluid matter, able to change its shape. They are at a distance from each other such that the ordinary gravitational effects of the one on the other are inappreciably small.

Each of these bodies is at equilibrium due to the force of gravitation of its own parts on each other, and the remaining physical forces (of an electromagnetic nature, for example), so that no relative motion of the parts of one with respect to the other occurs naturally. Still, each is forced to execute a relative motion of uniform rotation with a constant angular velocity around the line connecting their centers. This signifies that an observer on body S_1 notes a uniform rotation of body S_2, with respect to his own system of reference and vice-versa. Now, suppose that each of these observers determines the shape of the body on which he stands. Suppose they both found that S_1 is a nice sphere, and that S_2 is flattened by the effects of centrifugal forces. Newtonian mechanics would conclude that the difference in shape of the two bodies is caused by the fact that S_1 is at rest in "absolute space," and that S_2 executes an "absolute rotation." As S_1 cannot be considered responsible for the flattening of S_2, it is clear that absolute space has been introduced as a fictitious cause. In any case, this does not satisfy our logic with regard to causality.

As Born says: "The concept of absolute space is almost spiritualist in character." We might say Platonic.

Thus, we reject the idea of absolute space as the cause of the different shapes of S_1 and S_2. Let us assume that no other material body is present in space. The different shapes would then be completely inexplicable. But is this different shape an empirical fact? Of course not. We have never observed two bodies alone in the Universe! The only possible cause, actually, of the different behavior of S_1 and S_2 is the presence of very distant masses. This is indeed the Machian point of view. The new requirement, therefore, should be that the laws of mechanics, and of course of physics in general, involve only the *relative* positions and motions of bodies, if only because we do not know the Universe as a whole, and we cannot define the Machian system of reference. No system of reference can be a priori favored any more. Otherwise, absolute acceleration with respect to these favored systems of reference, and not only relative motions of bodies, would enter into the physical laws. This is a considerable extension of the principle of relativity, as expressed on p. 399. Let us remember that, by limiting the motions to uniform motions, Special Relativity ignored accelerations, as it ignored gravitation.

To fulfill this postulate, a reformulation of the *law of inertia* is necessary. Inertia cannot be an effect of absolute space, but of all other bodies, known and unknown, in the Universe. We know, however, one interaction, and only one, between material bodies at the scale of the Universe, namely gravitation. We also know (Oetvös, Dicke) the equality between gravitational mass and inertial mass, which, although postulated a priori, never failed. This gives the gravitational constant G, measured by Cavendish, the status of a fundamental constant of physics (just as we have already established in the case of c). So it seems that inertia and attraction must have a common root. Einstein proposed that there is no way to distinguish between a motion due to some gravitation-identified effect and a motion due to some other kind of acceleration, such as centrifugal force, for example. This is not as obvious as it may seem to us now! We can conceive, for example (Fig. 8.2), that in an accelerated rocket, the acceleration appears to all inhabitants of the rocket as if it were due to the attraction of the gravitational field of some heavy body located under the rocket (such as the Earth, for a rocket at rest).

Now we are confronted, however, with a basic difficulty. What geometry shall we use? This is where the "invariance principle," already expressed, will enter the picture. In a 4-dimensional space-time system of

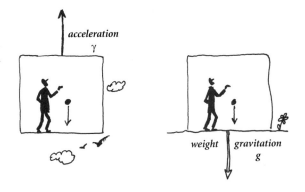

Fig. 8.2 *The equivalence principle.* A man in an elevator is subject to the acceleration γ of the elevator. He is subject to the same feelings and phenomena as a man in a non-moving elevator in the field of gravitation *g* of the Earth, provided that, numerically, $g = \gamma$.

coordinates, a *world point* represents an "event," and a *world line* is the graphic picture of a moving point, i.e. the representation of the evolution of the "event." This is a notation introduced in 1908 by Minkowski. It gives the word "event" a limited meaning (see Fig. 7.17, p. 403). The world line imposed by inertia is only a straight line in Euclidian space; but as we can see (Figs 7.5, p. 380, and 8.3), lines that are straight in an inertial system may be curved in an absolute system, if we follow the classical terminology. In the new mechanics, this should not occur. Therefore, a new geometry has to be adopted. It cannot be classical Euclidian geometry.

We must consider the possibility that space is curved, as we have already said in chap. 7. To operate correctly in that type of geometry, we must develop four-coordinate systems of reference, and a means for measuring the distance between two world-points, the length of a world-line. The ds^2 already introduced defines the *line element,* or *metric* of the space. Knowledge of the $g_{\mu\nu}$ really defines the way to determine an "interval" (the word "interval" being less misleading than the often-used word "distance") in space-time, the tensor $g_{\mu\nu}$ actually defining the "geometry." There are no straight world-lines in a curved space-time (except very locally); but between two world-points, there are, if not straight, more "direct" lines, the "geodetic" lines, for which the interval ds between two world-points is minimal.

The *curvature* changes from point to point in space-time. We can illustrate this notion (very qualitatively!) by representing 4-dimensional space-time by either a 3-dimensional space-time (*x, y, t*) or even by a 2-dimensional space (*x, t*). Without embarking on a lengthy mathematical excursion, we can refer to Fig. 7.16, p. 402, which shows various aspects

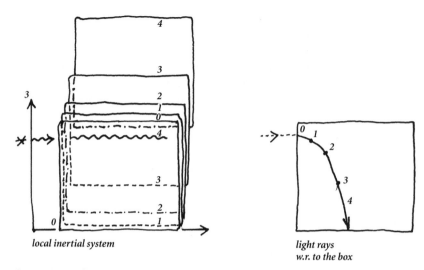

local inertial system

light rays
w.r. to the box

Fig. 8.3 *Inertial motion.* In a local inertial system, a system of reference (say: a box) occupies positions 0, 1, 2, 3, 4, at successive time intervals. A ray of light is sent through that box throughout its motion. With respect to the moving box, the light ray appears bent, its shape being determined by the motion of the box.

of the geometry we have just outlined, based on the principle of invariance and the principle of equivalence – in other words a "physical geometry." One important factor is part of the geometry, and indeed essential, to Einstein's concepts: *the distribution of mass is closely linked to the geometry of space-time, in a very strong bond.* The presence of an important mass creates in itself, or, better, is in essence identical to, an important curvature (or a small radius of curvature). The distribution of curvature in 4-dimensional space-time is described by another tensor, the *Riemann-Christoffel* tensor $R_{\mu\nu\varrho\lambda}$, or in its condensed form, the Ricci tensor, written $R_{\mu\nu}$. In other words, the metric tensor $g_{\mu\nu}$ and the tensor $R_{\mu\nu}$ reflect the mass distribution. In a flat space, empty, i.e. with no mass, the appropriate metric tensor is the one we have written above, completely diagonalized, and the Ricci tensor is equal to zero. The *energy-impulsion tensor,* another table of 16 terms, represents the aggregate of causes (mass distribution, energy distribution, momentum distribution) which act upon the geometry of the space. It is designated by the symbol $T_{\mu\nu}$.

It is clear now that the usual equations of classical Galilean-Newtonian mechanics, (a) $F = m\gamma$ (the equation for the dynamics of a moving point), and (b) $F = G \, Mm/r^2$ (universal attraction) have to be replaced.

In essence, they associate the value of the force linked to the gravitational field, (b), and the motion given to a body by this force (a). Instead of these local equations of motion, there is a preference in classical mechanics for a global formulation which defines the fields of force and their effects at the same time. The classical formulaton of the field equation is the *Poisson equation*, widely used since the beginning of the 19th century: (c) $\Delta\Phi = -4\pi\varrho$, where ϱ is the density of matter, at any point, $\Delta\Phi$ the *Laplacian operator*, $d^2\Phi/dx^2 + d^2\Phi/dy^2 + d^2\Phi/dz,^2$, and Φ the gravitational potential function from which the gravitational field can be computed (by the relation (d: $F = -$ grad Φ). In this classical vectorial representation, only vectors are used. The relativistic equivalent of the Poisson equation is much more complicated. It involves tensors instead of vectors:

$$R_{\mu\nu} - (1/2)\,\mathcal{R}g_{\mu\nu} = -\,G\,T_{\mu\nu}.$$

The Einsteinian equivalent of the equations of motion (a) and (b) are now the equation of *geodesics* (i.e. of the lines representing the motion of particles in space-time).

Here the term $\mathcal{R}$ represents an invariant expression derived from the Riemann-Christoffel tensor, or in other words, the local curvature of space-time, a physical quantity. This equation may look rather simple, but it is not. It implies as many equations as there are terms in the tensors involved.

For reasons which we will revisit below, Einstein added a term $\Lambda g_{\mu\nu}$ to the first term, where Λ is the so-called *cosmological constant*.

This equation is indeed a tensorial equation, therefore equivalent to 10 equations (not 16, because of some relations between $g_{\mu\nu}$, $R_{\mu\nu}$, and $T_{\mu\nu}$). Still, it is simplified, as the more general form of the Riemann-Christoffel tensor is much more complicated by far!

It is no wonder that in cosmological applications the various authors, starting with Einstein, have tried to simplify this tensorial equation, as we shall see in due course. Naturally, all these mathematical developments may be physically wrong, or over-simplified; but they are quite rigorous from the mathematical point of view. They satisfy very obvious requirements, as follows.

When still further simplified: (1) If $G=0$, the tensorial equation practically amounts to ordinary Special Relativity. This is no longer an equation concerning the gravitational field, but only a geometrical construction permitting us to change coordinates. (2) If $c = \infty$ (as c enters in $g_{\mu\nu}$), one is left with Newtonian physics.

General Relativity appears then as a very general theory, embracing classical Newtonian physics as well as Maxwell's electrodynamics, in one single (and simple) tensorial equation.

Hereafter, we shall not explicitly use the Einstein equations of motion except in a significantly simplified form. A great deal of ingenuity has gone into providing these simplified solutions for application in some specific cases, as Einstein's formulations constitute really complicated systems of equations, when written in their more general form.

8.1.3
The Einsteinian Cosmological-Local Facts

On a small scale, space is flat (as a plane is tangential to a curved surface in ordinary space). The masses of the planets and of the Sun are superimposed in a dominating way on the gravitational field of the Universe and the "flat Universe" is locally a good approximation, just as Newtonian mechanics is an excellent approximation of relativistic dynamics.

Ordinary Newtonian mechanics is valid then, except when we apply it to motions in which relative velocities are close to the velocity of light (for example, in modern particle accelerators).

There are still, however, phenomena, observable on a small scale on Earth and in the solar system, which are contradictory to Newtonian mechanics.

(i) The advance of Mercury's perihelion.
We have already mentioned the behavior of the perihelion of Mercury, which turned out to be different from the prediction based on Newtonian theory (Fig. 7.1, p. 373, and Table 7.1, p. 374). We have not yet mentioned two other facts predicted by the theory, the deflection of light in the vicinity of a massive body, and the redshift imposed upon spectral lines in a gravitational field.

These three phenomena have been considered the key observational facts supporting the Special and General Theories of Relativity.

In Einstein's conception, the motion of a planet around the Sun is indeed a geodetic world-line in space-time. The calculations (solutions of equations of motion, sufficiently and conveniently simplified) show that the excess of 43", displayed by the observed motion in longitude of Mercury's perihelion with respect to the complete Newtonian theory (as shown by LeVerrier), can be obtained by the GR, which finds $43.03" \pm 0.03"$. This confirmation has been reinforced by the careful study of artificial satellites. In fact, at the level of accuracy at which astrometric observations in the solar system are undertaken at the end of the 20th century, it is necessary to use GR calculations in all cases of planetary motions within the solar system.

(ii) The gravitational redshift.

We know few phenomena which displace a spectral line, towards either the longer wavelengths and shorter frequencies (redshifts), or towards the shorter wavelengths and higher frequencies (blueshifts).

Very well-known is the Doppler-Fizeau effect, which applies to all waves, be they light waves or sound waves, of which the emitter at rest has a given frequency v_0, and a given wavelength λ_0. Fig. 7.24, p. 416, makes perfectly clear that, when a source of light is approaching the observer, there is a blueshift; when the source of light is receding away from the observer, there is a redshift. If v is the observed frequency of the source, and $\lambda = c/v$ its wavelength, it is easy to show the modified frequency in the observer system. Noting positively the velocity V away from the observer and the displacement of wavelength towards the higher wavelengths (redshifts) (a standard convention in an astronomical context), one finds that

$$v = v_0 \, / \, (1 + V/c) \qquad \text{and} \qquad \lambda = \lambda_0 \, (1 + V/c).$$

Einstein's theory predicts that a redshift of spectral lines must also be observed in a gravitational field. On the surface of a star, there is indeed a strong field of gravitation, because of the large mass of the star and its relatively small radius. The result of this strong gravitational field is to affect the metric, and to slow a clock with respect to what it is on Earth where the field is much smaller. It happens that there are some clocks on the stellar and solar surfaces. They are precisely the atoms and molecules of gases responsible for the absorption lines of the solar (or stellar) atmosphere.

Let us consider one clock K in a field-less region of space, at a distance h from observer O; T_0 is its common characteristic period, $v_0 = 2\pi/T_0$ its characteristic frequency. The clock K is subjected to a constant acceleration with respect to O, γ, in the direction of the line OK (Fig. 8.4).

Fig. 8.4 *The slowing of the clocks.* Clock K is accelerated with respect to observer O. It is slowed down accordingly (see the text). Its signal is "redshifted" by the acceleration.

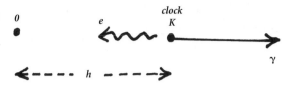

At the end of each period , the clock K sends a signal to O, which takes the time $t = h/c$ to reach O, at a state of rest. But during time t, the clock K is not at rest. It will have changed its velocity with respect to its initial velocity: $V = \gamma t$. Then, the ordinary Doppler effect would result in a shift in the observed time-lag, and the apparent period of clock K would be (with the same sign convention as above):

$$T = T_0 (1 - V/c) = T_0 (1 - \gamma h/c^2) \qquad \text{(a redshift)},$$

a formula which links the change in velocity to an acceleration γ instead of to a velocity V.

Let us now apply the principle of equivalence. The same phenomenon is observed in a gravitational field. For an observer located at the center of the Sun, the clock on the surface of the Sun, even at rest, is accelerated towards the observer. The acceleration is GM_0/R_0^2 (based on the equations of classical Newtonian mechanics: $F = m\,\gamma = G\,mM_0/R_0^2$), where M_0 and R_0 are the mass and the radius of the Sun. But if the distance h is equal to R_0, as the observer is at the center of the Sun, the "shift" becomes

$$\frac{T - T_0}{T_0} = -\frac{V}{c} = -\frac{\gamma h}{c^2} = -\frac{G(M_0/R_0)}{c^2}.$$

or, in wavelengths

$$\frac{d\lambda}{\lambda_0} = +\frac{G(M_0/R_0)}{c^2} \qquad \text{(a redshift)}.$$

On the Sun, this dimensionless ratio can be computed as equal to 2.12×10^{-6}. This was observed at the solar limb in the solar spectrum during the years 1940–60, first by Adam, and finally by Roddier who compared solar wavelengths of selected well-defined lines with Earth-based spectrograph spectral lines, using the resonance line of strontium as a scanning device of the solar spectrum.

On Earth, the ratio has also been measured by observing an emitter-clock put either at the top or the bottom of a tower. This is an extremely difficult measurement to make, as spectral lines are not narrow enough in the solar spectrum. On Earth, the effect is much smaller than it is on the Sun.

We cannot forget the existence of the Doppler effect, just because we are operating within the "new universe" of relativity physics. The new effect is, in essence, a second order one, linked to acceleration (or to gravitation, as implied by the principle of equivalence), but not to velocity.

It is well to remember in this context that the Doppler formula was written for weak velocities, or weak accelerations. It is possible to write a new formula at a much greater level of generality, which of course is a function of the metrics adapted to any specific problem. We shall not, however, present the generalized Doppler formula here.

At this stage, we should say that there are no other well-accepted explanations for redshifts except the Doppler effect, gravitational effect, and, in a different physical context, the Compton effect. We shall come back to this much debated question in sections 8.2.3 and 8.3.1.

(iii) The deviation of light by massive bodies.

Even in the Newtonian case, one can predict a deviation of light rays. This can be observed, for example during an eclipse, near the limb of the Sun. Let us assume a particle travelling at a uniform speed V in the direction of the y-axis, passes a mass M at a distance $x = h$ (Fig. 8.5).

The acceleration produced by the gravitational force of that mass M is (a very elementary and classical Newtonian expression) $\gamma = GM/(x^2 + y^2)^{1/2}$. This enables us to write and solve the equations of motion. For large values of y (long after the closest approach), one can write the trajectory easily. It amounts to a deviation; the angle of deviation is equal to $\theta = 2\,GM/c^2 h$. This value is independent of the mass of the moving particle; therefore, it applies to a photon. This remark was made as early as 1801 by Soldner, in the framework of Newtonian mechanics.

The relativistic computation is quite different. It works with equations similar to those used in the computation of the motion of the perihelion of Mercury; we shall not perform it here. The main difference is that the deviation computed according to General Relativity is *double* the Newtonian one, namely

$$\theta = 4\,GM/c^2 h.$$

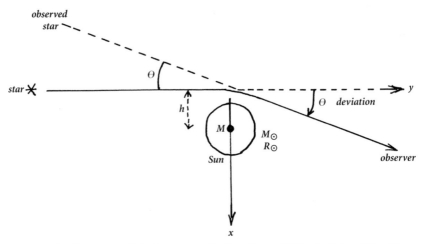

Fig. 8.5 *The deviation of light by massive bodies (Sun …).* The deviation θ is a function of the distance h of "closer approach" of the straight motion. If the light rays are deviated, the source appears to the observer as located in another direction. In the case of the Sun, and just near its limb, $h = R_0$, the deviation is $2\ GM_0/R_0c^2$ (in the Newtonian computation), and $4\ GM_0/R_0c^2$ in the relativistic computation.

One way to check this is by comparing the positions of stars near the Sun (as they are observed during a total eclipse), and the positions of the same stars six months later, when the Sun is in the opposite part of the sky. This is quite a difficult measurement, as the two observations are performed one in daylight, the other at night. The temperatures will be different and the photographic determinations depend upon temperature, as the instrument itself changes (slightly, of course, but the measureable effect is indeed very small!).

Actually, this test has been done. On May 29, 1919, during the total eclipse on the west coast of Africa, and in northern Brazil, a British expedition on the initiative of Eddington made the measurements. In November 1919, the final result was proclaimed, and gave a displacement, at the limb of the Sun ($h = R$), of 1.75 seconds of arc, as predicted by Einstein. This was the triumph of Einstein's theory, a historical signpost quite comparable to the return of Halley's comet, and later confirmed by many other eclipse expeditions (Van Biesbroeck). From that date on, only a very few scientists continued to doubt the Einsteinian concepts.

Let us note that the two relativistic effects *(ii)* and *(iii)* are linked one with the other, such that:

$$\theta = 4\ (G/c^2)M_0/R_0 \quad \text{and} \quad d\lambda/\lambda_0 = G(M_0/R_0)/c^2.$$

A single parameter M_0/R_0 determines the importance of both effects. This may have applications in the case of cosmological masses of simple shapes and structures.

8.2
The Cosmological Solutions of General Relativity

It is interesting to note that, when moving from the physics or astronomy of close-by phenomena to considerations of the *Universe as a whole*, we are crossing a sort of epistemological frontier, entering, so to speak, the "Terra Incognita" of the World. (We use here the old word of Newton and Laplace, the common terminology before modern astronomers substituted for the word "World," the word "Universe" to designate the totality of that which exists. Scientific professionals generally disdain the use of the word "cosmos," which has several mystical connotations.)

No doubt, the concepts of Plato, Ptolemy, Copernicus, and Kepler, were "cosmological" in nature, concerning every part of the Universe then conceivable. On the other hand, Aristotle, staying close to the observed facts, just laid down principles and constructions, as did Newton (although both were interested and seemed involved in some general view of the Universe). Tycho and Herschel, as well as many others, more or less limited their work to the investigation of the observable part of the Universe, without speculating too much about what might lie beyond the observed part.

So far, we have seen Einstein acting in a perfect Aristotelian spirit, trying to limit himself to the observed phenomena, or the observable ones, as possible checks, and basing his philosophy on the famous "σώζειν τὰ φαινόμενα," "*save the phenomena*," which we have shown to underlie the Aristotelian attitude.

Now we shall see Einstein and his successors entering into the field of speculation, emerging from Plato's cavern, and constructing systems of the World, views of the Universe, with either a Platonic tendency (Newtonian, or big bang?) or a Pythagorian tendency (some of the modern cosmologies?). It seems to me a very interesting shift in the philosophy of scientists, perhaps a shift which exists in the mind of every scientist. Perhaps! We must remember Kepler, who, following the opposite path from everyone else and even from some of his own earlier ideas, rejected the Pythagorian point of view of his youth and adhered to the principle of "saving the phenomena" alone without exception.

Newtonian cosmology elevated gravitation (universal attraction) to the highest place in its construction. But observations were limited. The Newtonian extrapolation, hardly expressed in the *Principia*, is that of an eternal, unchangeable world, because of the periodic regularity of all astral motions. This is also a world where God is an essential part; but let us quote Newton himself:

"Since every particle of space is *always*, and every indivisible moment of duration is *everywhere*, certainly the Maker and Lord of all things cannot be *never* and *nowhere*."

Let us note that there is a problem here. How are we to conceive of the Creation by Newton's God? Has eternal time a beginning?

Einstein started from such concepts, at least the concept of infinite duration of the Universe. This was Newtonian and Aristotelian too. It was also in keeping with Einstein's tendencies, already affirmed in the two steps of relativity theory, SR and GR. Entering the field of cosmology, of course, implied, as always, an extrapolation. Basing a cosmology on GR is only a natural construct. It means basing a cosmology on the best theory to explain the nearby part of the Universe, the only one which could describe some newly discovered phenomena. Because of the independence of physical laws, hence of cosmology, with respect to the system of reference, and because of its symmetry, GR is also the most satisfactory theory from an aesthetic point of view. This last point puts great weight on the aesthetic character of the reasoning to be made.

Considering the whole Universe, it will be essential to work within one system of reference, the Machian system, related to the real masses distributed in the Universe. We have to take into consideration one single time, universal time, or cosmic time, valid in this system. We will not need to consider the role of the propagation of gravitation in this space. The geometrical nature of space-time permits us to avoid the problem.

However, as suitable as it may be in principle to treat the cosmology of the Universe as a whole, it is clear that the GR, with its many equations (implicit in equations written above, p. 443, and in the Sect. 8.8) is extremely difficult to handle. Therefore, simplifications will have to be introduced. But how do we know, having observed so little about the real Universe, whether these simplifications are suitable? We know the *local* character of the GR equations well. But any cosmology has to have a *global* character! Some scientists considered that it was quite possible that mathematical integrations and derivations, operations in permanent use in the solution of any system of cosmological equations, are physically valid at all scales. For that reason, and others, many authors have used quite different ways to solve the GR equations, trying to eliminate some of these limitations. True enough, as we shall see, many of the simplifications commonly used were perhaps, and still are, quite unjustified. In other words, we shall see that Occam's razor has cut the hair of Samson, depriving him of all his potential strength.

Moreover the authors, up to the early fifties, assumed the GR equations to be valid whatever the physical conditions, even at very high densities, for example, or very high temperatures. This is far from certain. Other forces might well then overtake the gravitational forces.

Therefore, the ideas we will discuss in this section are only "models." They have only a suggestive value, and they should certainly not be considered either as complete or as final, or even as first approximations. They are only a few simple solutions, out of the infinity of possible models, which are solutions to the complete GR equations.

8.2.1
The Cosmological Principles

The first simplification will be to assume that the Universe has a spherical symmetry. This leads either to a symmetrical treatment of Cartesian coordinates x, y, z, or, even better perhaps, to the use of spherical coordinates, θ, π, r, as described in Sect. 8.8. The time t is not affected. It is "cosmic" time. The "simplified equations", written in spherical coordinates, are explicated in Sect. 8.8. They are already difficult enough, and complex enough to give rise to a multiplicity of cosmologies, those which appeared between 1917 and, roughly, 1950. They imply the recognition of the so-called *"absolute cosmological principles,"* a very misleading name indeed, as these "principles" are only a way to simplify highly complex equations. Their solution also implies the choice of a sufficient number of *"initial conditions."* Some are indeed given by the observations. (The actual average density ϱ_0 of the Universe is an example, although there are even doubts regarding the validity of this notion; the age of the Universe is another example, for which we have at least lower limits.)

The principles can be enunciated as follows:

The Universe is spatially homogeneous.
The Universe is spatially isotropic.

The fact that there is a *unique cosmic time* is inherent in the very concept of cosmology. It is not one of the "principles."

We have considered the introduction of "cosmological principles" merely as a way to simplify the equations. We know quite well that in our vicinity, we are very far from homogeneity and from isotropy. The density in the center of a *neutron star* is 10^{13} g cm^{-3}, whereas the density in the intergalactic medium is 10^{-29} g cm^{-3}, or even less. As far as isotropy is concerned, at the scale of our Galaxy, it is not present. We are eccentrically located with respect to it and the Galaxy has the shape (roughly) of a flattened disk.

One might, however, also consider the introduction of "cosmological principles" as a tendency to assert that the Universe must be simple, a sort of application of Occam's razor, or again, a reversion to the Pythagorian temptation. There is little doubt that Einstein wanted to stay close to the observations, and therefore was considering the principles as necessary simplifications. However, many of his followers were acting out of a more idealistic view of the principles, which they were generally unwilling to discuss. The latter have argued that, while it is true that locally the Universe is non-homogeneous and anisotropic, it is not less true that *at a sufficiently larger scale*, not yet observable (except perhaps in the so-called cosmological background radiation, see pp. 459–460), *the Universe is indeed homogeneous and isotropic*. This is a very daring thing to say, since it is based on no real knowledge.

8.2.2
The Static Universes; Einstein's Cosmological Constant

The Einstein model envisioned a spherical 3-dimensional Universe of radius R, completely static. This means that $R = $ const; and that, in the space-time representation, the Einsteinian Universe is *cylindrical*.

Tolman has shown that there are only three possible solutions to the simplified equations. One is the SR model, that of a *flat Euclidian spacetime*, "Minkowskian" space. The other two were extensively discussed in 1917 by Einstein and by de Sitter.

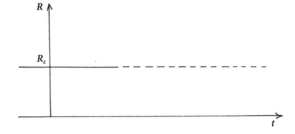

Fig. 8.6 *Stable cosmological models.* The Einstein-de Sitter solution: the value R_E is a function of Λ and ϱ_0 (the density of the present-day Universe).

(i) The Einsteinian Universe.
One model was proposed by Einstein (Fig. 8.6). It assumes first that space is filled with distributed matter, perfect fluid, with a pressure p_0, and a constant density ϱ_0. It further assumes a simplified form for the line element (see Sect. 8.8):

$$ds^2 = - dr^2 \frac{1-r^2}{R^2} - r^2 \, d\theta^2 - r^2 \sin^2\theta \, d\phi^2 + dt^2.$$

This has to be compared with the corresponding SR line element, in a flat Euclidian space-time Universe:

$$ds^2 = - dr^2 - r^2 \, d\theta^2 - r^2 \sin^2 (\theta) \, d\phi^2 + dt^2.$$

The radius R, which now enters the picture, is constant, and linked to the cosmological constant by the relation:

$$R^2 = \frac{1}{\Lambda - 8\pi p_0} = \frac{3}{\Lambda - 8\rho_0}.$$

This model, the simplest one indeed, has been criticized on several occasions and on several grounds. For non-vanishing values of pressure and density, Λ has to be largely positive for R to be real.

First of all, the model was considered *unstable*. The demonstration was given by Eddington and Lemaître. It rests upon the hypothesis of homogeneity; when r_0 changes *everywhere*, the radius changes. The instability results from the fact that if an expanding perturbation is applied (diminution of ϱ_0), a further expansion (increase of R) of the Universe occurs. This is obvious from the equations above. We should make clear that we do not consider that demonstration as valid, except for a strictly uniform Einsteinian Universe. A local perturbation of the density would not have that effect; but then the model would be inappropriate, being no longer strictly homogeneous. It is possible to conceive of quasi-static models and we shall return to that subject later in this chapter.

The other criticism one could make of Einstein's model is the *arbitrary* need for him to introduce the *cosmological constant* Λ. Without it, the radius of the Universe could not be equated to a constant. It would not be static! The cosmological constant has the effect of a sort of stabilizing force, opposing gravitation, and acting mostly at large distances, i.e. only on the cosmological scale. It is a sort of repelling term. It might, however, have another meaning, and again we shall come back to it (see p. 492).

It is interesting to compare Einstein's Universe with the observations, at least those of the quantities which enter his equations. Tolman did just that in the thirties. Hubble had estimated the density ρ_0 to be (1.3 to 1.6) $\times 10^{-30}$ g cm^{-3}. Leaving aside the mass of the unseen matter, and the pressure and density of radiation, and also leaving aside the material pressure p_0, Tolman's value for ρ_0 of 10^{-30} g cm^{-3} leads to a value for Λ of 9.3×10^{-58} cm^{-2} and $R = 3.3 \times 10^{28}$ cm $= 3.5 \times 10^{10}$ light-years. These values are then a lower limit for Λ and an upper limit for R, the assumed value for ρ_0 being considered by Tolman as a lower limit.

The Einstein model does not, however, explicitly predict any shift of spectral lines in the distant galaxies, as observed in the teens and early twenties of this century by Slipher and Hubble. This is probably the reason for its final rejection in the late twenties.

(ii) The de Sitter model.
This model differs from Einstein's in the way that it achieves the static solution. The line element of the de Sitter model is

$$ds^2 = -\frac{dr^2}{1-r^2/R^2} - r^2\,d\theta^2 - r^2\sin^2(\theta)\,d\phi^2 + \frac{1-r^2}{R^2}\,dt^2.$$

The construction of the de Sitter Universe implies strictly that: $p_0 + \varrho_0 = 0$. A fluid, any fluid, having a non-zero density would then exert a negative pressure. This is hard to accept and, as noted by Tolman, leads to the conclusion that de Sitter's model is empty.

One can then show easily that the line element of de Sitter corresponds to the very simple relation:

$$R = (3\,/\,\Lambda)^{1/2}.$$

As in the case of Einstein's Universe, this is an apparently static Universe, cylindrical in 4-dimensional space. Its curvature is only linked to the value and the sign of Λ.

The Tolman comparison of the de Sitter Universe with the measured data assumes the density and pressure to be vanishing. It leads to a value of $R = 1.75 \times 10^9$ light-years and $\Lambda = 1.08 \times 10^{-54}$ cm^{-2}. The value of R, as that of a boundary, means that we should not observe events that are more distant than 1.75×10^9 light-years, or older than 1.75×10^9 years. This puts a severe limit on the age of the "observed" Universe. In fact, we have by now penetrated to greater distances in our observed Universe of galaxies.

There is actually no serious argument against the use of a cosmological constant $\Lambda \neq 0$. If one wanted to construct the Einstein equations from some minimum complexity principle, one would indeed have to equate G to zero (at the first order) while Λ (at the zero order) could remain different from zero. Indeed, one could even say that both G and Λ are "non-Occam quantities!" The only real criterion we have is the following: would the Einstein or de Sitter value found for Λ affect motions in our nearby Universe? Actually, in both cases, Λ is not large enough to influence planetary motions.

8.2.3
Non-static Universes: The Meaning of the Singularities

There are several reasons to doubt the "strictly static" Universe, as described by either Einstein or de Sitter.

First, one might consider that these models impose the *non-necessary limitation* of $R = const$ (keeping Λ at a non-zero value does not force $R = const$!). To quote Tolman (who here takes a clearly anti-Occam attitude, that of a pure Aristotelian) "The history of human endeavors to understand the Universe would certainly indicate no a priori right to demand mathematical simplicity of nature."

Second, the Einstein Universe *does not seem to be stable* (a good argument if we limit ourselves to homogeneous models, but it does not rule out quasi-static models, and inhomogeneous structure).

Third, the *redshift of spectra of galaxies* indicates (independently of some general expansion) motions which at least suggest the possibility of such an expansion. This must be taken into account, even though the expansion has been labelled first by Hubble himself, then by Tolman and many others: *"apparent expansion."*

Finally, *the Doppler effects seem to be the only way to explain the redshift*, in spite of Hubble and Tolman's use of the qualifying term "apparent expansion." All other redshifting effects of classical physics are obviously too small. We shall return to this question (see p. 510).

If, however, we abandon the hypothesis of a static behavior, we find that an enormous number of solutions are possible.

It is not possible here to reproduce the details of the derivation of the line element (however essential!) of the solutions of even the simplified equations. These solutions were discussed extensively by Friedmann (1922), and accommodated to the "real" Universe by Lemaître (1927). We shall first examine the time behavior of some of these solutions. Some models have been constructed in which the amount of matter is constant (closed models).

The generalized solution, however, depends mostly upon the "initial conditions." It also depends upon the value of Λ. Friedmann, considering Einstein's reason for introducing it not completely rational (based, as it was, not only on observations, but on a possible metaphysical subtext and the desire to satisfy Mach's principle), decided it was probably an unnecessary hypothesis, and assumed instead $\Lambda = 0$. Figs. 8.6, 8.7, and 8.8, show different cases. The main parameter is now density.

We shall return to this subject, but let us note here that, between the Friedmann hypothesis: "$\Lambda = 0$" and the hypothesis: "Λ is undetermined, to be computed to fit the observations" (as in the static models), $\Lambda = 0$ is a more arbitrary hypothesis, and a more restrictive one, than any other. This may be a poor use of Occam's razor. Assuming any quantity to be either zero or infinity, in order to limit complexity, to avoid any unnecessary device, is indeed very restrictive. Of course, it allows one to drop the very symbol of that quantity from the equations; but indeed, what is "simplicity?" Where is the "minimum" hypothesis? So far as we are concerned, we think that choosing $\Lambda = 0$ is far more restrictive than leaving these two quantities undetermined, and to be determined only with the use of the observed data.

The de Sitter-Lemaître models which have an initially static state are represented in Figs. 8.7 and 8.8. These models start with either radiation without matter or matter without radiation. Eddington criticized them on the basis of the instability which could have, or which should have, occurred at any time in the infinite past.

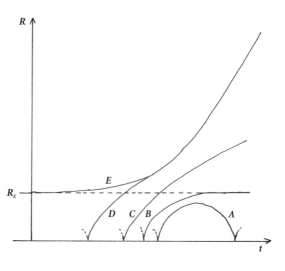

Fig. 8.7 *Expanding models* (except for C, these models assume that $\Lambda = 0$). From a value of R equal to 0, dense models (A) will not reach the value R_E, and will finally collapse. Models having the critical density will expand forever, but in a "quiet" way (C). Some, having a critical Λ-value tend to a limit (B). Models having a lower density will expand in a "catastrophic" way (D). Models starting from an Einstein-de Sitter stable model will ultimately expand (E).

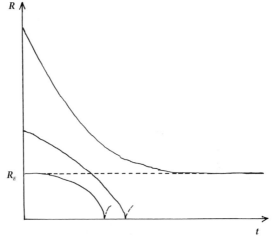

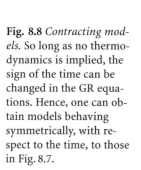

Fig. 8.8 *Contracting models.* So long as no thermodynamics is implied, the sign of the time can be changed in the GR equations. Hence, one can obtain models behaving symmetrically, with respect to the time, to those in Fig. 8.7.

Other models start from a Universe reduced to zero (R (t=0) = 0), as suggested by the extrapolation of Hubble's law, with various values for the parameter Λ, including $\Lambda = 0$ (Lemaître). Some others start from a very diluted model of a contracting Universe. They all finish with an expansion, some tending asymptotically towards a zero density de Sitter non-static model.

A third type of solution (Fig. 8.8) is in contraction from $R = \infty$ with $\Lambda = 0$, towards a limit value R_E. So long as the equations only involve the variable t (time), and no thermodynamic consideration, time might indeed run both ways. The "arrow of time" does not exist. Any solution found to the equations has another solution, which is symmetrical and where it is sufficient to change the direction of time. These solutions are indeed symmetrical to one of the solutions of Fig. 8.7, including the Eddington-Lemaître one.

A fourth type of solution implies *oscillating* models. These are the only solutions for closed models, where $\Lambda = 0$. The extreme case of an expanding model starts from a value $R = 0$, and reaches a value R_E (as in Fig. 8.7). Other models, still more "close" or denser, reach a maximum value R_{max} (curve A in Fig. 8.7). In view of thermodynamic considerations (second principle) applied to the Universe as a whole (a daring concept, actually!), two successive cycles reach different, increasing values for the radius (Fig. 8.9).

How can we determine which of these models best represents the actual Universe? First, by observing the present rate of the expansion (assuming it to be real), and the rate of change of this rate (or the acceleration of expansion). Second, by measuring the actual average density of the Universe, which is a parameter of the solutions. Third, by finding a way to estimate values for Λ and R at the present time from observed data. Fourth, by taking note of the various remnants, in today's Universe, of the physical state of the Universe near its "initial state," such as the chemical composition of the residual matter, and the quantity and properties of the residual radiation. Problematically, we see that, in all these models, the initial conditions are not clearly defined. A radius equal to

Fig. 8.9 *Oscillatory models.*
Models of type A in
Fig. 8.7 are "oscillating
models." If the entropy of
the Universe is constant,
these models will behave
as in A. However, if we
consider the Universe to
be an isolated system, entropy is increasing. Therefore, the behavior of the
models will be as in B.

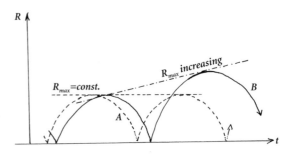

zero? A density equal to infinity? This is indeed insufficient for formulating a conclusion. Moreover, is it not almost obvious that near $t = 0$, the density is so great that matter cannot possibly behave like it does in our Universe? Forces other than gravitation (the intra-nuclear forces that we now know about) come into play then. Would the GR equations continue to be valid then? Probably not. We are no longer in the GR Universe, with only the force of gravitation. We are in a different Universe, where different forces and different equations may apply.

This idea is the origin of several categories of remarks, made by various astronomers in the 1950s:

Some, such as Hoyle, Bondi, Gold, and others, rejected any initial infinitely condensed Universe and tried to substitute a *static universe*, accounting for its *expansion* through the process of *continuous creation of matter*.

Some discussed the nature of the physics functioning in these early instants, that would explain *the actual composition of the Universe, and the remnants of its early life*. This was primarily the focus of Gamow and the theory of the "big bang." (We will refer to this *"standard" theory of the big bang expanding universe* as the "old big bang," implying an initial singularity, an initial infinite density.) This is still the basis of what most cosmologists currently believe.

Others (Hawking, Penrose) went a step further. Having studied the physics of black holes, hyperdense regions of space resulting from a violent collapse of a stellar or galactic mass, they noted the symmetrical way time enters into the equations. They concluded that it was more or less a necessity for one or several singularities to have existed in the past Universe. This theorem is valid even in a non-homogeneous Universe. It implies a general statement about the instability of static universes, be they inhomogeneous or anisotropic!

A fourth tendency, which has given rise to the "new big bang," is to reject the very first moments of the theoretical solutions, and to replace them with an elaborate theory, implying forces other than gravitation, even aiming at a unification of all forces, and implying, in addition, the possibility of a "quantum" Universe, of infinite duration, or, better, in which the very notion of time loses its meaning.

Still others, following in the same line of thought as Hoyle and others, namely a rejection of the classical big bang (Zwicky, Born, Findlay-Freundlich, and later Segal, and Pecker and Vigier) have defended static or almost-static models, and looked for causes other than expansion to explain the observed redshifts.

At this point, it is clear that it will be necessary to look more closely at the astrophysical arguments, and at the physics of hyperdense matter. True enough, observational arguments have been piling up, since the twenties. We can no longer discuss cosmology solely as a problem in pure mathematics. We must act as physicists and astrophysicists as well, and we have to move on to an "observational cosmology," a very Aristotelian approach indeed, at least in principle. We will illustrate this approach in the next sections of this chapter.

8.3
New Cosmological Facts

We have enumerated those facts which indeed were known before Einstein (Olbers' paradox for example), and those which were known to Einstein, i.e. the first cosmological facts, or, if you will, the first cosmological tests.

We must be very careful with this nomenclature. A "fact" may not be a "test," in the context of different cosmologies. Therefore, the status of cosmological facts varies greatly, from one author to another. There are some facts that are of some cosmological significance but function poorly as tests, for example the age of the solar system. This datum certainly shows that the *minimum* age of the Universe is 4.6 billion years, but all cosmologies (except "modern creationism," to which I would without hesitation deny the distinction of a scientific theory) lead to ages of the Universe much larger than that value!

8.3.1
The Redshift of Galaxies; The Hubble Ratio

During the first years of this century, it became possible to determine the spectral position of some well-known lines of the spectra of galaxies, such as the H and K lines of ionized calcium, and the Hβ line of hydrogen. Vesto Slipher, in the years 1915–1920, showed, without ambiguity, that these lines were displaced, when compared to lines of the same elements in the solar spectrum. In most of the cases, the displacement was to the red; hence the name *redshift* given to this effect. At the present time, the word redshift is still in use in a much broader sense: "displacement of lines towards longer wavelengths." It can even be a displacement from the X-ray domain to the ultraviolet! Obviously, we will have to use it in its broader sense, remembering that the word must not be taken literally. In cosmologically meaningful observations, radio lines are not shifted "towards the red", but in the opposite direction, towards the less energetic radio waves of longer wavelengths.

The redshift, a difference in wavelength, $\Delta\lambda$, could be caused, as in the motion of double stars, by the classical Doppler-Fizeau effect. For this reason, it is generally expressed in km/sec. Or, better, it is expressed by the non-dimensional number $\Delta\lambda / \lambda_0 = z$. A positive value of z indicates a displacement towards longer wavelengths; and the negative value a displacement towards the shorter wavelengths.

The z was obviously larger when the magnitude of the galaxy was larger, i.e. when the galaxy was further away from us. However, a relation between z and the distance D of galaxies is not easy to establish. It depends on the calibration of distances of galaxies. How can we determine such a distance?

To answer this question, we have to consider the idea of "distance calibrators," objects which, because of some observable character, allow us to determine some absolute brightness. By comparing the absolute brightness with the apparent brightness, we can obtain a measurement of the distance. The distance calibrators then give an estimate of the distance of the galaxy in which they are observed.

For years, the best-known of the calibrators has been the *period-luminosity relation* established during the years 1908–1912 by Henrietta Leavitt. As the period is determined, unambiguously, whatever the distance, this relation gives the absolute magnitude M. Observations give the apparent magnitude m. The difference $m - M$ permits a determination of the distance. The period-luminosity relation had been established in the Small Magellanic Cloud, a neighboring galaxy. It was calibrated in 1918 by Harlow Shapley by using similar stars found in globular clusters of our own Galaxy. This relation is a function of a type of the Cepheid. It differs from Cepheids to RR Lyrae stars, for example, hence its application has not always been easy. The first to make extensive use of this calibrator was Edwin Hubble, at the Mount Wilson 254-cm telescope. He established firmly that the galaxies were definitely located outside our own Galaxy.

Little by little, beginning in the 1920s, Hubble, and his colleagues, students and successors at Mount Wilson and at the Palomar Mountain observatories established a whole set of calibrators. Humason, Mayall, Sandage and Tammann must be mentioned in this connection. First, they used the brightest stars of the galaxies, which always seemed (in the beginning) to have the same average absolute brightness. Then, they used the diameters of planetary nebulae, which have a relatively constant absolute value. The novae were also used as distance indicators, but these were in essence "primary indicators," based upon our knowledge of objects belonging to our Galaxy. They are not very bright and are unobservable in distant galaxies. They allow us to determine distances up to a few dozen million light-years. In other words, they enable us to determine the distances of the nearest galaxies, those of the "local" group of galaxies, roughly speaking (Fig. 8.10).

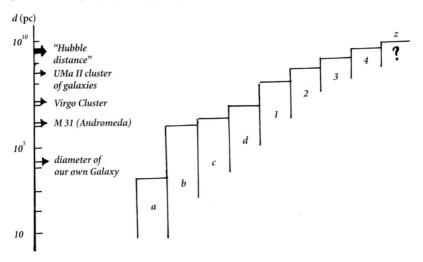

Fig. 8.10 *The cosmic distance indicators.* The letters a, b, c, d label, in essence, the calibrators used before Hubble or by Hubble to scale our own Galaxy. a: kinematic methods; b: brightness of supergiants; c: RR Lyrae period-luminosity law; d: Cepheid period-luminosity law. By these methods, the distances to the Andromeda Nebula (M31) and the Magellanic Clouds can be reached. Later indicators of a primary nature are: (1) diameter of planetary nebulae, maximum brightness of novae and of bright stars; of a secondary nature: (2) bright supernovae, and diameter of bright H II regions; of a tertiary nature (3): diameters of galaxies of a given morphology; and of a quaternary nature: (4) brightness of the brightest galaxies in clusters of galaxies. To go further, the redshift z has actually been used, but this last distance criterion is subject to criticism.

"Secondary indicators" soon had to be introduced. Their calibrations were based on the distances of the galaxies which had been determined using the primary indicators. In some way, they constitute a reasonable linear extrapolation of the laws established using only our Galaxy. Such were the bright supernovae, the diameter of HII spheroidal regions, and the same type of supergiants which can be observed in distant galaxies, but cannot be calibrated in our own, because of the small number of them observable there. The distance that can be determined with secondary calibrators is about fifty million light-years. One can reach ten times as distant galaxies, by using "tertiary indicators," the properties of which have been established with galaxies whose distance was determined with primary or secondary indicators.

The bases of these tertiary determinations are generally criteria of a global nature: total luminosity and the diameter of a galaxy of a given morphology (hence, its apparent magnitude as the angular diameter of such a galaxy functions as a measure of its distance), or again the amplitude of the rotation of the galaxy. A fourth scale is reached by using the average brightness of the brightest galaxies in clusters, apparently a rather well defined quantity.

The primary indicators allowed Hubble to establish his linear law (Fig. 8.11a):

$$cz = H\,D.$$

He used only 24 galaxies in his initial work and succeeded in determining the *Hubble ratio*: $H = 500$ km/sec per Megaparsec (a strange unit, but easy to understand). Using secondary indicators, and correcting the initial calibrations after criticizing the earlier statistical methods, led Hubble's modern followers (Fig. 8.11b, c) to the value $H = 50$ km/sec per Mpc (Sandage and Tammann, 1974–76).

This determination has been criticized, several times, but without substantially changing the faith of the astronomical establishment in Hubble's linear law.

A first criticism (with cosmological implications) of the value of Hubble's ratio H, came as early as the twenties. Öpik determined a significantly larger distance for the well-known galaxy M31 (Andromeda) than that of Hubble. Moreover, in the forties, Baade, and Mineur, independently, questioned the Shapley calibration, on the basis of obscuring matter in our own Galaxy, and some dynamical considerations. At the

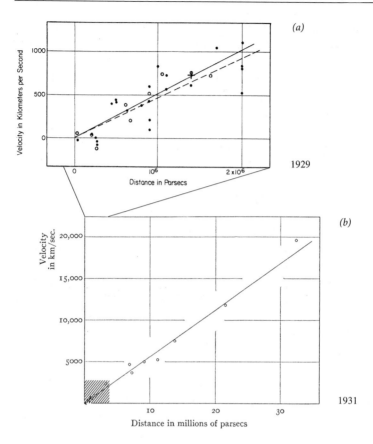

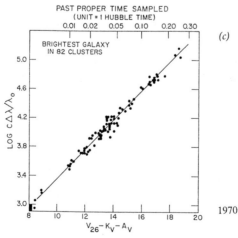

present time, the radio-astronomical data (Bottinelli, Gouguenheim) and the critical study of the óptical calibrators used by de Vaucouleurs, has led these investigators to oppose the 50 km s^{-1} Mpc^{-1} value of Tammann and Sandage with a larger value of 100 km s^{-1} Mpc^{-1}. The true value (if the law is indeed linear) is probably between these two values, but there are still astronomers deriving smaller values than 50 and some larger values than 100 kms^{-1} Mpc^{-1}. Very recent data seem to lead to a value near 70 kms^{-1} Mpc^{-1} ± 10 kms^{-1} Mpc^{-1}, according several observers. Indeed, the classical method implies that each time we change scale, we have to adapt the loosely defined end of the law applied to the closest galaxies to the more fully defined end of the relation between z and D for the more distant objects. This accommodation is hazardous and can lead to errors (Figs. 8.12a and b).

The scales z and D are not really appropriate to a large range of variation of the distances. We would do better to use logarithmic scales log z vs log D. Since log D is obviously linked to the *apparent magnitude m* of any object ($m = M + 5$log $D - 5$, M being the *absolute magnitude* of the star, i.e. the magnitude of the star were it located at a distance of 10 parsecs), the "Hubble diagram" is most often replaced by a log z, m diagram. This diagram is easier to build, as well as to read, since the slope has to be 1 for a *linear* Hubble diagram (Fig. 8.11c).

A second group of criticisms concerns the linear character of the law. Already Lundmark, in the twenties, even before Hubble's publication, had suggested a quadratic law, using the same data, concerning 24 nearby galaxies. Using only galaxies determined with primary indicators, Segal and Nicoll are now also advocating a quadratic law ($z = kD^2$), which agrees with Segal's chronogeometric cosmological theory. Their arguments are strong, but the linear character of the law has now been adopted by the establishment, almost as a dogma. For this reason, one almost never sees citations of Segal and Nicoll's paper, nor Lundmark's.

←

Fig. 8.11a–c *Hubble's linear law, in three steps.* (a) Hubble's law as observed in 1929. Circles represent averages, small black points are individual galaxies. On the abscissa, the distance expressed in parsecs. (b) According to Hubble and Humason (1931): the black square at the left- hand side represents the totality of Fig. 8.11a. The abscissa is as in Fig. 8.11a. (c). According to Sandage (1970), using the brightest galaxies in clusters of galaxies. The black rectangle at the left-hand side represents the totality of Fig. 8.11a. Note that here, the scales are logarithmic, and the magnitude, not the distance, is put on the abscissa. The linearity of Hubble's law is indicated by the slope (= unity) of the straight line in these ad hoc units.

Fig. 8.12a,b
(a) *The Malmqvist bias*
(schematic). If there is a
certain dispersion of the
galaxies of a given sample,
the limitation in magni-
tude may lead to some
overestimation of the
Hubble constant. Observa-
tional limitations on the z
– value, when it is large,
may also induce biases.

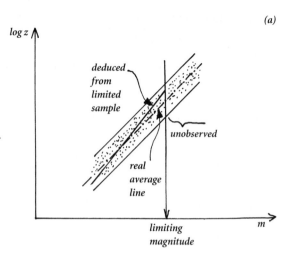

(b) *The difficulty of ac-*
commodating the log z-m
law (schematic). When
one uses different samples,
obtained with different
distance indicators, one
may commit errors in the
fitting of each of them on
the others, because of the
bias described in part (a).
This may lead to a bias in
the assessment of the linearity of Hubble's law, or to
biased values of the so-
called Hubble constant.

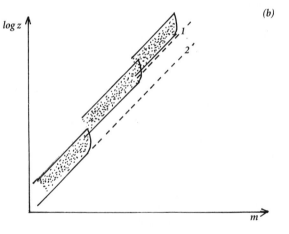

From a third perspective, note that H can be measured for only one galaxy G_i (whatever the distance calibrator used), and written as H_i. What is determined is the average value of H_i, i.e.: $H = \langle H_i \rangle$. The averaged character of H should not be forgotten. After all, the dispersion may be real, due to various causes. It may seem that large values of H correspond to nearby samples, small values to the samples including very distant galaxies or clusters of galaxies. This would imply a relation between H and the local conditions. It may mean that H is a local property of 4-dimensional space, not a general property of that space. Again this remark bears upon the real nature of H, to which we shall return.

Can this remark be the source of the different values found for H by different observers? Only in part, as in many cases the same galaxies are used.

We have said the the observed redshift is expressed in km/sec, as if it were due to a velocity of recession. Hubble himself, and Tolman, up to the forties and even later, constantly used the words "apparent velocity" and "apparent recession," as we have already said. Their followers went further in their interpretations. Comparing Hubble's finding and Friedmann's models, Georges Lemaître remarked that they are perfectly compatible with each other, the Hubble ratio measuring then a *real* recession of the galaxies, or, in the more classical expression in use, a *real "expansion" of the Universe*. Indeed, if z measures a velocity of recession, the Hubble ratio, its constancy ("Hubble's law") shows then that any galaxy G_a recedes from *us* at a velocity which is proportional to the distance of that galaxy to us. This is the same for any other galaxy G_b. Therefore, a simple construction (Fig. 8.13) shows that G_a recedes from G_b with a velocity proportional to their mutual distance.

Hence, accepting the Lemaître conclusion, the Universe appears to be expanding, in a quasi-uniform way. The observed inhomogeneity of the H_i value may lead us to believe, alternatively, that the Universe is expanding slower in non-dense regions, faster in more dense regions (clusters).

The *Hubble age* is the inverse of H (in proper units). It would measure the age of the Universe, if the expansion had been uniform since the origin of time. At the two extreme values given here above, the Hubble age is 9.6×10^9 years ($H \approx 100$) to 19.2×10^9 years ($H \approx 50$). If we compare these ages with the age of globular clusters (said sometimes to reach 18 to 20×10^9 years) (see hereafter), we see why the dispute between these two values of H, or more precisely, between their aggressive proponents, is important! However, it seems now (1998) that a convergence towards $75 \text{ km s}^{-1} \text{ Mpc}^{-1}$ may be reached. The possibility of this agreement arises out of the discovery of the importance of metallicity in determining the period-luminosity relation of Cepheids. The Hubble age would then be $(14 \text{ to } 15) \times 10^9$ years.

Homogeneous expansion is the basis of the classical "big bang" cosmologies, often called "standard" cosmologies.

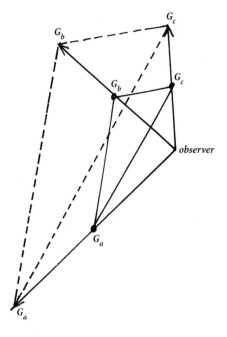

Fig. 8.13 The recession of galaxies according to Hubble's law means the expansion of the Universe (see text)

8.3.2

The Universal Background Radiation

The universal background radiation was discovered in 1964 by Penzias and Wilson (although some credit should be given to Peebles and Dicke, who publicized the importance of this discovery).

Actually, the history of the discovery is strange. Its classical description goes as follows. In the years 1949–54, George Gamow, on one side, Ralph Alpher and Robert Herman on the other side, then the three together, and, once, with Hans Bethe, defended the idea that, if Lemaître's cosmology, supported by Hubble's so-called law, were right, then the Universe must have experienced a state of extreme density several billion years ago. (The association Alpher-Bethe-Gamow became a joke, because of its similarity to the first three letters of Greek alphabet, α-alpha, β-beta, γ-gamma. Bethe did not play an essential role in the writing of this paper!). Gamow and the other protagonists of this work, assuming that it must have also implied a very high temperature, wondered whether the present composition of the Universe in light elements as well as in heavy elements, could not be explained by an equilibrium situation

Fig. 8.14 *The Background Radiation (BR) of* 2.735°. Observations of the sky background are possible in the millimetric, and centimetric wavelengths. At longer wavelengths, the Galaxy dominates the radiation field. At shorter wavelengths, the atmospheric and galactic infrared radiations dominate.

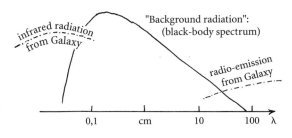

reached at high temperature between the different elements. The elements would have been linked by nuclear equilibrium reactions, followed by a quick cooling. A quick "tempering" (as in steel metallurgy) would have retained this equilibrium composition in the Universe, as is now well-known.

Assuming for a moment the elemental composition predicted by the theories of Gamow and others, note that they predicted, as a by-product, that the radiation filling the Universe had at some stage decoupled from matter and, due to expansion, has now reached a temperature of only a few Kelvins (Fig. 8.14). (The two dozen papers then published on this subject did not agree on the predicted value for that residual temperature of the radiation distributed in the Universe.)

It is interesting to note that in 1954–55 Findlay-Freundlich and Max Born, both friends of Einstein, both exiled in 1933 from Germany to Scotland, published in British journals an alternative explanation of the redshift. For them, redshift was not the measurement of a velocity, but a consequence of some reddening (loss of energy) of the travelling photons, having interacted during their long trip with space and its various contents of fields and particles. This *tired-light mechanism* also predicted that one should observe around us some undiluted radiation of about 2 K, therefore observable only in centimetric, millimetric, and submillimetric wavelengths. These papers passed unnoticed.

Note also that, many years before, on the basis of an equilibrium static Universe, Guillaume, then Regener or Nernst, then Eddington (!) also predicted values very close to 2.7 K, even closer than those of Gamow and his coworkers ...

At about the same time, Emile Leroux, at the Nançay station of the Paris-Meudon Observatory, detected some signal that he interpreted as a sort of background radiation. He did not identify it with the predictions. He did not assert that it was black body radiation. Still, his measurements, as we revisit them today, were undoubtedly the Penzias and Wilson radiation (often called the CBR, "cosmological background radiation," although it is cosmological only by inference and comparison with predictive theories; to avoid any premature conclusion, as it is predicted by non-cosmological phenomena as well, we shall designate it only by the letters BR, for "background radiation").

Note also that McKellar, from studies of interstellar molecular lines, derived a very similar value to that of Penzias & Wilson, – but in the early forties! …

Penzias and Wilson got the Nobel prize for this discovery. Since 1964, observations of this 3 K radiation have been pouring in. In particular the COBE satellite, launched for that very purpose by NASA, has made a few things clear, and we can now describe the background radiation with relatively great assurance, but still a few question marks, as follows:

The BR is the combination of three radiation fields, easily visible in the data obtained by the COBE satellite (Fig. 8.15).

On the one hand, BR is mainly composed of an isotropic field of radiation, that of a strict black body of 2.735 K $\pm$ 0.01 K (Fig. 8.14).

To this basic component, whatever its origin or localization, whether cosmological or not, is added the effect of the motion of the Sun with respect to the isotropic source of this black body, a velocity of 350 km/sec towards a well-defined apex. The location of the apex, the value of the velocity, results from the vectorial combination of several motions: the motion of the Earth around the Sun (which changes slightly with time in the course of a year), the motion of the Sun around the center of the Galaxy (220 km/s), the motion of the Galaxy with respect to the local group of galaxies (80 km/s), that of the local group with respect to the local supercluster of galaxies (250 km/s), and finally that of the motion of the local supercluster with respect to the isotropic source of the 3 K BR. It gives the sky map an apparent general distortion, a bipolar anisotropy, to be more precise, when observed at a submillimetric wavelength.

Finally, a third and very small component of the BR, made of fluctuations of about 0.001 K amplitude (or of that order), is observed to be not randomly distributed, but organized on large filaments. Unfortunately, not much is known concerning the spectrum of these filaments. Do these also have a black-body spectrum? At what temperature? If so, is it a diluted spectrum, or not? Do they correspond to some black body dis-

Fig. 8.15a–c *The COBE observations.* The NASA satellite COBE has measured the BR with the best accuracy obtained so far. It has measured **(a)** the temperature of the blackbody spectrum to be 2.735 K, at 10^{-4} accuracy, and it has shown: **(b)** the bipolar character of the radiation field (due to the solar motion), and its fluctuations **(c)** which may be indicative of the history of the structure of the Universe.

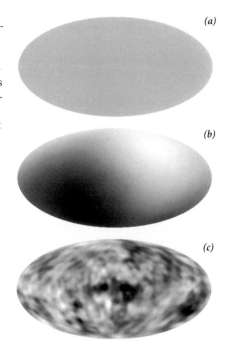

placed towards the long wavelength in one way or another, by some z-effect? These questions are so far answered only partially by COBE, but later observations may soon provide answers.

8.3.3
The Abundance of Light Elements

It became clear very soon after the publication of Gamow's papers, that the heavy elements (carbon and heavier) were produced continuously inside the stars through thermonuclear reactions occurring at moderate temperatures (from 10 million K to a few billion K). These reactions, however, cannot explain the respective proportions of light elements, i.e., ^{1}H (hydrogen), ^{2}H (or ^{2}D, deuterium), ^{3}He, ^{4}He (isotopes of helium), ^{6}Li and ^{7}Li (isotopes of lithium).

The relative abundances of light elements are measured primarily by spectroscopic means, in stars and galaxies. It would not be possible to infer these abundances from chemical analysis of terrestrial materials, lunar samples, or meteorites, as is currently done for heavy elements, be-

Table 8.1. *Cosmic abundances of*
light elements

H		0.92–0.93
^{2}D		$(2\text{–}5) \times 10^{-5}$
	(& QSO)	2×10^{-4}
^{4}He		0.07–0.08
^{7}Li		$(1\text{–}4) \times 10^{-10}$

cause of the extremely uneven composition of these objects, and the fact
that light elements, precisely because they are light, departed from these
objects long before they became solid. From spectral analysis, however,
one does not measure abundances in very distant objects. The spectral
lines involved are difficult to observe in any case, except those of hydrogen. For deuterium, the lines Hα and Lyα, in the wing of the H lines, are
used; for helium the lines (He I) at 447 nm and 1083 nm, and (He II) at
454 nm; for lithium the line at 671 nm. Are these results reliable? Certainly they are, but they are not characteristic of the composition of the
Universe, only of that of the relatively nearby Universe. The data are reproduced in Table 8.1.

Do the light elements have "cosmological significance"? Classical big
bang cosmology enables us to understand the observed proportions and
we know that temperatures within stars (even in the case of stars in their
terminal evolutionary stages, such as supernovae) are too low (no more
than 10^{10} K) to modify the abundances of light elements in a noticeable
way, during their "short" lifetime (20×10^9 years at the most according
to the standard big bang). Let us note that some cosmologies (the
steady-state or quasi-steady-state cosmologies, for example) do not impose a "short" lifetime to the galaxies. They could well be old enough for
the light elements to have been produced in the observed quantities.

8.3.4
The Inhomogeneous Structure of the Universe

One of the a priori principles of mathematical cosmology is the homogeneity of the Universe. The evidence is opposite to that principle. In the
intergalactic vacuum, the density is as low as 10^{-30} g cm^{-3}, in the neutron
stars, as high as 10^{13} g cm^{-3}. These differences cannot, obviously, be considered "fluctuations!" (Fig. 8.16) The obvious question is therefore: in
spite of this huge range of densities could the Universe be homogeneous

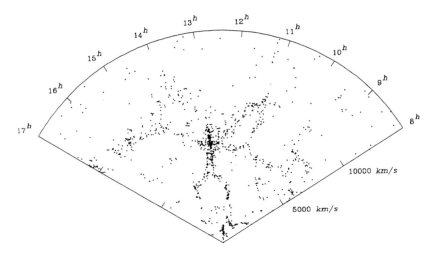

Fig. 8.16 *The large-scale distribution of galaxies.* Representation of the distribution of galaxies in some slice of our Universe. Note *The Great Wall* of galaxies, one of the largest structures in the Universe, here extending horizontally over the entire map.

perhaps, at a sufficiently large scale, much larger than the observable universe of galaxies? Could we then conceive of an "average density" of the Universe, which would then be a meaningful quantity?

A partial reply to this question was given by de Vaucouleurs. It consists of a "hierarchical" and "fractal" description of the distribution of matter in the Universe. Structures exist that are similar to that of Russian dolls, called matriotchkas: quarks, electrons, protons, neutrons, other particles, atoms, molecules, asteroids, planets, stars of the main sequence, supergiants, dwarf galaxies, galaxies, clusters of galaxies, and larger scale structures, (often called supergalaxies). The density ϱ decreases from one of these structures to the next larger. Actually, the larger the volume in which density is measured, the smaller is that density. This remarkable statement is exemplified by Fig. 8.17. It is valid from the scale of elementary particles up to the scale of the Universe explored so far. The question is now put in a different way. If, as shown by de Vaucouleurs, the *fractal dimension* of the distribution of matter in the Universe is 1.3, which express the fact that the mass or the density can be expressed as a function of the size D of the structures:

$$\varrho = D^{-1.7}, \text{ or } M \approx D^{+1.3}$$

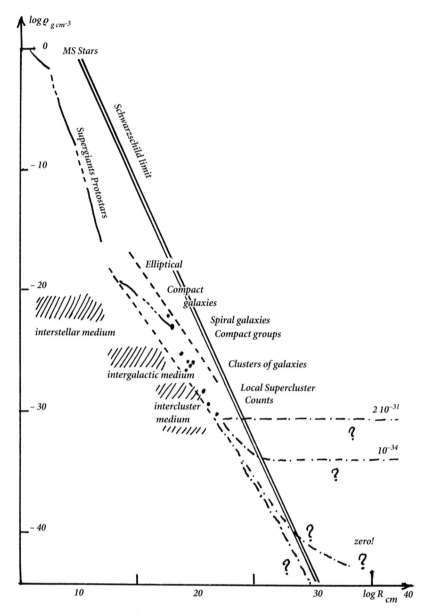

Fig. 8.17 *The hierarchical distribution of matter in the Universe* (similar to Fig. 7.21, p. 412, and from same sources) From stars to galaxy counts, the observations show a fractal structure, as noted by de Vaucouleurs. The density is decreasing as a power law of the radius. An open question is of course: for larger structures, will the density

(instead of ϱ = const, or $M \approx \pi \varrho D^3$, which would be observed if the Universe were homogeneous). The fractal measure of this hierarchical Universe is defined as is 3–1.7 = 1.3.

This observed property leads to a first conclusion: the "average density of the Universe" ϱ_0 is a quantity which depends upon the degree of penetration of observation, or upon the size of the regions in which it is measured. Hence, the usual value of this density has to be looked at with the utmost care and suspicion.

The second conclusion is that one has to modify the solution of the GR equations assuming a uniform average density and find solutions for a hierarchical Universe.

In any case, a question remains: As the exploration of the Universe progresses, what would then be the average density? Would the fractal distribution continue, with a fractal index 1.3? Would this index change? Would it change sufficiently to reach the zero value, i.e. to allow the representation, at large, of the Universe, as an empty Universe? Or would it change to reach a single non-zero average density (smaller than the usual standard density of the standard models), i.e. a homogeneous Universe?

Even if this were the case, are we justified in applying the "homogeneous" solutions of the GR equations, with a single density ϱ_0, even though we now know that this density results from averaging over a very large range of densities?

We shall return to these questions in our discussion of the modern cosmologies.

←

reach a finite limit? Will it decrease much more, giving rise to the idea of a Universe in which the visible matter would be only a very tiny fraction? Or will it be an empty Universe, asymptotic to a de Sitter Universe for example? The *Schwarzschild limit* defines a mass having a density sufficient, for a given radius, to become a black hole. One sees that the density of inter-structure matter is decreasing at a lesser rate than that of the density of structures. If the two curves intersect (where? for which very low density?), the Universe might become homogeneous, but still fractal.

8.3.5
Alleged Abnormal Redshifts

In the years 1960–1963, the existence of quasars was discovered (San-dage, Schmitt). *Quasar* is an acronym meaning: *quasi stellar astronomical radio sources*. (Strictly speaking, there should be an "*s*" in the singular, but we will follow the common usage, hereafter.) Quasars are often divided into two classes, the QSO, and the QSS, O for Optical, S for radio source, distinguishing those objects that are more conspicuous in the visible domain, from those radiating primarily in the radio domain. We should also add that many AGNs (*active galactic nuclei*) have been discovered. They are often considered a sort of evolutionary phase between quasars and galaxies, although no proof has ever been given of this evolutionary link, except perhaps a certain continuity in morphology, and the assumption that all quasars seem to be surrounded by an associated galaxy, generally not observable.

Quasars are affected by a very large redshift $z = \Delta\lambda/\lambda_0$ often comparable to that of galaxies (up to $z \sim 1$), or even much larger (up to $z \sim 5$).

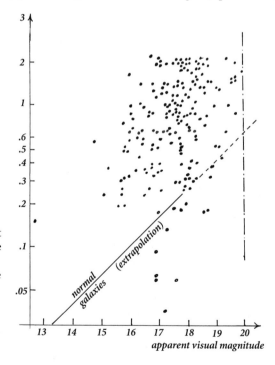

Fig. 8.18 *Magnitude-redshift diagram for the quasars* (schematical). Note the huge dispersion of data, indicating either that the z is not linked with the distance-commanded z of the Hubble law, or that the dispersion of the absolute magnitudes of the quasars is so large as to rule out considering them as a single group of objects.

They are so different in structure from galaxies, looking almost like stars, point-sources in other words, that no other distance indicator is usable except either their magnitude m or their z-value.

Using both in z-m diagrams leads to a strange result: The distribution of the z for a large value of m (less bright quasars) shows a very large accumulation of points in a band corresponding to the larger m, according to the data catalogued by Burbidge *et al.* This contrasts dramatically with the behavior of galaxies, for which the z-m relation is relatively well-defined, with a seemingly small dispersion (Fig. 8.18).

On the condition of assuming that the z is indeed a measure of distance, according to Hubble's law, this diagram shows us that whereas the less bright quasars are not much brighter than ordinary galaxies, the brightest are intrinsically much brighter than ordinary galaxies.

This result, the dispersion in quasars of absolute brightness compared with galaxies of the same apparent magnitude, has led several authors to deny the z of quasars the same significance as the z of galaxies. For galaxies, z indeed reflects the distance (whatever the meaning of Hubble's law). For the quasars, it would reflect a local velocity, a very large one, often not far from the velocity of light.

Halton Arp has attached his name to the hypothesis, resting upon a large number of individual cases, that quasars are small dense objects, essentially young, produced inside galaxies and later ejected by the galaxies. Then, the redshift z is "local," not "cosmological" as it is for galaxies. Often the grouping of quasars around galaxies seems to preferentially indicate polar ejection by galaxies displaying some active nucleus. Some cases are represented in Figs. 8.19a–c.

There has been active opposition to this result, as it does not allow us to consider active galaxies or quasars as cosmologically meaningful objects (thus reducing the range of explored z). One of the main arguments against Arp's conclusion was the discovery of gravitational lensing (Fig. 8.20), which allows us to see several images of a single quasar in the vicinity of a perturbing but much closer galaxy.

Moreover, the number of quasars in the vicinity of a galaxy (and located much beyond that galaxy, according to these critics) is increased with respect to what it is in reality, because of the "intensification" of apparent brightness associated with the gravitational lensing effect. We will not explain in this book the theory of gravitational lensing and this gravitational intensification. We already know that any light ray is deviated by a gravitational field, just as a light ray is deviated in a refracting med-

Fig. 8.19a–c. *A few cases of abnormal redshifts.* In a group of galaxies, the "companion" galaxies (fainter) are affected systematically by a larger redshift than the central galaxy. Other authors have found similar results. Often an excess of redshift is also associated with an excess of ultraviolet emission. Typical groupings of low-z galaxies (striped in the figures) with high-z quasars. Note the alignment of quasars with the polar axis, suggesting the hypothesis according which the quasars have been ejected by the galaxy. **(a)** NGC 3384, **(b)** NGC 3516, **(c)** NGC 5985. (All according to Arp and colleagues)

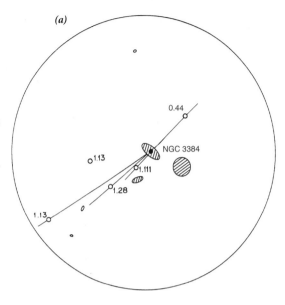

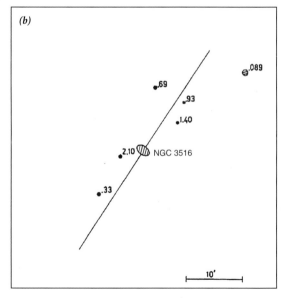

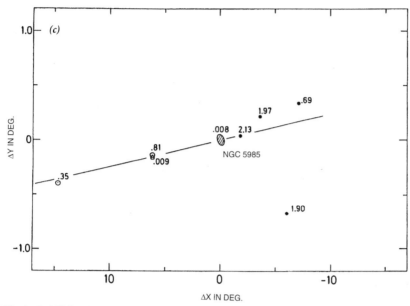

Fig. 8.19c NGC 5985.

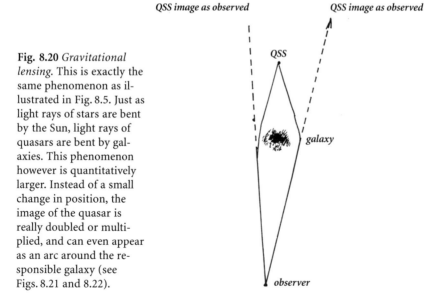

Fig. 8.20 *Gravitational lensing.* This is exactly the same phenomenon as illustrated in Fig. 8.5. Just as light rays of stars are bent by the Sun, light rays of quasars are bent by galaxies. This phenomenon however is quantitatively larger. Instead of a small change in position, the image of the quasar is really doubled or multiplied, and can even appear as an arc around the responsible galaxy (see Figs. 8.21 and 8.22).

Fig. 8.21 *A classical case of gravitational lensing (after Stockton): the double quasar Q 0957+561.* The two images on left are in appearance two almost identical quasars. Assuming they are a double image of one single quasar, one can subtract the image on top from the bottom left image. This results in the figure at the right. At the bottom right appears the image of the brightest galaxy of the deflecting cluster.

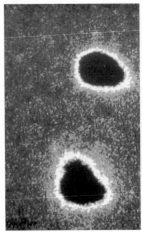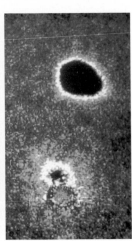

ium, glass or water. A gravitational lens acts much like optical lenses. Fig. 8.21 shows a clear case of gravitational lensing, that of the "double quasar" 0957+561.

The problem is very much disputed. A large number of arcs of high or very high z have already been identified, with a high degree of certainty, as being the result of gravitational lensing effects. Some other cases, however, often used against Arp's claims, suggest instead morphological differences between the several observed apparent quasars, indicating that they are not the images of one single quasar, thus giving some weight to Arp's allegations. Such is the so-called Einstein-Cross (Fig. 8.22).

A conservative attitude, which I will follow here, is to say: there are gravitational lensing effects, and images of real quasars given by this mechanism. There are, however, other quasars which cannot be accounted for by any effect of that sort. A whole range of objects affected by some "not simply cosmological" z exists, from the ordinary galaxy for which the z is indeed a distance indicator, to the quasars of large z, or "abnormal redshift," for which the z results from the addition of the distance calibrator z_{dist} to some intrinsic z_{intr} due to causes that are still hypothetical. We will have to come back to the hypotheses that might account for them.

A recent group of data is, in this sense, quite disturbing. I refer to the periodicity discovered in the quasars' z-distribution by Burbidge *et al.*, by Carlson, by Arp, and by Fang *et al.* This discovery first suggested a

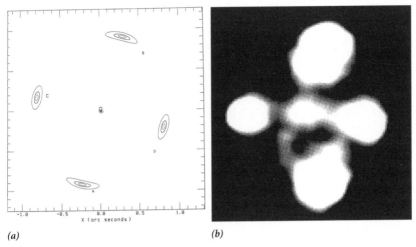

(a)   *(b)*

Fig. 8.22a,b *An ambiguous case: the Einstein cross.* When discovered, the Einstein cross, a configuration of four quasars having the same large redshift z around a central galaxy having a much lower redshift appeared to be a splendid confirmation of gravitational lensing, the four quasars being only images of one single quasar located at a very great distance behind the galaxy. The computation of the gravitational lensing responsible gave rise to the configuration illustrated in **(a)**. But in the image observed later **(b)** with the Hubble Space Telescope, Arp noted the bridges of light (or matter?) between the five objects, their shape quite different from the arc-shaped images of the computed configuration, and he concluded that the five objects are all real and physically associated quasars being perhaps ejected by the galaxy, as in Fig. 8.19b. The field is open for more observations and theories!

rather large periodicity (of Δ $(1+z)$ = 0.089, according to a very careful statistical analysis by Depaquit *et al*). The analyses by Tifft of the Coma and Virgo clusters of galaxies, and a more recent analysis (Fig. 8.23) by Napier and Guthrie suggested a much smaller periodicity of 70–75 km/ sec and/or 37.2 km/sec. One of these effects does not contradict the other; several periodicities may be superimposed, as in the case of several physical oscillatory phenomena. Again, the questions are: why? and how? There are some answers, but they are not yet accepted very widely.

Are abnormal redshifts artifacts, as suggested by the adherents of the standard cosmology? Or are they real? Clearly, as soon as some observational fact is suggested, great care should be taken in confirming or falsifying it. It is unfortunate that, whereas Arp and some of his colleagues have been working constantly at finding new "abnormal" redshifts, the adherents of the "classical standard cosmology" have never made a real

Fig. 8.23 *(After Napier and Guthrie) Periodicity of 37.5 km/s in the redshifts of galaxies of the Local Supercluster of galaxies.* When analyzed by the statistical method using the correlation function displayed in the figure, this sample of 261 redshifts has shown a very pronounced periodicity. What could be the cause of this periodicity? Is it real, or spurious? In the Virgo cluster, and in other samples, redshifts had shown a periodicity of 75 km/s. The Napier and Guthrie results confirm fully the effect discovered by Tifft in the seventies, from the analysis of binary stars, and within the Virgo cluster.

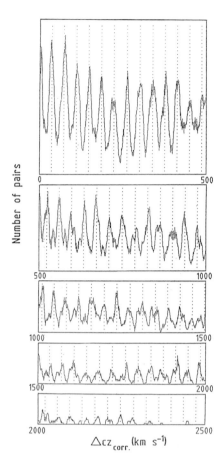

effort to remeasure the data pertaining to these findings. We feel that, at the present time, the evidence for abnormal redshifts, whatever their cause, is truly convincing.

8.3.6
The Distribution of Quasars in Space

If one assumes z to be a measure of the quasar distance (contrary to Arp's evidence and convictions), the periodicity of their z indicates that the Universe has a peculiar structure, stratified in a few layers. As the Earth is obviously not at the center of the Universe, this peculiar distri-

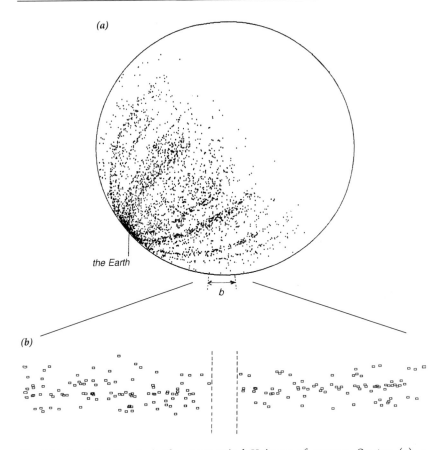

Fig. 8.24a,b *(After Souriau) The symmetrical Universe of quasars.* On top **(a)**, a spherical projection of the Universe, and of all the quasars known up till now, assuming them to be at the distance fixed by their *z*-value and Hubble's law. The location of the Earth is indicated. At the bottom, part b of the same figure is "compressed," in order to demonstrate a zone empty of quasars at the equator of the sphere. The Souriau theory (see page 504) assumes that this equator divides the Universe in two equal parts. On the side of the Earth, there is "matter." On the other side, there is "antimatter." At the equator, the annihilation of matter by antimatter is the cause of the absence of quasars, and the presence of a strong gamma-radiation.

bution, if corresponding to a real layered structure, is sort of blurred; and no effort, to our knowledge, has been made to "deblur" it.

However, another distribution effect has been extremely well-studied, from a statistical point of view. In a plot where 3-dimensional space is projected on one plane, provided the choice of the projection orientation is done properly, a zone void of quasars has been detected by Souriau that is statistically significant (Fig. 8.24). We shall return to Souriau's very peculiar and interesting cosmology.

8.3.7
Does the Observable Universe Show Traces of Evolution?

This question is an essential one; but so far replies to it have not been conclusive. It is essential because, if there is such an observable evolution, the stationary Universe advocated by Einstein and, later, by Hoyle, Bondi and Gold, (but not the Universe advocated in 1998 by Burbidge, Hoyle, and Narlikar which is only "quasi-stationary") can no longer be considered a valid cosmology. What evidence would indicate this evolutionary process? It would be necessary to count the galaxies of various morphological types, the radio-sources, and the quasars, and estimate whether the density (absolute and relative) of the number of these objects is a function of distance, or better a function of the z-measure.

The counting of radio-galaxies alone, in absolute values, gives us a reply (Fig. 8.25). If there has been a real expansion, the number of objects per volume unit "located" (in the standard interpretation of the redshift) at a large enough z, would be larger than the number of objects per volume unit located at a smaller z, near us (where $z=0$).

Such a count has actually been done, but the conclusions are not clear. They depend strongly upon the sampling of numbered objects.

They have led different authors to contradictory solutions. Quite recently, statistics concerning the distribution of galaxies of different morphological types up to $z \sim 4$, as obtained by the Hubble Space Telescope in the fields chosen for being best for extremely deep exploration, seem to indicate, at high z, an excess of elliptical galaxies, and a lack of spirals.

The relation between the apparent size of either optical or radio-sources and their z value is also an evolutionary index, for which we could reach similar conclusions. So is the variation with z of the ratio of large-disk galaxies to galaxies displaying a weak-disk structure. However, the statistical data are no more than suggestive. They are inconclusive in favor of some evolution, within the range of measured z.

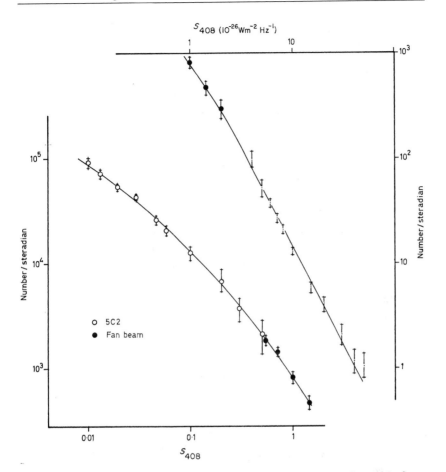

Fig. 8.25 *The counting of radio-sources as an indicator of expansion.* This figure shows the relation (as measured by Pooley and Ryle in 1968) between log$N$ (ordinates), N being the number of radio-sources per unit of solid angle, and log$S$, the flux density at a frequency of 408 MHz. The detailed analysis of the steepening of the slope of that curve (power of the order of –1.8), and the flattening for low brightness, suggests the reality of the evolution of the Universe from distant sources (observed now as they were long ago, hence statistically younger) to nearby ones.

8.3.8

The Age of Globular Clusters

Globular clusters, about one hundred in our own Galaxy, distributed according to a quasi-spherical distribution, are undoubtedly remains of the very early stage of the life of the Galaxy before it started to flatten in its collapse and rotation. What is their age? This age is a minimum value for the age of the Galaxy, hence for the age of the Universe.

The age of globular clusters is determined by comparison between the observed color-magnitude diagrams (Fig. 8.26) developed for each cluster and the theoretical curves constructed for a sequence of "theoretical" clusters of a wide range of ages. Theoretical and observed curves are based on an identical initial chemical composition (the "cosmic" composition, with its uncertainties, in particular with the ratio Z/H of the abundance of heavy elements to that of hydrogen), and an identical structure and set of evolutionary processes. This is a sequence of which the only free parameter is the mass.

Fig. 8.26a,b *The age of globular clusters.* In **(a)**, the relation between the surface temperature and the absolute magnitude of stars in clusters. The "turning point" where the stars of a cluster no longer belong to the "main sequence" of stars determines (scale at the right, which depends upon the state of the theory) their age. Open clusters (Pleiades, Hyades...) have ages on the order of 10^7–10^9 years. A globular cluster such as M67 has an age of about 10^{10} years. For **(b)** see next page.

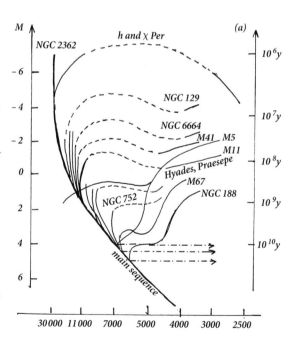

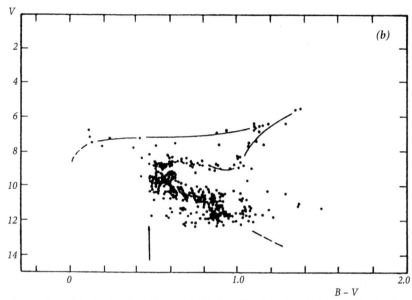

Fig. 8.26b In (b), the detailed diagram built for M67: the color index, on the abscissa, is a measure of the surface temperature of the star. The apparent magnitude is on the ordinate.

The results are of course rather dispersed. Not all globular clusters are born together. There is no clear reason a priori for that. What is important for us is the age of the oldest of them.

Even so, the difficulties are considerable. There is the difficulty concerning the abundance of heavy elements; but also the effects of physical diffusion processes which may contribute to the replacement of the central thermonuclear core by hydrogen having diffused from the outer regions. Not everything is known about stellar evolution, although this is perhaps the best established part of astrophysics in our time, the real "conquest" of 20th century astrophysics.

Several authors have worked in this field (see Table 8.2). As all of them take more or less similar models as comparison objects, they may all be affected by the same type of systematic error, an underestimation perhaps, if one assumes that the stellar core is constantly renewed with fresh hydrogen by some process (an idea for which there is some basis).

1.5×10^{10} years seems to be a reasonable estimate of the minimum age for the Galaxy, hence a minimum age for the Universe. No realistic evidence has been offered to suggest that clusters could belong to other

Table 8.2. *The age of some globular clusters (in years)*

M71	8×10^9
47 Tuc	10^{10}
M15	10^{10}
M13	10^{10}
M5	10^{10}
M92	12×10^9
M3	12×10^9
NGC 5466	12×10^9
ω Centauri	15×10^9

Note that these ages may be affected by systematic errors due to the difficulties of the theory. The accuracy of the determination is greatly affected by the dispersion of measured data.

galaxies so far as age is concerned; some has been presented to determine the helium over hydrogen ratio. It is important to carefully follow the development of theories of stellar evolution, based on discoveries concerning the solar neutrino flux, and solar seismology. It may be that this "minimum value" of the age of the Universe will change significantly.

8.3.9
Other "Real Cosmological Facts"
and "Cosmological Hypotheses" that Are Not Facts

It is often difficult to decide whether a given group of observations has or does not have cosmological significance. Obviously, as we have just seen, the study of the Sun has yielded new results which force us to change our theory of stellar evolution, and which may in turn have a bearing on cosmological theories. Obviously, the discovery of black holes, the discovery of gravitational lensing, and of gravitational waves are important. Are they important enough to allow us to distinguish between two basically different cosmological views? Not necessarily; or at least, it remains to be demonstrated. Therefore, all observed facts should be scrutinized with this question in mind; and perhaps, one should not oversimplify the global description of the Universe, as is often done, mostly by the adherents of the standard cosmologies.

A greater difficulty in cosmology is the temptation to fit cosmology to facts that are completely hypothetical. One classic example is the assumed stationary nature of the Universe, which led Einstein to introduce the cosmological constant Λ. Another example, more in fashion nowadays, is to assume that the average density is equal to the "critical density," at the limit between ever expanding models, and models ending in a final contraction. This puts a great emphasis on the quest for unobserved, or invisible mass (which should be of the order of 10 times the mass so far observed). There is, of course, nothing wrong in looking for heretofore unobserved objects, or dark matter. But why this factor 10? Why assume the density must be equal to the critical density? This is an hypothesis of an absolutely metaphysical nature.

There are many such examples of a priori ideas, imposed upon scientists by some cosmological preconception, which became the basis for a confirmation of that very same cosmological view!

* 8.4
The Standard "Big Bang" Cosmology

Let us define the standard "big bang" as a cosmology (i) which aims to account for the first three facts quoted above, in Sect. 8.1.3 and (ii) which satisfies the GR theory, and the Einsteinian cosmological principles (or hypotheses), of homogeneity and isotropy.

It begins with the Friedmann-Lemaître solution of the GR, which we have briefly described in Sect. 8.2.3 and goes on to add the Gamow conception of the hot big bang, from time $t=0$ of the Friedmann solutions.

8.4.1
The Geometric Evolution of Space-time

The metric to be used as in Sect. 8.2.2 is the metric adapted to the uniform-and-homogeneous structure of the Einstein Universe, i.e. the so-called Robertson-Walker metric, described in Chap. 8.8.2, p. 530, in its polar coordinate form. Of course, as suggested by Friedmann and Lemaître, there is no cosmological constant in the model – or, more precisely, the value of the cosmological constant Λ is assumed to be equal to zero.

Then there are only a few quantities that can be expressed as a function of time:

(1) *The geometrical space-time quantities*, i.e. the radius of curvature, or "scale factor," R of the 3-dimensional space projection of the Universe, and the sign of this curvature, k, which may be equated either to 1 (closed space, like a sphere), or −1 (open space, like a cylinder, or a hy-

perboloid); $k = 0$ (the "critical" model) corresponds to the strictly Euclidian Universe (see 8.8 about the definition of flatness, and of curvature, and 3-dimensional space and 4-dimensional space-time).

(2) *The physical quantities*, i.e. the density of radiation and of matter (the sum of which, ϱ, measures the density of energy, provided convenient units are used), the pressure p, the temperature T, and the respective proportions of the various elementary components of this Universe, from radiation to elementary particles, and to matter organized in stars and galaxies.

(3) *The cosmological quantities*, i.e. the Hubble ratio H, measuring the rate of the apparent expansion, and q, measuring the acceleration of the apparent expansion, i.e. the rate of variation of H, – the cosmological constant Λ being equated to zero (after Friedmann and Lemaître).

For each of these quantities, variable with time t, the index $_0$ indicates the present epoch ("now") of the evolution of the Universe (or a quantity as measured by the local observer).

We shall not describe the detailed calculations, and will give only the results. Still, there are several ways to build a cosmology, according to the choice of the constants intervening in the Robertson-Walker metric, k and R, essentially.

Without limiting ourselves to the singular points of the solutions, we must state that solutions maintained by the "standard cosmology" have a mathematical singular point at the "origin," at which the scale factor R is zero, the density infinite. They also all assume a pressure equal to zero in the present epoch of the expansion they describe, if not in the "primordial" phases of the Universe.

The flat model ($k=0$), in essence the Einstein-de Sitter model (1932) (which must not be confused with either the Einstein or the de Sitter models, which we have already described) is simple. In Fig. 8.27, we see:

$$R \propto t^{2/3}, \varrho \, \alpha \, 1/t^2, H \propto 2/3t, \varrho = \varrho_{\text{crit}} = 3H^2/8\pi G.$$

R is of course not the "radius" of the Universe (infinite, for $k = 0$!) but the scaling factor introduced in the line element (see p. 530).

For negative or positive curvature models, the figure sufficiently describes the behavior of R, in the three cases. Closed models and open models behave similarly to the case of $k = 0$ near the origin $t=0$. Closed models (where $\varrho > \varrho_{\text{crit}}$) give rise to a strong contraction at a large value of *t*, where *R* returns to the value 0. Bad neologisms, the "big crunch"or "big splash," are often used to designate such a contraction to a state of infinite density. On the other hand, flat ($\varrho = \varrho_{crit}$) and open models (where $\varrho < \varrho_{crit}$) expand forever (Fig. 8.27).

The existence of some initial singularity, complicated by the possibility of another one at the end in the case of the closed models, brings up an interesting problem, often discussed. There is nothing in the mathematical description of this problem which rules out a "splash" before the "bang," or a new "bang" to occur after the "splash." The two successive "cycles" may occur. Is this, however, thermodynamically correct? Actually, we know that during one cycle, the Universe, if considered as an isolated system (but have we the right to assume such a thing?) has gained entropy. It cannot be physically identical to the preceding and following cycles! We have discussed that point above, in Sect. 8.2.3, p. 448.

In essence, the increase of entropy from cycle to cycle is an increase of entropy between the successive states of the Universe at the time of its maximum extent. Therefore, successive cycles should appear as in the

Fig. 8.27 *The three main Friedmann-type models. Assuming $\Lambda = 0$, the only parameter is the curvature factor k, which can be negative, zero, or positive corresponding (resp.) to open, critical, and closed models, or (resp.) to q < 0.5, q = 0.5, q > 0.5. It has been shown that the "critical" value ϱ_{crit} of the density allowing the model to be "critical" is about 10 times the actual measured value ϱ_0. This may imply, if one insists on keeping the density equal to the critical value, that 90% of the mass of the Universe is not yet observed, and perhaps not observable (dust, white dwarfs, black holes, neutrinos, etc.?).*

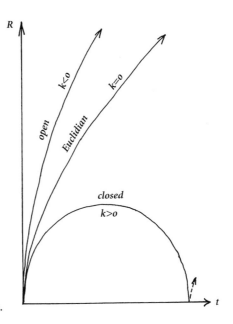

Fig. 8.28 *Hubble time.* The time $t_0 = 1/H_0$ deduced from the extrapolation of the Hubble rate of expansion is an overestimation of the "age of the Universe" elapsed from the time of the assumed big bang ($t=0$), as it results from Friedmann-type models.

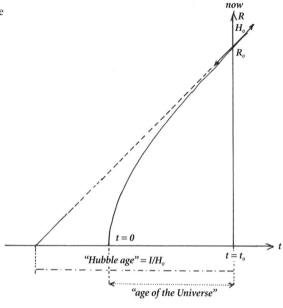

Fig. 8.9, p. 448, where the quantity measured on the ordinate is in essence the radius of curvature.

It is often thought that, as observations show that the observed density is smaller than the critical density (by a factor of 10 roughly – and a factor of 10 is, after all, rather small), the real Universe actually has the critical density, and that we must, therefore, look for some "hidden" mass. It is true that there is invisible or unobservable matter in the Universe, but why should the density be exactly equal to the "critical density?" This is, in our view, a highly idealistic point of view, essentially metaphysical. There is no good reason why the density should not be 5 times smaller, or perhaps 100 times larger than the density of the observed matter!

One should note that, in an open or closed model, as well as in the Einstein-de Sitter model, the curvature of the relation $R(t)$ is always negative, if, as assumed, $\Lambda = 0$, meaning that the quantity $1/H_0$ (Hubble's age of the Universe) is a maximum estimate for the real time elapsed between the time $t=0$ and the time $t=t_0$ of the present epoch. Actually, the ratio $W=\varrho/\varrho_{\mathrm{crit}}$ is critical in the evaluation of t_0, as shown in Fig. 8.28. It is linked to the parameter q, as follows in the Einstein-de Sitter model: $q=1/2$ for $k=0$; $q>1/2$ for $k>0$; $0<q<1/2$ for $k<0$.

8.4.2

The Hyperdense Hyperhot Big Bang (Standard Model)

What happens, physically, at the times very near $t=0$?

Near the origin of time, it is quite difficult, due to the high density, to accept easily that $p = 0$. The situation must be that of a hot gas, not a pure matter situation, but a pure radiation situation. The photon field exerts a pressure which is a third of the energy density of radiation (radiation pressure). This hyperdense medium is called "hyla" (ὕλη) by Gamow.

The radiation pressure decreases with time as R^{-4}, and t^{-2}. The density ϱ decreases as R^{-3}, and $t^{-2/3}$. Therefore, the densities of radiation and of matter do not behave in the same way. The earliest Universe is a radiation Universe. Later, matter takes over, and the density of energy is mostly the energy of matter. In fact, an essential item of the standard cosmologies is the fact that the equation of state of the Universe implies an adiabatic expansion, at constant entropy. It is only in the later phase, the phase where stars and galaxies, and life are forming, that entropy is created.

The observed values of H_0, of ϱ_0, and of q_0 (the latter is very difficult to assess!) must help us in selecting the constants necessary to adjust the solutions of the equations to these boundary conditions. A solution is represented in Fig. 8.29.

This solution can be described, schematically, as follows:

(a) $t=0$ (strictly speaking!) – *Infinite density, infinite temperature.* (Note that these very concepts are in essence meaningless. No physical measurement can give an infinite result. This is a purely mathematical notion.) The physics then is of course unknown, therefore unspecified, but essentially irrelevant at that stage of the model. At such high temperature, there is a pure thermodynamic equilibrium among all species (photons, neutrinos, protons, etc.) that are linked by reactions between particles of high energy, even among species that we have not yet taken into account. In such conditions, T is an increasing function of ϱ and decreases quickly as t decreases.

(b) $t < 1$s – *Neutrinos have decoupled* from the other particles because of their long lifetime. The Universe is transparent to these neutrinos, but all other particles are coupled, by virtue of thermodynamic equilibrium.

(c) $t < 10$ s – This *situation of equilibrium* lasts for 10 s. Then T has dropped to 10^{10} K and ϱ to 10^5 g cm^{-3}, values that are perfectly within

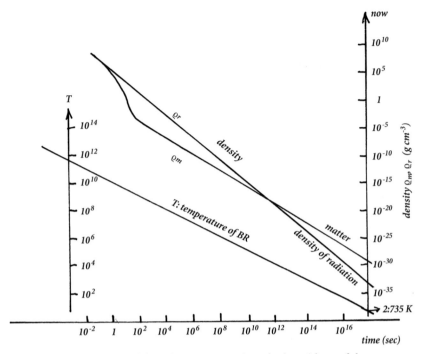

Fig. 8.29 *The standard model.* In this representation, the logarithms of the tempera-ture of the radiation, *T,* and the density (density of matter ϱ_{matter} and of radiation ϱ_{rad}) is on the ordinate. The time *t* is on the abscissa. Due to the deceleration of ex-pansion, this is not a strict linear scale, if one wants a representation in which the be-havior of the representative curves is linear with respect to the scaling factor.

the realm of manageable physics. Electron-positron pairs are recombin-ing. The Universe is made of photons, protons, neutrons, electrons, and light nuclei in thermodynamic equilibrium.

(d) 10 s < *t* <1000 s; *T* decreases as 1/*t* – *"Primordial" elements form from the basic particles.* In this short period, light elements are formed in the proportion we observe now. This was more or less shown by the earliest investigators, Gamow, Alpher and Herman, Fermi and Turke-vitch, and more recent authors have confirmed it. At the end of this peri-od, *T* is too small for the light elements to modify their abundance in any sensible way.

(e) 1000 s < *t* < 10^5 years – For a relatively long time, *the matter-ra-diation equilibrium continues to hold,* but temperatures are too low to modify the abundances. The *opacity of the Universe is still great.* The equilibrium is maintained by a constant transfer of energy from radiation to matter.

(f) $t = 10^5$ years – The less dense Universe now becomes *transparent*. *Radiation and matter decouple*, and the energy of photons no longer affects matter. The two "temperatures" evolve separately, as $1/t$ for T_{rad}, and $1/t^2$ for T_{matter}. The Universe then ceases to be radiation-dominated and is finally matter-dominated. Photons are not energetic enough to influence the energy of the matter. They will evolve separately, with little interaction.

(f1) 10^5 years $< t < t_0$ – Radiation density decreases as $1/t^3$. Radiation cools at the rate of $1/t$. The measurements of the background radiation (BR) allow us to fix the time scale of the successive steps described. "Now" (in the modern world), the temperature of radiation is about 3 K (a sort of "flag" temperature, symbolic so to speak), more precisely 2.735 K $\pm$ 0.0001. Its density is now 10^{-33} g /cm^{-3}, when expressed in mass units.

(f2.1) 10^5 years $< t < 10^8$ years – At about $t = 10^5$ years, the temperature has reached a few thousand degrees, the density $\varrho = 10^{-20}$ g cm^{-3}; *the plasma (protons and electrons) can recombine*, this process being essential in the decoupling of matter and radiation.

(f2.2) 10^8 years $< t < 10^9$ years – The photon pressure is still high. It prevents matter from condensing into stars and galaxies. *Small perturbations will start to develop* and form clusters of galaxies, superclusters etc., growing larger and larger, denser and denser.

(f2.3) Still 15 or so billion years to go! – *Stars, galaxies, clusters of galaxies, and perhaps quasars, form and evolve.* We are no longer investigating a cosmological problem!

This is, in a considerably abstracted form, the standard cosmology of the "big bang," according to most authors of the seventies. It explains correctly, in a relatively simple and straightforward way, with the same choice of parameters, three important observed facts: (i) the *rate of expansion*, as measured now, H_0; (ii) the *BR* as observed now, i.e. as a blackbody of $T_0 = 2.7$ K; (iii) the *relative abundances of light elements* H:D:^{3}He:^{4}He:^{7}Li, as measured now and on Earth. The theory fits the observations properly with a flat Universe ($W = 1$, $q = 1/2$), on the condition that $\varrho_0 = 2 \times 10^{-31}$ g cm^{-3}, the observations not enabling us to really determine the present acceleration of the expansion, i.e. q_0.

This success, with a relatively *simple* analysis, became a completely accepted truth from 1964 on with the discovery of the background radiation. Clearly, the assumed existence of an "initial singularity" led people to an obvious metaphysical interpretation. In 1951, Pope Pius XII went so far as to identify the *big bang* with the *fiat lux*. This very assimilation, on the other hand, has led many scientists to be very careful and suspicious of the a priori beliefs in the unconscious, or subconscious, minds of the authors and adherents of the standard big bang theory.

We shall see that there are other arguments against such a simplified version of the standard cosmology.

It is obvious these big bang views, defended with vigor, and even sometimes with alacrity, are linked to some Platonic view of the world. Quite clearly, the δημιουργός is back and there is a strong odor of spirituality entering the description of the physical Universe, notably in the first popular books devoted to the big bang, such as those by Reeves, Hawking, Weinberg, and others.   ✳

8.5
The "New Big Bang"

8.5.1
✳ The Need for "Inflationary" Models

If we consider one situation that is based on important cosmological facts, but which does not enter the standard cosmology, we face the contradiction between:

(i) the observed isotropy of the 3 K BR, described in Sect. 8.3.2 and Figs. 8.14 and 8.15; and
(ii) the observed strong anisotropy and non-homogeneity of the distribution of matter, described in Sect. 8.34.

Why is this so? Are these facts really contradictory? The isotropy of the 3 K BR is easy to understand, since the Universe was highly coupled before the time when z was about 10^3–10^4 (or roughly at $t = 10^6$ years). Let us use the concept of the light cone, which we introduced earlier (Sect. 7.7.4). The entire part of the cone before that period is obscured to our eyes, but the radiation we observe, which is isotropic, comes from that age. At that time, two very distant points in the Universe, such as A and B, should have the same past. This could explain the fact that the ra-

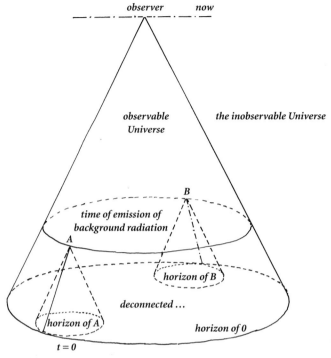

Fig. 8.30 *The observable Universe.* From the point of view of the observer, now, the past can be seen, inside the "light cone" (see Fig. 7.20). This cone, in the standard model, is limited at $t = 0$, by a certain space which, in essence, fixes our "horizon." Two events in the past such as A and B are completely disconnected, with different horizons. Why should they appear to suggest the isotropic background radiation?

diation emitted is the same, having the same spectrum (Fig. 8.30). However, the light cone of A and the light cone of B do not intersect, even at $t = 0$, at least in the standard cosmology. In other words, the "horizons" of A and B are so small and A and B so distant that one does not see how they could have intersected in the period prior to the decoupling of light and matter, but since the "big bang." The isotropy of the big bang cannot be explained by this model.

A way around this difficulty is the concept of "inflation," suggested in the eighties by Guth, Linde, and others. This implies that the light cones of the past are not formulated as in Fig. 8.30, but rather as in Fig. 8.31. Then, in the period $0 < t < 10^{-32}$ s, the expansion would have been much more rapid than in the standard model. Not only would the expansion

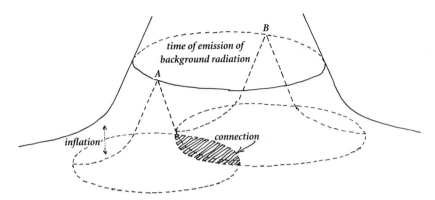

Fig. 8.31 *The light cone of the inflationary Universe.* Two events A and B are now connected, in that their horizon extends to the limits of our own horizon and beyond. Now there is no difficulty in reconciling an anisotropic distribution of matter, and an isotropic field of radiation coming from all points such as A, B, etc.

have been much quicker, but so rapid as to imply an increasing rate of expansion. This is the inflationary phase. Actually the expansion would have been 10^{28} times as large as the "standard" expansion during a phase 10^{-35}–10^{-32} s after the big bang. The volume of the Universe would have increased, during that short but decisive inflationary phase, by an enormous factor – on the order 10^{40}, or even more (Fig. 8.32)!

The light cone then has a different shape, as in Fig. 8.31, and the homogeneity of the 3 K BR as now observed is a reflection not only of the Universe at the time of decoupling, but of the Universe near $t = 0$. Of course, one might ask how the inflation can occur within the constraints of a GR model, as we have exhausted all mathematical models without encountering this inflationary behavior. We will note that the physics of GR is indeed questionable at these high densities and high temperatures During the inflation, the behavior of the Universe is not adiabatic, at is the "old big bang".

8.5.2
The Grand Unification Associated with Inflation

The response to this question lay in a very powerful idea which had been floating around more or less since Maxwell: the idea of the unification of forces in physics. We know of several kinds of forces in the universe of

Fig. 8.32 *Inflation.* At the top of this highly schematic sketch, the standard Universe, evolving adiabatically (RT = constant). At the bottom, the inflationary Universe. The adiabatic relation has been maintained for most of the time up to now (dotted parts of the curves indicate an enormous change of time scale). At a very early time, however, (between 10^{-34} and 10^{-32} s after the assumed big bang), there was a quick cooling due to some breaking of symmetry in the elementary interactions, followed by an increase in scale factor by a factor of 10^{40}, and a corresponding reheating of about 10^{40}, followed by a slower return to "normal."

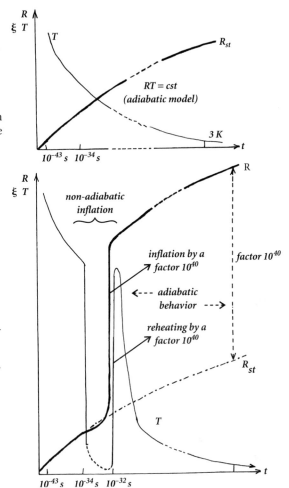

terrestrial physics: *The gravitational force* (in $1/r^2$) which exerts at the cosmological and cosmic scale, generally as it does on the laboratory scale. This is the only one so far known to act on the scale of the Universe. On a smaller scale, *electromagnetic forces* act, but as there are negative and positive particles (protons and electrons), the magnetic field acts only at the scale where matter is not globally neutral, i.e. on a relatively small scale – typically intra-atomic. In essence, Maxwell's theory was a unification of the electric and magnetic forces. Since then, there has naturally been a tendency towards a much more ambitious project:

unifying the electromagnetic forces and the gravitational forces in only one type of force. This unification has not yet been achieved.

On a still smaller scale, we now also know of the existence of forces acting between elementary particles, *weak interactions* (at small distances, inside the nucleus of an atom), and *strong interactions* (at still smaller distances, inside the components of the nucleus, typically quarks). We should note here that in spite of some attempts, it has never been possible to uphold any argument in favor of a fifth type of interaction, or a sixth. The natural trend is to aim at a complete unification of these forces. We shall describe in some detail the steps towards such a unification, but we should note that the idea of looking for it is basically a metaphysical idea. It aims at an aesthetic view of the Universe. It reflects more or less a typically Pythagorian philosophy. Is there a real need to unify the forces? Does the truth of the Universe imply this unification?

To clarify the ideas, and to facilitate the reading of the following paragraphs, Fig. 8.47 (p. 537) and Table 8.6 (p. 541) describe in a simplified way the now recognized particles of modern physics, and their interactions, implying "mediating" particles.

The first question is: what do we mean by unification? Can we really "unify" these four different types of interactions? We essentially mean that before the breaking of symmetry in the history of the Universe the four interactions were the same, identically the same. Although there is not much physical argumentation for that (at least for unifying the gravitational interaction with the others), speculations about it may appear to be promising.

It means that we have to start from a "symmetrical Universe." This notion is defined in Sect. 8.9. The breaking of symmetry is, of course the real trick (Fig. 8.33).

Once the energy of the medium is increased by going further back in time, on from the present the symmetries can be re-established in succession, and all the forces united into a single one, all coupling constants being then equal to each other. One should note that this implies a variation of fundamental constants with time. We shall discuss models with varying fundamental constants. Unfortunately, they are difficult to substantiate. This huge theoretical effort, based on the development of group theory applied to interactions, is the GUT, or "Grand Unification Theory" (i.e. the unification of electromagnetic, weak, and strong interactions). Beyond the GUT is the TOE (a very strange acronym standing

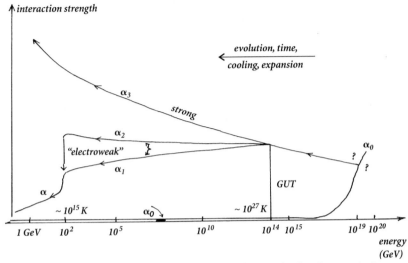

Fig. 8.33 *The Grand Unification Theory, GUT (schematic).* The three main interactions are characterized by coupling constants, as described in Section 8.9. The three constants α_1, α_2, α_3, have very different values in the present-day Universe. When the energy of the medium (or its temperature) increases, or when its interaction distance decreases, these three constants become (by steps ?) equal to each other. This is the GUT. The constant α_0, characteristic of the gravitational interaction, is equal to the three others only at very high energies ($> 10^{19}$ GeV, i.e. the *Planck energy*). The unification begins with the inflation and disappears then progressively between $t = 10^{-35}$ s and $t = 10^{-11}$ s. The evolution of the Universe is represented from right to left on this figure.

for "Theory Of Everything," i.e. the unification of the four fundamental interactions) for which we have no observational or experimental evidence (whereas the unification of the two weakest forms of interaction could be demonstrated in very large particle accelerators). It is not impossible that the increased power of terrestrial accelerators will succeed in this, but it is not at all certain. In any case, the accelerators must be regarded as instruments of high energy astrophysics, looking into the depths of the Universe. The observation of proton decay is, in that light, a very important experiment, being performed in several laboratories. For the time being, several predictions of GUT have failed to be upheld by the observations.

The successive breakings of symmetry in essence imply a change in the equation of state of the matter. This is the trick through which inflationary models could be solutions of the GR equations, in which we need, in any case, to re-establish a non-zero cosmological constant, which now has the meaning of the "energy of vacuum," and which therefore is variable with time – as it is also from place to place, according to the distribution of density. Still, it is difficult to find in the literature any detailed model satisfying the GR equations and describing an inflationary Universe using the GUT concept.

8.5.3
Superstring Theory

In this effort towards the unification of physics, string theory appeared at the end of the sixties, in an attempt to describe the strong interactions. The idea was essentially to identify particles such as neutrons and protons with waves on some (imaginary? real?) string. When two such strings meet through some intermediary string, the strong interactions are linked to that third intermediary string. In essence, the Feynmann symbolism gave rise to something which appeared as a real, not a symbolic figure. The strings had to have some elasticity, and be able to bear the weight of several tons. Geometrically, the new vision implies that strings replace the "point" particle, and therefore occupy a part of space. This suggested that space-time had to have many more dimensions than the four which are usually attributed to it. The strings, in space-time, appear as tubes of a possibly very complex structure.

An idea which emerged was that all types of symmetries should be verified initially (i.e. immediatlely after the "big bang"). We can see (see 8.9.2) that the T parity (as the arrow of time) was certainly non-symmetrical at some early moments of the big bang Universe. But very early (probably at the time when the radius of curvature would be the same size as the *Planck value*, i.e. 10^{-43} s), even the T parity should be verified. We must, therefore, assume that some "supersymmetry," abbreviated SUSY, existed, implying an exchange between particles (fermions) having a half-integer spin ($s = 1/2, 3/2,...$) and other particles (bosons) having an integer spin ($s = 1, 2, 3$). SUSY implies the existence of other intermediary particles (gravitinos, carriers of the photon-graviton interactions). SUSY is a basis for "superstring" theory, which requires fewer dimensions than the other string theories.

Before the time $t = 10^{-43}$ s, there was a quantum Universe, where the very idea of time is meaningless. We start from a supersymmetric state, but how does this Universe then evolve? If time does not exist, there is no bang. Why should the energy decrease in the direction of the successive breaking of symmetries?

These theories are often dismissed as mere abstractions that have little correspondence to physical reality. They are considered to be a little bit like the epicycles and other geometrical tricks of the Ptolemaic theories, as H. Alfvén once noted. Therefore, there is a tendency to reject them and wait for something better to come along. On the other hand, we could also say that the atomic structure of matter appeared in theory long before it was the only obvious solution, and a great length of time before the appearance of direct observations of individual atoms. It may very well be that, influenced by observations at very high energies, we

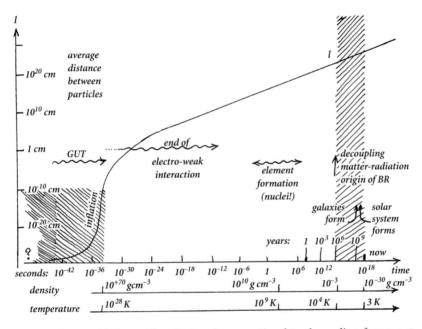

Fig. 8.34 *The new big bang.* The abscissa is proportional to the scaling factor, not represented here, and the time scale *t* is almost linear. In dark gray, the *quantum phase* of the Universe, about which nothing is really known. In light gray, at the left, the period of the Grand Unification (GUT), ending with the inflation. In light gray, at the right, the period of the observable Universe, beginning with the decoupling of matter from radiation, i.e. the emission of the background radiation (BR). Typical (rounded) values of density and temperature are indicated on the bottom scales. Note that Fig. 8.29 above occupies only the part of this diagram beginning (roughly) with element formation by nucleosynthesis. See also Fig. 8.35 on this point.

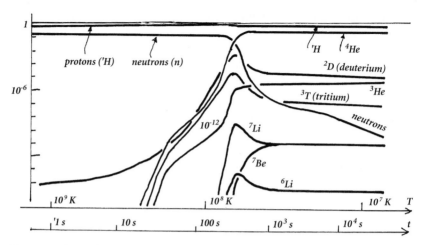

Fig. 8.35 *The nucleosynthesis of light elements.* According to both the old and new big bang models, the formation of elements has occupied only a short interval of time in the life of the Universe.

will feel forced, in due course, to accept the extreme views of the cosmo-physicists, whose efforts must certainly be considered very worthwhile.

Figs. 8.34 and 8.35 illustrate the evolution of the Universe as described by the "new big bang," with all its complexities.   ✳

8.5.4
The Quantum Universe

From what we have already written, it is clear that, whether or not one accepts the expansion, whether or not one accepts the inflationary phase and grand unification (GUT), there remains a basic question: what was the Universe before...? Before what? Before a certain time all these the-ories seem to fail, for very basic physical reasons. If scientists often quote the time $t_{Pl} = (Gh/c^5)^{1/2} = 5.4 \times 10^{-44}$ s (the so-called *Planck time*) as the time after which everything is "clear" (at least for the fans of the "new big bang"), and before which one just does not know (we do not even know what to know, or whether there is anything to know), we still have to ask the question: Could we know? Time, as a physical oriented quantity, as it enters the equation of General Relativity, has a clear mean-ing, but which one? Has this past (before t_{Pl}) left any observable relics, as may be detectable from the following period?

One obvious question, which we have already alluded to many times, is that a simple formula in which both macroscopic gravity (through G), and the quantities such as h, typical of the quantization at small scales of the Universe intervene, raises the problem of quantization of gravitational energy. That gravitational energy be transmitted through "gravitons" is the expression of a partial reply to this fundamental question.

Another investigation, which becomes crucial at this point is: what is the meaning of time? We are certainly aware of the meaning of the time of GR: a reference time, in an absolute reference frame, defined according to Mach, by the aggregate of all masses present at once in the Universe. Obviously, the GR equations imply a continuous time, which enters the equations, as t, and fits our notion of time as defined by the usual clocks, be they astronomical, or more precisely, atomic. There are no unexplained gravitational or astronomical phenomena which would impose a change in this point of view.

The quantization of gravitation, the introduction of *gravitons* in physics, is linked primarily to the need for insuring the unification of the four fundamental forces; the theory of the other three (GUT) (electromagnetic, weak, and strong) already being quantum theories. As clearly stated by Narlikar the Universe around the big bang epoch is so small as to make the usual description impossible, in terms of "action," the classical notion used for example in the (macroscopic) "principle of least action," or of "stationary action." One can actually demonstrate the GR from the macroscopic principle of stationary action, as shown by Hilbert as early as 1915. However, the fact that the classical laws of physics cannot be applied for $t < t_{\mathrm{Pl.}}$ also results from Hilberts's demonstration.

We are immediately burdened by the fact that in GR the force of gravitation is merely a geometrical effect. In quantum gravity, the very process of quantization must therefore affect the structure of space-time. In particular, the light cones could be distorted; and as shown in Fig. 8.36, this may distort the causality linking two events A and B.

An answer might lie in *conformal quantization*, which keeps the angles of the light cones unchanged. Without going into detail in this difficult theory, we will give the description of the hydrogen atom, as an indication of more general problems (always according Narlikar). In the classical (Newtonian!) description, the electron, massive, accelerated, loses energy (all accelerated particles radiate) and therefore spirals inwards and falls onto the proton, in a time on the order of 10^{-23} s. Obviously, this does not occur. By quantizing only r, the distance from

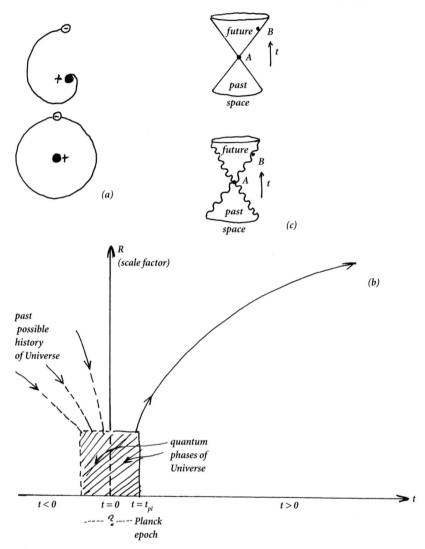

Fig. 8.36a–c *The quantum phase of the Universe.* **(a)** The electron and the proton: in the classical case, the electron winds its way into the nucleus of the hydrogen atom. In the quantum scenario, the electron remains stationary, at a distance from the nucleus. **(b)** In the Narlikar view of the quantum universe ($t < 10^{-45}$ s, in the usual units of cosmic time), the quantum phase of the Universe may have followed virtually any possible history. **(c)** Its duration is indeed not even defined, time not having a well defined meaning when distances between particles are shorter than the Planck distance. During that phase, the laws of causality are broken, as the geometry of the light cones is now affected by quantum transitions the "future" point B becomes out of the "future" come inaccessible.

proton to electron, the electron can exist for an extremely long time in a "stationary" orbit of radius $r = h^2/mc^2$, as shown in Fig. 8.36a. This could be done as well on the scale of the hypercondensed Universe of Friedmann's models, before the Planck epoch. We could introduce conformal transformations that would not satisfy the classical GR equations of Einstein. Narlikar concludes from this discussion that, generally, these new models do not have any singularity (Fig. 8.36b, see also Fig. 8.36c); the big bang models being actually extremely unlikely. So the question is: can stationary states thus exist for the Universe? It has been demonstrated that it is indeed the case, and that their characteristic scale is (not surprisingly) the one associated with the Planck time, i.e. the Planck length $l_{Pl} = (Gh/c^3)^{1/2} = 1.6 \times 10^{-33}$ cm. It is interesting that such models eliminate the need for inflation (in the sense that the horizon problem does not occur anymore). One may conclude that the pre-Planck era may be a very important phase in the history of the Universe, although it remains unknown.

⋆ 8.6
Big Bang or Not Big Bang? Alternative Cosmologies

It is clear that the standard "old" big bang was not satisfactory. Inflation is, to a large extent, ill-defined. The intrusion of modern physics is conjectural, and neither SUSY nor GUT, nor even the inflation, can be supported by very strong and final observational arguments.

Moreover, if the order in which the main elements of theory and the main arguments of observations were arrived at were changed, other theories would have flourished. We will look at some of them, from those closest to the big bang theories to those most distant, and even to the "heresies". At this point, however, a *very strong warning* is necessary. Almost none of the *working hypotheses* described in this section are really at the stage of *bona fide* theories. They cover mostly only part of the observed data, and they have been heavily criticized, sometimes on sound bases. However, we feel we must give a short survey of these heresies. Each of them may contain some element of truth, a seed that can in the future give rise to a real theory, complete and consistent with the observed data. Some of them are certainly more fully achieved than others, but the reader must be careful in this reading. We are now at the limits of explored physics, and sometimes much beyond those limits!

After all, just as we have seen for the heresies of the Middle Ages, the modern heresies could, later on, generate some important ideas about modern cosmology, the cosmology which will emerge, but perhaps not for another century! We should indeed notice, at this stage, as a follow-up to Chap. 6, that it seems that the 20th century has been marked by two profound advances in the knowledge of the Universe: our understanding of most of stellar evolution, and most of planetary evolution. Although we know that evolution is taking place everywhere, and at every scale, there is still little we can say about the evolution of galaxies, still less about the evolution of the Universe. On the other hand, we know little about the evolution of life, and nothing about the evolution from ordinary matter to living matter, i.e. about the origin of life. The future of the theories of universal evolution is indeed very wide open!

8.6.1
The GR Models Where Actual Expansion is Accepted

(i) Gödel's anisotropic model.
One limitation of the models of the Friedmann type is that they assume no rotation whatever in the description of the Universe, or, to put it another way, that they assume the general isotropy implied by Einstein's so-called cosmological principle (rotation in itself is a deviation from isotropy). A term involving angular momentum, which does not vanish in this Universe, has been introduced for example by Gödel. The implications of this model are manifold. The metrics is complex. The model implies terms containing exponential functions of the space-time coordinates. It also suggests that the cosmological constant is non-zero. Strangely enough, the trajectories (time-lines) are such that a person could travel in his own past! Gödel's model is in absolute rotation, hence test particles are influenced by it, just as the Foucault pendulum is influenced by the Earth's rotation.

We should note that Gödel's model is in obvious contradistinction to Mach's principle. In essence, Gödel assumes the Universe to have a relative rotation with respect to some absolute space. The consequence is that the Einstein field equations (on which Gödel's model is based) do not imply Mach's principle. This is perhaps the only lesson to be learned from it, as nothing has ever begun to serve as a justification of Gödel's model, which is otherwise very similar to the classical Friedmann model.

(ii) Steady-state models with expansion.

During the fifties, in a revival of older ideas, the possibility that it was not incompatible to accept the expansion of the Universe as a fact, while at the same time rejecting the 20 billion year scale for the lifetime of the Universe, came to some scientists reluctant to accept any big bang as the origin of this expansion. Their reasons were essentially philosophical, Aristotelian one might say. These scientists, Hoyle, Bondi, Gold, and later (1998) Burbidge, Hoyle, Narlikar suggested a "steady-state Universe" or a "quasi-steady-state Universe." To account for its permanent expansion, they suggested the *continuous creation* of matter, everywhere, at a rate sufficient to replace the matter subtracted in the expansion, in such a way as to maintain density at its constant value. The new "perfect cosmological principle" states that the Universe looks the same at all points, in all directions, and also at all times. One can formulate the model's space-time geometry, and specify its metrics, predict from it phenomena that can be observed, and even explain values of the measure of q_0.

The problem was, of course, to physically explain the continuous creation of matter. Galactic nuclei seemed to be likely natural candidates for the processes of continuous creation (Hoyle, Narlikar, Arp). The appearance of the BR does not seem possible in this model, according to Weinberg's criticism. The authors deny such a failure. The debate is now (2000) progressing very quickly. Hoyle and Narlikar have formulated a theory of action at a distance implying retarded effects, and this offers a contribution to the new model which must be scrutinized carefully.

The "quasi-steady state model" proposed by Burbidge, Hoyle, and Narlikar (hereafter BHN), which we shall now describe, deserves serious consideration. It is extremely complete and rigorous in its deductions.

(iii) The BHN "quasi-steady state cosmology" (QSSC).

This is probably the most coherent and successful attempt in the current literature to take all the cosmological facts into account without positing a big bang.

Without going into too much detail, we can say that the theory is a *bona fide* solution to the General Relativity equation. It allows for the creation of matter without violating mass-energy conservation. Mass is created in explosive events, which incidentally had been assumed much earlier by Ambartsumian as "explosive evolution of hyperdense matter." Mass is created in the form of "Planck masses" of the value $m_0 = 10^{-5}$ g. That quantity is linked by simple relation to the gravitational constant G.

This mass m_0, formed at any phase in the life of the Universe, decays into equal numbers of the 8 species of baryons, namely: p, n, Λ, Σ^+, Σ^-, Σ^0, Ξ_0, Ξ^-, with a negligible amount of other particles (see Chap. 8 p. 541). Then first, n and p combine into He; the rest decays into n and p. A crude estimate gives the ratio He/H=2/8=0.25; but some n and p (a computable fraction y) remains free at the end of the nucleosynthesis; hence one obtains He/H=0.229, remarkably close to the observed value. Other light elements can also be accounted for.

The BHN Universe oscillates and we are now in a phase of expansion; but expansions follow contractions (Fig. 8.37).

Of course, as there is no "initial singularity," the temperature does not go as high as in the new big bang. Nevertheless, there is relic starlight from many previous cycles that has to be thermalized, after periods of time counted in millions of millions of years. A precise estimate of this shows that this equilibrium radiation has a value of 2.7 K. The spectrum is indeed fully thermalized by multiple scatterings of light by iron-graphite intergalactic"whiskers" of dust, formed in supernovae explosions, or through continuous ejection of matter by stars.

It is interesting to note that the BHN model can be worked out to conform to actual predictions. One of them is the fractal distribution of mat-

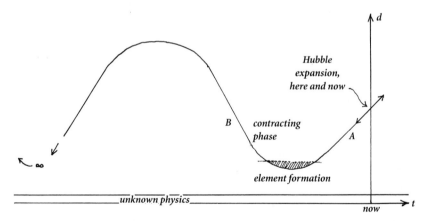

Fig. 8.37 *The oscillating quasi-steady state Universe.* According to Burbidge, Hoyle, and Narlikar, it is possible to conceive of a Universe which oscillates. Galaxies A are redshifted; the background radiation results from a quasi-permanent equilibrium between matter and radiation, in an infinite duration universe; light elements were formed during the phases of highest density. Galaxies B are blueshifted. Their observation could be an essential test of this theory.

ter with an index close to 1.2, close to that implied from observed data by de Vaucouleurs (see p. 464). Another, which could be tested in the future (Fig. 8.37), is that galaxies belonging to a preceding cycle are indeed blue shifted. The relation between shift and magnitude, even before the observation of blue shifts, is quite different than it is in standard new big bang cosmology. In order to perform that test, it would be necessary to clearly identify very distant faint galaxies and determine their spectral shift. It will be necessary to employ "very large next generation", or "out – of the-ecliptic", space telescopes to perform that test – so we will have to wait for verification of the theory!

(iv) Remarkable relations between fundamental constants of nature. Models with variable G.

The Pythagorian temptation, taken up quite seriously by the young Kepler, has indeed flourished ever since in many papers of little value. The best example is the so-called Titius-Bode law, and all laws which have been suggested to improve upon it. Although the distribution of planetary distances results from well-defined physical processes, their expression by a "simple" numerical relation is somewhat arbitrary. Most scientists have been very suspicious of this type of thinking. One of the *bona fide* scientists who nevertheless dealt with this type of numerical relation was Dirac (1937, 1938).

According to Dirac, one can easily construct several simple non-dimensional relations between the numerical values of c, h, e, m_e. Some of these non-dimensional numbers are "very large." Eddington (1932, 1948) reached similar conclusions. The Dirac interpretation is cosmological: are the fundamental "constants" of physics, which intervene in cosmological theories, constant? Are they the same now as they were in the first minutes after the so-called big-bang? Wouldn't their large value represent an increasing complexity in the Universe, resulting from the change with time of some of these so-called constants? Several of these non-dimensional ratios have a value of 10^{39} or 10^{78}. The age of the Universe, expressed in some "convenient" unit, would be 10^{39} units. But since when? Actually, arguments in this debate depend largely upon the time-scale on which this variation is discussed!

As we will see (p. 506), it is impossible to detect in our human lifetime any change of G with time in spite of some hopes expressed a few years ago by van Flandern, based on the study of the Sun-Moon-Earth system. It is still more difficult to assume any variation of the fine-structure constant in the time spent since light left the more distant radiogalaxies known ($z = 4$ or so). This is easily shown by measuring, in the spectrum of the same object, the difference in frequency between the three hydrogen lines, Hα, Lyα, and the 21-cm radio line. Their difference is indeed linked to *the fine-structure constant α*; and this difference is strictly constant over the entire range of z explored. Of course, the fine-structure constant combines three fundamental constants. It would not be logical that one of them or even all three are varying, while α is constant!

Still, α does not contain the gravitational constant G. Can G vary?

It is remarkable to note that indeed the gravitational forces are very weak with respect to electric forces. Between an electron and a proton, this force is $G\, m_p m_e / r^2$, whereas the Coulomb force is e^2 / r^2. The ratio (independent of the distance r) is easy to compute and equal to 4.4×10^{-40} (one of Dirac's "large numbers"). Of course, the gravitational force has a large range of action, whereas electric forces do not act at a distance because of the global neutrality of matter in the Universe at large scales.

The equalities implying the constants of physics have been studied extensively, as we have said, by Eddington, and by Dirac. How can we be sure that any of them has a cosmological importance? In particular, why would such a weak force as G have cosmological significance? Part of the answer has been given: no screening effect diminishes the gravitational force at large distances. Another part lies in the possibility of constructing a mass on the order of a pion, using a simple combination of only fundamental constants:

$$(H_0 h^2 \,/\, Gc)^{1/3} = m_\pi.$$

Of course, one could also construct other simple combinations, such as $\alpha = hc/e^2 = 1/137$ (a number not so different from unity!), which will permit the introduction of some other masses, smaller than that of the pion. Actually, the range of masses of elementary particles allows for much ingenuity and many possibilities in finding new relations – and one can find them. Dirac finds that the ratio of the "mass of the (finite) Universe" to that of the proton is 10^{78}. The significance of such ratios is slightly more questionable. We shall return to this question when discussing the "anthropic model." One such ratio is:

$$(e^2/c)\, H_0\, Gc = e^2 H_0\, G = m_e.$$

Such "coincidences" led to the assumption that one cosmological parameter H_0, and the physical "constant" G are, so to speak, playing a similar role in the structure of the cosmological Universe.

The problem is that H, in Friedmann-type cosmologies (when $k \neq 0$), is dependent upon time. This leads to the idea of replacing H in equations with some quantity of comparable magnitude, but which would be constant. This idea, however, does not lead to much in the way of reasonable solutions.

Another idea is that G (or e, or h or c) should vary with time. Dirac developed the consequences of this idea. He admitted the possibility that G could be the variable which is not "constant." Then, in the models:

$$G \propto (dR/dt)/R$$

and this equation remains valid over time. As $r \propto R^{-3}$, one has:

$$GR^{-3} \propto (dR/dt)^2/R^2$$

and, eliminating G between these two relations:

$$dR/dt \propto R^{-2}, \text{ with the solution } R \propto t^{1/3}, G \propto 1/t.$$

The reason for the small value, $\approx 10^{-39}$, of ratios such as those given in examples above, is therefore, according to Dirac, purely circumstantial. It means essentially that the Universe is old. Of course, Dirac's cosmology has many other consequences. The GR has to be replaced by some other, broader, field theory of gravitation. Actually, Dirac did so, in 1975, but some consequences of Dirac's cosmology, as limited as it was, contradicted some of the observed facts, within the general framework of standard cosmologies. Still, some authors, such as Jordan, and Brans and Dicke, tried to formulate field theories of gravitation that were actually extensions of the GR.

(v) Brans-Dicke models.

Brans-Dicke models also imply a varying G, but the consequences fit the observations better than Dirac's. They indeed contain the Friedmann models, of which they are a generalization. Gravitation is not represented by the constant G, but by a more complex quantity. In essence, Brans-Dicke theory generalizes the Einstein equations by adding an additional term, a function of an arbitrary constant ω, and reduces to the Einstein equations for a particular value of this constant, $\omega = \infty$, for which value we find the Friedmann models.

Brans-Dicke cosmology has several strange aspects. It implies that an increased abundance of helium formed in the earliest Universe (the abundance of the "primordial" helium can indeed be double that of the standard model!). It also implies a decrease of the constant G with time, the rate of decrease being a function of H_0, r_0, and ω. This could be tested in the solar system. Although some authors have found such a change through the analysis of observations, their work is highly disputable. There is indeed no convincing indication of such a behavior. However, models with variable G can still be considered, and discussed, as we shall see. Another consequence of this decrease of G is that the Sun, formed billions of years ago, should be flattened. Again, this effect is beyond detection, in spite of some efforts. The difficulty is that a flattening can easily be confused with a latitude-dependent change in the limb-darkening of the solar surface.

(vi) The symmetrical Universe.
One point which we have not discussed so far is the extreme dominance, in the part of the Universe we observe, of ordinary matter over antimatter. Antimatter is, in essence, produced as individual antiparticles in accelerators. These antiparticles tend to be annihilated very quickly by collision with their corresponding particles.

Why is the observed Universe so asymetrical?

Several attempts were made to explain that, since the big bang, there has been a separation of matter and antimatter, but in a sort of random chaotic way. Nothing has really confirmed these attempts. From the opposite perspective, Souriau's model has an interesting observational test. He assumes that, beginning with the big bang epoch, the Universe was divided in two equal parts, one of matter, the other of antimatter. In the present epoch, the Universe is still divided in two equal parts of matter and antimatter. At the border between the two parts, matter annihilates antimatter. This leaves: (a) a region void of galaxies and quasars (for which the z is a good distance indicator, as it is for Sandage and other defenders of the standard cosmology); (b) a flux of gamma rays coming from that annihilation region.

By projecting the position of all the quasars known up to now on a two-dimensional map (with the arbitrary choice of the direction of projection), Souriau finds for one of these possible directions of projection, a region clearly empty of quasars. This is the "equator" of the Universe (Fig. 8.24a,b, p. 473). Is the "equator" found by Souriau from the data, but using a pre-conceived theory as a guide, completely convincing? In any case, according to Souriau, the known Universe is almost entirely on one side of this equator. The equator's location fixes the location of the Sun in the Universe, and establishes the radius of the Universe and other cosmologically meaningful quantities. Souriau suggests that some groupings of quasars and even galaxies are oriented in the same way, their plane being originally parallel to the equator. Their local evolution may distort them somewhat, but there are remnants of this parallelism. However, this is not entirely convincing, the statistical sample being perhaps too small.

This cosmology does not deal with the physical nature of the big bang. It remains essentially a Friedmann cosmology, with a non-zero value for Λ. The Souriau Universe is closed.

The gamma-ray radiation from the zone of absence of quasars might also be identified with the observed gamma radiation, but this is a less obvious test. At least, the zone of absence of quasars is, seemingly, a very convincing argument. However, an effort to explain this absence could be made in any type of cosmology, not necessarily implying either a big bang or expansion, nor necessarily implying that quasars are at their cosmological distance.

The idea of the Universe being divided into matter and antimatter, in equal quantities, is not particular to Souriau. Alfvén (1977), for example, presented such a cosmological point of view, called a "symmetric" cosmology, after Klein's work in 1966. Alfvén's model is otherwise quite comparable to the oscillating models described in Fig. 8.36. He rejects the "bang," and emphasizes the importance of inhomogeneities. His criticisms of the classical (old) and new big bang are similar to ours.

8.6.2
Models Without Expansion

The very fact that this group of theories is completely outside the main current of thought justifies our presenting only a very short abstract of each point of view. Nevertheless, we should not put them aside, just because they are not in the mainstream. It is quite possible that the trends of future research will demonstrate the qualities of these models, as incomplete as they may appear now, on the basis of future determinations of the structure of the distant Universe, based on observations. There are actually several other suggestions and discussions, even well-elaborated "models." We apologize to those we have not cited, having limited ourselves to the demonstration that the "standard" cosmology is not unique, and that there are not only many variations of it, but also some radically different alternative solutions, without the big bang or even any expansion.

(i) Extreme variation of G.

When it was thought that the variation of G had actually been measured in the solar system, this variation was shown to be a good alternative to the expanding model. It was even possible to compute H_0 from the variation of G and a correct value was found, notably by Hoyle and Narlikar, and by van Flandern. However, these measurements were later shown to be wrong and the idea has not been further developed.

(ii) Chronogeometry.

Segal's theory is highly mathematical, and we shall not provide much detail about it. It is the only one which accommodates a square law relation between redshift and distance, which seems to be closer to the observed data (Fig. 8.38 a,b).

Segal's theory describes a Universe which could be seen as still more relativistic than that of Einstein. The very notion of simultaneity has to be entirely ruled out. There is everywhere some local time; and there is a universal time, which can be linked to the local time only through local characteristics of the four-coordinates geometry of space-time around that point. Only local and stationary observers are involved. Causality, is of course, strictly respected. Homogeneity and isotropy are assumed, as in the Einsteinian and Friedmannian cosmologies, but the Universe is curved, although anthropomorphically local measurements are in essence flat (as if they were made in a flat Universe, tangential to the real Universe). From the mathematical study of this curved space-time, the notion arises that redshift is indeed a property of space, not due to any expansion, or to the relative motion of distant galaxies. The z-r law is quadratic, not linear.

The model claims to explain the BR, as the others do. It does not contain, to our knowledge, any attempts to explain the abundances of light elements.

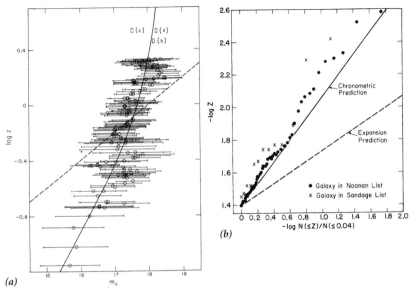

Fig. 8.38a,b *Segal's Universe: observable consequences.* (a) The relation between magnitude and log z for quasars. The data are compared with the standard expanding model prediction (dotted line) and with chronogeometry, which predicts not a linear law but a square law for the relation between the redshift z and the distance (full line). (b) Comparison between the actual counting rates (log N, on the abscissa) and log z, for the brightest galaxies in clusters. On the solid line, the chronogeometric prediction; on the dotted line, the linear Hubble expansion prediction. Clearly, Segal's arguments are very strong. Still, they have never been taken seriously enough (perhaps because they imply no global evolution of the Universe).

It is remarkable to note that Segal's model accounts for the z-m relation, the N-z counting relations, and the angular diameter z-relation rather better than the standard big bang cosmology accounts for these relations. Strangely enough, the discussion of Segal's work, although it has all the characteristics of a very strict and rigorous theory, has never really convinced other scientists, because of its disagreements with most of the observations. Most of the commentators have actually disregarded or even despised Segal's theory, in spite of some very interesting aspects.

(iii) The new steady state theory of abnormal redshifts.
Arp, who has been the main provider of observations in support of the reality of non-cosmological (or "abnormal") redshifts, has constructed a theory to account for them. For him, the essential cause of redshift (be it

normal or abnormal) in extragalactic objects can be attributed to some intrinsic property of matter in these objects, not to the intervening material.

The assumptions at the basis of Arp's theory are: (i) that there is a continuous creation of matter, suggested by the many ejections we see at work; (ii) that all redshifts are of the same nature, measuring the age of created matter; there is therefore no expansion (unlike in the Hoyle *et al.* stationary model); (iii) that the mass of elementary particles changes with time.

They do so because of the number of "gravitons" they may interact with, within their individual 3-dimensional light sphere (leading to $m_p \propto t^2$). Then the photon energy varies as m_p; and the redshift is a function of the time elapsed since the creation of the matter in question $(1 + z \propto t^2)$. This simplified concept can be expressed in a form which corresponds well with the GR field equations. This theory leads to remarkable predictions, such as that of the Hubble constant, from the observed age of our Galaxy. In this theory, H diminishes from distant galaxies (observed in numbers at $z = 1$ to $z = 5$ or more!) to nearby galaxies $(z \sim 0)$.

The stationary Arp model (Fig. 8.39) is simple. From the observer on Earth, we penetrate the Universe only to a certain radius $R = ct_0$ (t_0 being the age of the Galaxy). The outer parts of the Universe are now invisible. They will become observable progressively, at a very slow rate. The BR is not cosmological in essence, but formed in the static intergalactic space. The periodicity in redshift distribution has to be explained by properties of elementary particles, and their formation within galaxies. The abundance of light elements is not discussed by Arp. It can be explained, however, naturally within Jaakkola's so-called *equilibrium cosmology*, or even from local considerations.

Arp advocates (as do Burbidge, Hoyle, and Narlikar) a continuous ejection of matter by galaxies, forming quasars from "young matter" (the hyperdense Ambartsumian's matter? Or the Gamow's hyla?). The quantized properties of the redshifts are linked with the properties of young matter as it is expelled by old dense matter from galaxies.

(iv) Equilibrium cosmology.
This model is not an expanding Universe either. Jaakkola accepts the cosmological principle at a large scale, as did Einstein, Friedmann and others. The model assumes (as does Arp's) that different types of gal-

Fig. 8.39 *Arp's non-expanding model.* The background radiation has been, in essence, in equilibrium for billions and billions of years. Stars, galaxies, and clusters are constantly forming, evolving and dying.

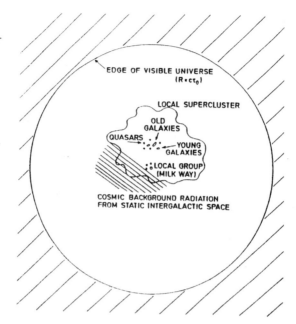

axies have different z (when located hypothetically at the same distance). The quantity H/c (which links z to distance) is proportional to the square root of the local density, which fact explains the observed differences in determinations of H (p. 455). As far as the BR is concerned, the perfect Planckian shape of the 3 K radiation is typical of an equilibrium Universe, *of infinite age*.

Coupling is necessary to maintain an equilibrium in this Universe which implies a large scale balance between the number of baryons (b), leptons (such as photons (γ), etc.), and gravitons (g). This coupling is called the *electro-gravitational coupling*. Of course, the details are mathematically formulated. It is interesting to note that the quantum behavior of photons and gravitons permits the prediction of the quantified fine-structure of redshifts. It also predicts the equilibrium of light elements.

It seems to the author that this theory is really aiming at some kind of Grand Unification of interactions, but acting at all scales, in the present Universe. Its weak point is perhaps the fact that it does not justify many of its logical consequences numerically (in spite of the correct prediction of some others, such as the 3 K BR). It suggests many tests that would promote the selection of the equilibrium theory as against standard big bang cosmology.

(v) Tired-light theories.

Unlike the preceding theories, the tired-light mechanism theories do not present any mathematical model of the Universe. They attempt only one thing: to account, with one single hypothesis, for all abnormal redshift observations, allowing the "normal redshifts," those which follow Hubble's law, to be interpreted in any possible way. According to these theories, one may consider the actual redshift z to result from the addition of a local, intrinsic redshift to the Hubble redshift. Both are linked to the tired-light mechanism, and these authors generally exclude any expansion.

The idea is simple, and has been expressed, implicitly or explicitly, in many ways, by many authors. It assumes in essence that the photon has a non-zero rest mass. Therefore, when travelling through space, through encounters (of any kind, with fields, or particles) it loses energy, at a rate proportional to the distance covered. The proportionality constant depends of course on the elementary processes involved. The idea has been considered by several authors with very different processes in mind, so that there are many different tired-light theories. The criticisms that can be made of some do not necessarily apply to others.

The pioneer in the field may well have been the physicist Walter Nernst, well-known for his contribution to thermodynamics (he discovered the "third law of thermodynamics"). Nernst assumed a decay of the photon energy through some interaction with ether (1935). Among the first to develop the same idea independently of Nernst was Fritz Zwicky (1929), who has not always been recognized as one of the major astronomers of this century.

In 1954–1955, Findlay-Freundlich, and Max Born, looking for an alternative to the GR, proposed an interpretation of all redshifts in the solar spectrum based on some tired-light mechanism, more specifically a photon-photon interaction. Indeed, the excess of redshift over the Einstein redshift at the solar limb is sufficiently large to justify that attempt. The authors predicted quite accurately (more accurately than Gamow and the big bang theories of the time) the existence and magnitude of the 3 K BR. Still, it was possible to show that the photon-photon interaction would not really work. Later, Vigier and Pecker, together with several co-authors, reviewed a large fan of facts they could explain by the interaction of photons (having a non-zero rest mass) with a so-far unknown particle, for which they defined the properties. Some might say, however, that assuming the rest-mass of the photon equal to zero

($m_\gamma = 0$) is an hypothesis as restrictive as Friedmann's hypothesis $\Lambda = 0$. Probably, it is indeed a logical hypothesis, as $m_\gamma \neq 0$ would violate the conservation of electric charge, which is observed at a high accuracy, whereas observations tend to impose a non-zero value of Λ. However, the conservation of charge may well not be satisfied at a much higher accuracy than now achieved. Other physical principles, based on the theory of elementary processes, also lead to $m_\gamma = 0$. Still, this may be (and it has been) questioned, in spite of its real strength.

This theory has been heavily criticized on many bases. It does not, according to the criticisms, predict a blackbody but only a combination of blackbodies. It would blur the image of distant galaxies as the loss of energy of the photons from the source is slightly deviated. The authors have not replied to these criticisms. We will say here only (a) that, in an equilibrium situation, the photons are thermalized by the bath of interactions in which they are immersed (as in Jaakkola's model); and (b) that the deviation can be very small in some processes; and therefore that the blurring can start to be observable at $z = 5$ or perhaps $z = 10$ or more. A search in very distant parts of the Universe to occur in coming years would perhaps shed some light on this effect. In any case, gravitational fields distributed in the Universe also produce a blurring, which is easily observable. A more serious criticism arises from the assertion that the photon rest-mass cannot be other than zero from the point of view of theoretical physics based on group theory. A careful study of classical experiments, such as Morley, Miller, and Sagnac's (Vigier, 1998), have demonstrated the contrary to be true. According to Vigier, the rest mass of the photon is not zero.

(vi) Partial interpretations of some observed facts.
Several authors have tried to find an answer to some particular questions posed by observations that might possibly be cosmologically significant.

We have seen for example, that the 3 K BR is often assumed, in an infinite-eternal Universe, to result from an equilibrium between photons and the rest. However, even before any big bang ideas, the 3 K was predicted, for example by Eddington (1926), and by Regener (1933), as being the equilibrium temperature of intergalactic space. This interpretation is similar to that of Burbidge, Hoyle, and Narlikar, as well as that of Arp.

Another example is the explanation of Olbers' paradox. Charlier, after Fournier d'Albe, expressed the view that a hierarchical Universe (for which the density diminishes as the size of the volume in which this density is measured increases) answers all of the questions raised by Olbers' paradox and its gravitational counterpart, Seeliger's paradox (although they did not realize that the two paradoxes are essentially different – see p. 411). Neither Charlier nor Fournier d'Albe intended to explain facts they were not yet aware of, such as the redshifts.

The abundances of light elements are regarded, by the standard cosmology, as a fixed composition resulting from processes in the very hot, very dense, early Universe. Those who prefer to accept the view of an eternal and infinite Universe must find other explanations, local perhaps, for the abundances of light elements. One could think of young galaxies as resulting from the condensation of diluted matter already enriched by the evolution of stars belonging to long-vanished galaxies. Moreover, the chemical composition may be strongly affected by other non-nuclear phenomena, such as the diffusion of elements, a known phenomenon, in gravitational and radiation fields. It is well known that galaxies are often characterized by polar ejections following a central explosion. Dozens of radio galaxies are double sources, and the two (or more) lobes are located symmetrically with respect to the exploded galaxy, the remnant of which is observable. Where is the explosion coming from? It might come from a very hot temperature in the central regions of the exploding galaxy. The radiation pressure (from the Lyα line of hydrogen) may expel hydrogen whereas the Lyα line of helium only ionizes hydrogen. Therefore, hydrogen is expelled, and the helium abundance is increased within the galaxy. This effect has been computed by Pecker (1974), in first order of magnitude computations. The enrichment can reach an order of magnitude. Precise calculations are difficult to make. They are, however, necessary; and these attempts can only be considered working hypotheses. Recent developments in BHN theory (1998) give some new answers to these questions.

8.6.3
Some Discussions with Possible Cosmological Consequences

(i) Hierarchical structure of the Universe.
Earlier, we have described the basic observations and expressed some doubt as to their extrapolation to larger distances than observed at present. A qualitative model can be built of course, such as the one constructed by Fournier d'Albe in the last century or Charlier (1908, 1922).

These, however, were essentially only intended to explain Olbers' paradox. If we could build relativistic models, would the density be assumed to be fractal, as observed, and as noted most emphatically by de Vaucouleurs in 1970? True enough, inhomogeneous models have been considered, by Heckmann and Schücking, and intended for application to the general Einsteinian field equations. However, the solutions were not numerically expressed. Actually, even if it were possible that the cosmological principle of homogeneity and isotropy could be applied to the very early Universe, it could not be applied at the scale, say, of our Galaxy, of galaxies, or clusters of galaxies. One could say, in a very qualitative manner, that the higher the density the faster the expansion; therefore the larger the structure, the slower the rate of expansion. Adding that a principle of continuity must be maintained, we reach a model that looks like the one in Fig. 8.40, where in essence condensations and explosions appear only on a small scale, and where at a large scale there is from time to time a highly condensed state, lasting a short time. In its entirety, it seems then to be neither expanding nor in steady state, but a fluctuating Universe.

This intuitive and not-rigorous solution could well be a suitable framework for tired-light mechanisms, and/or for matter creation mechanisms. The recent BHN mechanism might well prove very convenient, as we have seen.

(ii) Anthropic principles.

The existence of life, in particular of human life on Earth is, in itself, a surprising thing. The appearance of life out of some inanimate medium has been taking place for some time but by processes we do not understand at all, and of which the probability, which we cannot properly estimate, is in any case very small. Where did it take place first? In the interstellar medium, where many heavy molecules, precisely those of the Earth's carbon-based organic chemistry, have been found? On Earth or other solid bodies? Has life been transferred from one place to another through the "panspermy" suggested by Svante Arrhenius in the last century, or by cometary tails, as suggested recently by Hoyle and Wickramasinghe?

In any case … life exists – as many improbable things, such as hitting the winning number in some lottery, do indeed exist! The next question then becomes: does it exist only here, around us? Or has it appeared in many other places?

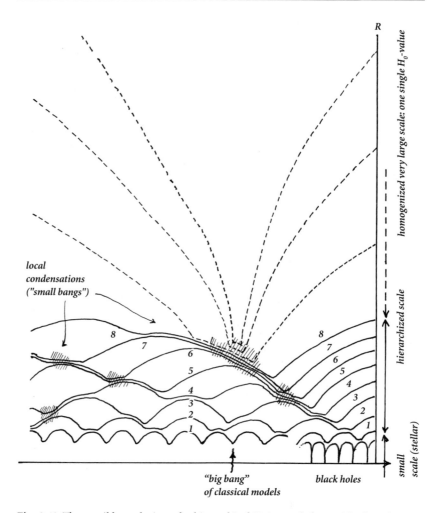

Fig. 8.40 *The possible evolution of a hierarchical Universe (schematic).* If, at time $t = t_0$, the Universe is hierarchical, the smaller the distance, the larger the local curvature, the quicker the expansion induced by that curvature. Here, figures 1, 2, 3, ..., 8 represent successive elements of the hierarchy, from the smallest ones to the largest ones. Different curves represent the distance from any point to the observer. Assuming that the continuity equation is verified, these curves should not intersect. The drawing here presented by the author represents the phases of matter accumulation, which are, in essence, local bangs of small importance. The Universe may have been behaving that way for billions and billions of years.

The Christian tradition teaches that Man is a unique creation, made in the image of God. Genesis is clear. Even the Mormon Church, which otherwise does not invest the old world with a monopoly on apparitions of Christ, or witnessing by apostles, assumes one single Creation. According to the Mormon Scriptures, some Jews left sinful Egypt at the time of Moses' flight to the promised land, and went to the West, crossing Africa and the Atlantic, to create a civilization of their own in America. In order to "save the Scriptures" (remember the Church Fathers, p. 132 ff.), many modern physicists have proposed rather far-fetched new ideas. Some people say that these attempts really do belong to cosmology. In fact they do. They present the most extreme case of spiritualist (Platonic?) cosmology.

The "anthropic principle," formulated by Brandon Carter (1973) in its more limited form, then extended by others to a less acceptable form, dates from the seventies. It can be expressed as follows: as the Universe (the Earth) is a home for human beings, the various constants which command the evolution of the Universe *must* have the value they actually have. If one changed the present measured values of G, or c, or k, or h, by an order of magnitude, the lifetime of the Universe would be much too short for the stars to have appeared, and its evolution much too slow to permit our human species to have remained in it. The extended principle goes one (large) step further, by asserting that the Universe must be subjected to some "fine tuning" *in order* for the observers – us ! – to have appeared in it. Needless to say, this very strong "finalist" attitude is adopted by only a very few scientists. It goes much beyond even the traditional Platonism of the big bang theorists, quite close to the attitude of the American creationists who assign a Scripture-based age (rounded) of 6000 years to our Earth, i.e. to the Universe.

The anthropic principle has been combined, more or less, with Dirac's remarks about the occurrence of simple dimensionless ratios of fundamental constants of nature as "large numbers" on the order of 10^{39} or 10^{78}. We have mentioned the ratio of the electrostatic force to the gravitational force ($\alpha\ 10^{39}$), the ratio of the mass of the Universe to the mass of the proton ($\alpha\ 10^{78}$). We can also note the numbers:

$N_1 = hc/Gm_p^2 = 10^{39}$ (inverse value of the gravitational fine-structure constant);

$N_2 = (c/H_0)(h/m_p c)$ $(R_0 = c/H_0)$, (i.e. the "size" of the Universe, over the Compton length) $= 10^{40}$;

N_3 (Eddington number) $= (4\pi/3)r_0 R_0{}^3/m_p)$ (mass of the Universe to proton mass) $= 10^{80}$;

Dirac takes out of these coincidences $(N_1{}^2 = N_2{}^2 = N_3)$ the basis of his cosmology, as we have seen. Nonetheless, Dicke, for example, asserts that these coincidences are a consequence of the existence of human beings. Carr and Rees have explored this question by introducing the ratios of the sizes of various structures in the Universe, including man. In the sequence of large numbers, they introduce intermediary steps. They even predict a priori relations, which they refuse to consider as mere coincidences. Such are the relations:

man $= (\text{planet} \times \text{atom})^{1/2}$;
planet $= (\text{Universe} \times \text{atom})^{1/2}$,

for the sizes, and for the masses the relations:

Planck's mass $= (\text{exploding black hole} \times \text{proton})^{1/2}$,
exploding black hole $= (\text{Universe} \times \text{proton})^{1/2}$,
man $= (\text{planet} \times \text{proton})^{1/2}$.

One should, I think, consider these relations with extreme skepticism, even with a smile as one might smile at a pleasant joke. First, it is easy to build many simple combinations, from any set of three of four numbers, with two or three significant figures. Second, the continuum of physical dimensions gives to the Carr and Rees relations a truly anthropomorphic character. Why man, and not, say, a flee or an elephant? Why not substitute examples of relations between quarks, whales, and quasars? There are as many relations of this type as structures identifiable in the Universe. However, there may be some rationale for looking for the meaning of these relations, which are linked to a change of scale. Scale invariant properties of the Universe are perhaps involved here; and we might (in spite of the fantasies mentioned above) gain some insight from a rational search for the physical meaning of some of these relations.

Others, at the opposite end of logic, are looking for intelligent life elsewhere, with numerous instruments, trying to catch some "message".

I feel skeptical, in the present epoch, of the realism of such attempts. However, in the future, they may carry more weight than they do now. Whatever their successes or failures may be, life may perhaps exist in a form with which it is impossible to communicate in many places in the Universe. After all, we do not communicate to any great extent with an ant or a jelly-fish, still less with a beautiful tulip! Yet, they are living beings.

In any case, these attempts involve us in a problem which cannot be escaped. Consider the Earth. It formed out of a cloud of some molecular and atomic gaseous medium, billions of years ago. Life and man have appeared more recently. How? This is still a very important astronomical question. Billions of years from now, life will disappear. Earth will eventually be swallowed up by the Sun, which will become a large supergiant, before a final collapse. The matter that now makes up the Earth will then return to its initial condition, and cool back again, although perhaps transformed by other processes in a different way. This brings us to a final problem: that of the arrow of time, that of the second principle of thermodynamics.

(iii) The Universe and the "second principle" of thermodynamics.
We have already evoked the problem of the second principle of thermodynamics in discussing oscillatory models of the Friedmannian family, but the problem appears in other models as well.

What, strictly speaking, does the the second principle, or (more emphatically!) the "Second Law," of thermodynamics say? "In an isolated system, a closed box, from which no energy goes out, which no energy goes into, the entropy is always increasing." We will not give here the precise definition of entropy, one of the important thermodynamic quantities. We will say only that entropy is a measure of the probability of a given physical state of matter and radiation. When we say that entropy is increasing, it means that the matter and radiation in the closed system are moving towards a state of maximum probability. This discussion led, at the end of the last century, to endless discussions about the "heat death" of the Earth (it was more or less assumed that the ensemble of Earth and Sun constituted an isolated system). The heat death was said to occur when everything became uniform, when no evolution could take place anymore, and when life itself was inconceivable.

If the Universe as a whole were considered as an isolated system, this situation would need to be reconsidered. Therefore, the Second Law has

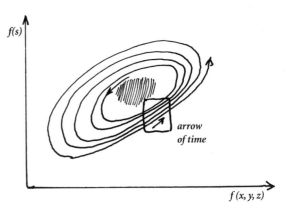

Fig. 8.41 *A thermodynamic principle and the topology (schematic).* On the abscissa, some parameter characterizing the position in space of an isolated system, which can be the Universe. On the ordinate, some function of its entropy. The arrow of time is well defined at a small scale, but not at a much larger scale. On this 2-dimensional graph, an area represents a volume of the 4-coordinate space-time. Each point in this volume represents a phase in the life of the Universe. There may be some states, grouped in space-time, that the Universe cannot occupy through this evolution. On this diagram, they are represented by a gray area.

been used more recently, for example by Reeves, as an argument against the steady state theories. In an eternal Universe, with all thermonuclear processes progressing from hydrogen to heavier elements, we should not find any more hydrogen in the Universe. But we do. Hence, the past Universe must have had a finite existence. Of course, all processes involved in nuclear reactions are not going in that same direction. For example, there are reactions that produce protons. Taking this into account, the Reeves argument, although rather strong, should be looked at more carefully.

On the other hand, there is no closed system which can be considered isolated, given that gravitational energy can leave or reach the matter and radiation supposedly isolated. Quite clearly, one essential principle, always implicit however, could be that "there is no screen against gravitation." Then what is the status of the second principle? Of course, it can be expressed without the artifact of an isolated imaginary box. It can be formulated locally, in any part of the Universe, subject to any influence, where the entropy can be defined quite independently of the isolation of matter and radiation in a sort of imaginary box. It is not impossible that locally entropy is always increasing, and that the "arrow of time," therefore, is always well defined, but that the topology on the scale of the Universe is such that entropy can disappear by some processes. The appearance of life indeed destroys entropy. The schematic drawing (Fig. 8.41)

Fig. 8.42 *The Novikov mini-universes.* Time is on the ordinates. In **A** one mini-universe collapses, and separates from the neighboring ones. In **B**, another mini-universe is going through a bang, seen as a "big bang" by its inhabitants.

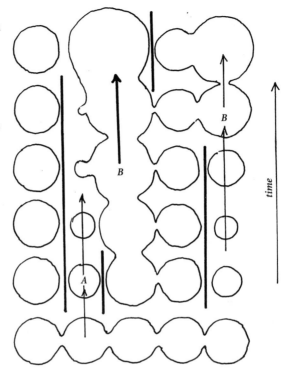

shows how this could be conceived, but it requires special topologies of space-time.

A way to turn this question around is also to imagine that the Universe is not an isolated system. Then we are in the realm of views expressed by Igor Novikov (1987) and schematized in Fig. 8.42. Novikov accepts the fact that we live in a universe (which he calls a "mini-universe," – we shall use the word "universe" – with an u, not an U –), affected by the big bang and now independent of any other universes; but it has not always been independent. A "Mega-Universe" contains our universe. At a certain time, our universe was linked to other parts of the "Mega-Universe." Fluctuations occur. Here and there, there is a big bang. In other places, a collapse takes place.

The inflationary phase of the bang keeps the volume in which it takes place independent of other volumes. Later on, a collapse will come, or perhaps the local expansion will continue. What we call the universe is only a local part of a Mega-Universe, of which we know nothing! At least we know that our universe, if now isolated, has not always been so, and that the entropy problem does not hold. On a larger scale, that of the Mega-Universe, one may say that the Mega-Universe lasts forever, with local fluctuations which give rise to a big bang such as the one in our universe. In a way, we face a model where the ideas of the steady-state cosmologies and big-bang explosion-expansion cosmologies are made compatible. Of course, there is no verifiable argument to be made in favor of this model, as other parts of the Mega-Universe are not accessible to us. Therefore, Novikov's attention has ultimately focused on the standard cosmology, and on the important stage of galaxy formation, for which tests have been proposed that may be performed in the coming years.

(iv) Topological properties of the Universe.
The Einstein Universe, and the Friedmann Universes, have more or less the same topology, a simple one, an ordinary space-time, with the big bang as its only singular point. However, in order to accommodate the SUSY theory strange topologies have emerged, some envisioning a Universe looking like a bunch of inextricable bundles of noodles (Figs. 8.43a–c).

Black holes, sinks for matter and energy, are conceived of as tunnels opening on some other universe, in which they are white holes, sources of matter and energy (Fig. 8.44). Some authors go so far as to propose "twin" Universes, everywhere in two different types of matter.

We have seen that some thermodynamic paradoxes may lead to geometries involving singular regions of the volume S (entropy) x, y, z, t avoided by the evolution of the Universe.

Other authors go in another direction by assigning a fractal structure to space-time geometry, involving such constructs as "fractal derivatives" (Nottale). This theory permits the explanation of some observed phenomena, which have not been explained by any other theory, such as a "mysterious" feature of the cosmic-ray spectrum.

(v) Scale-relativistic cosmology.
Nottale's theory comes out of a tradition of thinking about the Universe that holds that physical conditions, and their mathematical expressions, are different from one scale to another (the scales of the quantum domain, of the classical domain, of the cosmological domain). It seems clear also that the Universe is inhomogeneous at many scales, in spite of

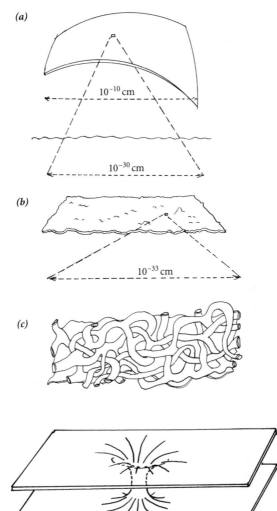

Fig. 8.43a–c. *Exotic topologies that could exist ...* (schematic). (**a**) The local space may be curved, but is rather smooth at the scale of say 10^{-10} cm. (**b**) At a 30 orders of magnitude smaller scale, the space (here we see a two-dimensional representation of it) may be corrugated. (**c**) At a still much smaller scale, the quantum universe may be (?) of a very high complexity, like a bunch of noodles, according to some commentators. Note that this is again a two-dimensional representation of the three-dimensional space. The surface drawn on the figure has two dimensions, but its imbrications give it the appearance of a three-dimensional object.

Fig. 8.44 *From the black hole to another universe (schematic).* The double-sheet model (first proposed by Wheeler, and again notably by J.-P. Petit, in the frame of reference of his concept of twin-universes) is a way to build a universe linked to ours through black holes. It solves some of the thermodynamic problems already mentioned.

the classical cosmological principles, and usual assumptions. Is scale really the determining factor in physics? Nottale's theory, its physics, its mathematical formulation assumes on the contrary that there is a clear unity in physical laws, whatever the scale, and that non-homogeneous structure is an essential ingredient of the observable fractal Universe (Fig. 8.17).

Nottale's theory is founded on an extension of Einstein's principle of relativity (previously applied to motion transformations) to scale transformations. It proceeds along the following lines. One first gives up the arbitrary hypothesis of the "differentiability" of space-time coordinates, while keeping their continuity. One can then demonstrate that such a non-differentiable space-time continuum must be "fractal," i.e. must be explicitly resolution-dependent. The space-time resolutions become inherent in the physical description, and they are re-defined as essential variables that characterize the "state of scale" of the reference system (in the same way as velocity characterizes its state of motion). Using the fact that resolutions can never be defined in an absolute way (only a ratio of space and time intervals has physical meaning), one can establish a "principle of scale relativity" according to which *the laws of nature apply whatever the state of scale of the reference system.* Its mathematical translation consists in writing the equations of physics in a scale-covariant way.

Motion-relativity evolved from Galilean relativity to Einsteinian Special, then General, Relativity. The same is true for the scale-relativistic laws of scale, whose construction involves several levels. The simplest scale laws can be given the structure of the Galileo group: they correspond to standard fractal power laws (with constant fractal dimension). The main axioms of quantum mechanics can be recovered from such scale laws. In other words, one can demonstrate that the quantum behavior is a manifestation of the non-differentiable and fractal geometry of micro-space-time, in the same way that gravitation is a manifestation of its Riemannian geometry at a large scale.

Actually, the usual (Galilean) scale laws are only a very particular case of the general scale laws which satisfy the principle of scale-relativity in the special (linear) case. These laws of transformation between scales have the structure of the Lorentz group of transformations. In this framework (that can be called "special-scale-relativity"), the fractal dimension is no longer constant but varies with scale. The effect of two successive dilations is, therefore, no longer their direct product. There appear minimal and maximal scales, invariant under dilations, that replace the zero and the infinite while keeping their physical properties. This is similar to the character of the velocity of light in special motion-relativity, which is impassable but has the physical properties of infinite Galilean velocity. Even more general scale laws can be considered if one also gives up linearity.

Various cosmological consequences of scale-relativity arise from these various levels of the theory. The present classical theory is ill-adapted to the description of physics at a large scale, and must be replaced by a statistical, quantum-like theory. The new theory, by a unique formalism and in terms of a unique fundamental constant, accounts for the structures observed in our own solar system (Titius-Bode-like laws). It also applies to recently discovered extra-solar planetary systems, for the Tifft effects of redshift quantization (see p. 471) and for several gravitational structures observed in a range of scales to 10^{15}.

A second kind of cosmological consequence applies to the primeval Universe. The *minimal, impassable scale* is naturally identified with the Planck length-and time-scale. In the special scale-relativistic framework, this scale is invariant under dilations, so that the whole of the Universe is connected at the Planck epoch. This solves the horizon/causality problem (see p. 486 ff.). The question of the initial singularity is also asked in different terms, since the expansion now starts asymptotically from the Planck length-scale. Inflation is no longer needed. The theory predicts the formation of scale-invariant gravitational structures, even in the absence of "initial" fluctuations.

The third class of consequences arises from the suggestion of the existence of a maximal scale, also invariant under dilations. Such a scale is naturally identified with the cosmological constant scale L (recall that the cosmological constant Λ is the inverse of the square of a length). This suggestion provides a meaning for the cosmological constant, implies that it is non-zero, and yields several new ways to measure it (galaxy correlation function, vacuum energy density, age of globular clusters). The resulting values are consistent (the reduced constant is found to be $0.36\,h^{-2}$) and solve the age problem. The ratio of the two minimal and maximal fundamental scales is then found to be about 5×10^{60} from several different tests and yields perhaps a basis for a physical understanding of Dirac's large number coincidences (see p. 500 ff.). Moreover, as in the microphysical domain, the fractal dimension is expected to vary with scale. This allows us both to recover the observed value of the fractal dimension of the distribution of matter at scales 10 kpc to 100 Mpc (it is related in scale-relativity to the cosmological constant), and to predict a transition to uniformity on a scale of about 1 Gpc.   ∗

8.7
Conclusions: What Is the Current State of Cosmology?
What Experiments Could Be Performed to Improve the Situation?

Within the limited scope of this book, it has been difficult to do justice to both the cosmologies which are credited extensively in the literature and the newer ideas. Many ideas with possible cosmological significance have no doubt been omitted.

Still, it is clear that several conclusions can already be drawn.

(1) We have reached a state of affairs where everyone, or almost everyone, is satisfied with the standard cosmology. Although the original big bang, which was so attractive in its simplicity, is now seen to be seriously flawed, physicists have managed to repair it, more or less, with important and promising ideas such as inflation, GUT, and SUSY. However, these ideas are not confirmed by any astronomical observations. In addition, the big bang cosmologies have many drawbacks, one of which is the obvious inhomogeneity. The hierarchical structure has not been taken into account in the mathematical models. Another problem is the existence (in my opinion quite well documented) of the various kinds of abnormal redshifts.

(2) On the other hand, the steady-state theories imply physical processes that are very unclear, and certainly no better confirmed by observations (including laboratory experiments) than those marshalled in support of the newest forms of the big bang theories.

However, progress has been made towards a solution to the abundance problem. The abundances of light elements observed have been accounted for theoretically within the steady-state frame of reference.

(3) The 3 K background radiation, although a very important and precise observation, can be accounted for, mathematically, by several cosmologies, from the new big bang to some quite unconventional ones. We could consider it as "cosmologically insignificant." Its fluctuations, which are small but real, should of course be studied, and analyzed, but without casting them in the light of any particular preconceived ideas about cosmology.

Although the standard cosmology is widely adopted, we feel that the door is still wide open to a multiplicity of criticisms, and discussions, and to "heretic" cosmologies. The general black-out of these heresies by the science media is actually quite shocking. It is shocking to see that some channels of publication seem to publish holy words, whereas others are written as if by the filthy hands of Satan, in unknown publications. This is certainly quite unhealthy. The tendency to stick to *Ap. J, Rev. Mod. Phys.*, and *Phys. Rev.*, is a misguided one. Everything that appears in these journals is not necessarily gospel truth. It is always surprising to see the silence which has obstructed the important contributions of Nernst, Born, Zwicky, and Arp, to mention only a few of the cosmological heretics. There is undoubtedly a sort of intellectual dictatorship of the cosmological establishment, which has its Gospels, its Paradise, and its Hell. But, in such a field, we are only at the boundary of the huge domains of undisputed science. Hence there should be some place here for controversy; but do not generalize hazily!

Another question we may ask now is, of course, the one asked by the Aristotelian physicists. They wanted to be able to "σώζειν τὰ φαινόμενα?" All existing cosmologies attempt to do that as well as to introduce new observations. How can we enlarge the field of cosmologically significant observations? We have mentioned, here and there, some possibilities in this respect. Let us, just briefly, mention again a few points.

Through space research, and instruments orders of magnitude more powerful than the Hubble satellite, we could reach much fainter galaxies and clusters of galaxies, therefore much more distant objects. Then the question to ask becomes: within that explored part of the world, can we see any convincing sign of evolution from the more distant to the closer objects? Examples could be: the proportion of radio sources, of AGNs properly identified, of large core galaxies, etc. We could also investigate the relative abundances of hydrogen, helium, deuterium, and lithium. The Hubble Space Telescope seems (1996) to indicate to some of the analysts such an evolution. We have to perform more tests, more statistical studies of this effect. The next generation telescopes and the out-of-the-ecliptic telescopes may provide some indisputable tests. But when will these telescopes be in use?

Even in the presently explored Universe, accurate measurements of helium abundances should be undertaken, notably in the lobes of the double radio sources, seemingly ejected by the poles of some galaxy, and discovered through their synchrotron radiation or through the 21 cm radiation, displaced towards longer wavelengths.

Laboratory experiments at very high energy should be developed to ascertain the processes involved in neutrino evolution, such as neutrino "oscillations," as well as to demonstrate the validity of the GUT ideas, and perhaps of the SUSY ideas. The measurement of the lifetime of the proton is important. Any attempts at measuring the rest-mass of the photon (now known only, from laboratory determinations, to be smaller than 10^{-54} g) are worthwhile.

The theory of the evolution of galaxy groupings, their observation and the data pertaining to the fractal nature of mass distribution in the Universe and its evolution all need further study. The formation and disappearances of galaxies are also fertile areas of future research.

It is still necessary to measure and discuss all the known abnormal redshifts in an "open debate," and to study the new cases of this phenomenon. More cases should be found. The assignment of those to gravitational refraction effects must not be based on only qualitative arguments. These cases should be discussed and measured quantitatively, as precisely as possible. Although cosmological, but not "cosmologically significant" in essence, one has to fully account for the 3K radiation and its fluctuations. Are these fluctuations galaxies in formation? If that were the case, the spectrum of each feature, each filament, should indicate the time, in the Universe's life, at which it is observed. This would not necessarily be the same time as that at which the 3 K is observed in the standard cosmology. We must face the fact that the obvious differences in morphology between the fluctuations, as seen at different wavelengths by COBE is not extremely encouraging. It should be properly explained.

* 8.8

Appendix I: Tensors, Line Elements

This appendix is not intended to explain the entire field of tensorial analysis, which has filled books and books. We have only grouped together here a few equations and relations that are used in the text, in order to make the reading of Chap. 8 an easier task.

8.8.1
Tensors

A *tensor* $g_{\mu\nu}$ (that which enters the definition of the line element) of order 4 is a collection of 16 coefficients (as is any tensor in the 4-dimensional geometry we are now using) and can be represented by a square table, which indicates the way the metric element (see text, page 428) ds^2 can be computed. This computation is that of a *determinant,* similar to those encountered in the classical solution of four ordinary algebraic equations of degree 1. We will not go into the details of this computation, unnecessary in the present context.

$$g_{\mu\nu} = \begin{pmatrix} g_{11} & g_{12} & g_{13} & g_{14} \\ g_{21} & g_{22} & g_{23} & g_{24} \\ g_{31} & g_{32} & g_{33} & g_{34} \\ g_{41} & g_{42} & g_{43} & g_{44} \end{pmatrix},$$

which, in the Euclidian case, becomes the "diagonal" tensor (where the terms not on the diagonal are equal to zero):

$$g_{\mu\nu} = \begin{pmatrix} -1 & 0 & 0 & 0 \\ 0 & -1 & 0 & 0 \\ 0 & 0 & -1 & 0 \\ 0 & 0 & 0 & c^2 \end{pmatrix},$$

In non-Euclidian geometry, tensors are not reduced to their diagonal. Actually, the choice of the metrics imposes a form to the tensors to be used.

This choice also indicates the way the interval element ds^2 is transformed by a change of coordinates. A general theorem, valid for all tensors similar to $g_{\mu\nu}$, states that this transformation can be done according to the following formulae, easy to demonstrate:

$$ds^2 = g^*{}_{\mu\alpha\nu}\, dx^{*\mu}\, dx^{*\nu} = g_{\alpha\beta}\, dx^\alpha\, dx^\beta,$$

where α and β, as μ and ν, may have four values, and characterize the four coordinates x of a given coordinate system $S^*(\mu,\ \nu,\ \dots)$, different from $S\ (\alpha,\beta,\ \dots)$. g and g^* are related as follows:

$$g^*_{\mu\nu} = \frac{\delta x^\alpha}{\delta x^{*\mu}}\frac{\delta x^\beta}{\delta x^{*\nu}}\, g_{\alpha\beta}\ ,$$

where the partial derivatives which characterize the change of coordinates enter into the equation. In these transformations, the ds^2 is *invariant*. The g are not invariant. They are said to be *covariant*.

The tensorial forms permit the definition of covariant quantities, and the expression of invariant laws. In particular, they allow us to use non-euclidian spaces. They could, of course, be used in an n-dimensional space; but the space-time of GR is 4-dimensional, therefore the tensors entering its laws will be tensors of order 4. This is not the case in some of the more recent cosmological attempts, which use higher degree spaces. The diagonal tensor is quite similar to a vector, in 3-dimensional Euclidian space, but the most general tensor implies physical quantities, in any space.

8.8.2
The Metrics

We can write the line element (the ds^2) implying a certain form for the $g\mu\nu$ which indeed *defines the geometry*. A rather general form, symmetrical with respect to x, y, z, is

$$ds^2 = -\, e^\mu\, (dx^2 + dy^2 + dz^2) + e^\nu\, c^2 dt^2. \qquad (1a)$$

The exponents are just functions of $r = (x^2 + y^2 + z^2)^{1/2}$, and t. They are introduced in that form for mathematical convenience, useful in the case of spherical symmetry.

In a Euclidian geometry, the line element can be expressed in its very simple Minkowskian form:

$$ds^2 = c^2 dt^2 - dx^2 - dy^2 - dz^2. \qquad (2a)$$

Fig. 8.45

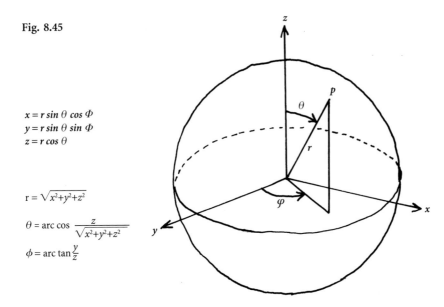

$x = r \sin \theta \cos \Phi$
$y = r \sin \theta \sin \Phi$
$z = r \cos \theta$

$r = \sqrt{x^2 + y^2 + z^2}$

$\theta = \arccos \dfrac{z}{\sqrt{x^2 + y^2 + z^2}}$

$\phi = \arctan \dfrac{y}{z}$

One is not obliged to use Cartesian coordinates. Polar coordinates are better adapted to cosmological problems. One can go from one to the other by using the well-known correspondence between coordinates (Fig. 8.45):

Then, the line element (1a) can be written

$$ds^2 = e^\nu c^2 dt^2 - e^\mu (dr^2 + r^2 d\theta^2 + r^2 \sin^2 \theta \, d\phi^2). \tag{1b}$$

and the Minkowskian line element (2) would become

$$ds^2 = c^2 \, dt^2 - dr^2 - r^2 d\theta^2 - r^2 \sin^2 \theta \, d\phi^2. \tag{2b}$$

A still more general form of the expression is:

$$ds^2 = e^\nu c^2 dt^2 - e^\lambda \, dr^2 - e^\mu (r^2 d\theta^2 + r^2 \sin^2\theta \, d\phi^2), \tag{1c}$$

where μ is also, here, a function of θ, ϕ, r, and which is therefore useful when we consider non-spherically-symmetrical systems.

Often, in these equations, the constant c is avoided and the time is $\tau = ct$, the unit of time being the time that light takes to travel one unit of length. Note that one can also use an imaginary transform $\tau = ict$, using the imaginary number $i = \sqrt{-1}$. This would result in $\tau^2 = -c^2t^2$, and all the expressions for the line element would look like the usual Pythagorian theorem in 3-dimensional space: $ds^2 = dx^2 + dy^2 + dz^2 + d\tau^2$. For reasons of easier physical intuition, we shall always use the negative sign in the expression of the ds^2.

It has been assumed (Karl Schwarzschild), that to arrive at the solution of the GR equations inside a 3-dimensional sphere of radius R, that: (a) the matter is an incompressible fluid of uniform proper density ϱ_0, where pressure is p_0; (b) boundary values have to be adopted at the surface of this sphere, namely $p_0 = 0$, and $r = R$.

A particular line element has been used to solve the GR equations in a spherically symmetrical geometry. It was introduced as:

$$ds^2 = \left[A - B\left(\frac{1-r^2}{R^2}\right)^{1/2}\right]^2 d\tau^2 - \frac{dr^2}{1-r^2/R^2} - r^2 d\theta^2 - r^2\sin^2\theta d\phi^2, \tag{3a}$$

where the λ and the ν are now expressed as functions of r only. This metric element can be compared to that actually used by Einstein and de Sitter in their static models (pp. 443 ff.). This is a metric valid at a certain distance from a massive point.

The *Robertson-Walker metric*, another metric, similar to it in some ways, is used more currently. It can be written, as a function of the dimensionless factor k):

$$ds^2 = d\tau^2 - R^2(\tau)\frac{dq^2}{1-kq^2} + q^2 d\theta^2 + q^2\sin^2\theta\, d\phi^2. \tag{3b}$$

This line element implies in particular that ν is constant. Here, R is the "scaling factor," and q a dimensionless quantity ($q=r/R$), different from the r of equation (1). The Minkowski metric, valid for a flat space-time, pseudo-euclidian, assumes $k=0$. This does not imply that the scaling factor R is infinite. It is important to be very clear on that point.

Note. A subtle question arises when one speaks on the one hand of the "scaling factor" R (to avoid the confusion, some call it a instead of R), which is also the radius of curvature of the 3-dimensional space at a given time t or τ; and on the other hand, of models with $k=0$, in a "Euclidian space," i.e. a flat space. This requires further explanation.

The word *"3-space"* should be used to designate ordinary 3-dimensional space, in which the location of a point is specified by three coordinates, Cartesian: x, y, z, or polar: r, θ, ϕ, often expressed in the neutral way x^1, x^2, x^3.

The word *"space-time"* (the words "hyperspace" or "superspace" are banned here, although often used in the literature, and reserved for larger orders of space, using 5 or more coordinates) should be used to designate 4-dimensional space. This is the "space-time" as such in use in SR as well as in GR, where the location of a point is specified by four coordinates, Cartesian: x, y, z, t, or polar: r, θ, ϕ, τ, or better (in order to define the line element in a better way, as seen above) x, y, z, τ, and r, θ, ϕ, τ – or again the neutral system x^1, x^2, x^3, x^4.

We should keep in mind the fact that *the line-element ds^2 has a cosmological meaning,* but part of it, i.e.: $dx^2 + dy^2 + dz^2$ is a local line-element, in the 3-space. Then, any of the constants implied in that 3-space part, are, in essence, local. When the same ds^2 is used to construct a cosmology, i.e. to account for the whole Universe, these parameters attain a universal value. Strictly speaking, *k is a measurement of the local importance of curvature of the 3-space* (it fixes its sign). On the other hand, *R is a quantity which allows to measure in space-time, the rate of expansion of the Universe,* more or less. It is a cosmological quantity, a function of time of course. This is why it is often called the *"radius of the Universe,"* even though the local curvature of the 3-space may be zero.

8.8.3
Equations of General Relativity Cosmology

Beginning with the rather general line element, written above (1c), and introducing, in the Einstein equation, the isotropic and homogeneous hypotheses, the Einstein tensorial equation takes the form:

$$8\pi p_0 = e^{-\lambda}\left(\frac{\nu'}{r} + \frac{1}{r^2}\right) - \frac{1}{r^2} + \Lambda,$$

$$8\pi p_0 = e^{-\lambda}\left(\frac{\lambda'}{r} - \frac{1}{r^2}\right) + \frac{1}{r^2} - \Lambda,$$

$$\frac{dp_0}{dr} = -\frac{(p_0+p_0)\nu'}{2},$$

where the accents on λ and ν mean a differentiation with respect to the radial coordinate r, to which the line element expression (1c) must be added. As the Robertson–Walker line element is simpler, when using the Robertson-Walker metrics, these equations reduce to two (the last one is identically zero, p_0 is uniform), the two following equations, the first one being the *dynamical equation* (where Λ may be a function of r and t) and the second the initial value equation.

$$8\pi p = 2\frac{R''}{R} - \frac{R'^2}{R^2} - \frac{k}{R^2} + \Lambda$$

and

$$\frac{8\pi\rho}{3} = \frac{R'^2}{R^2} + \frac{k}{R^2} - \frac{\Lambda}{3},$$

where the signs ' and " represent simple and double differentiation with respect to time.

These commonly used equations are the ones solved by various authors, notably (with $\Lambda = 0$) Friedmann. Some solutions are described in Figs. 8.6, 7 and 8. Naturally, the Hubble rate of expansion, at time t (or τ), is given by R'/R. As a consequence of our remark at the end of the last paragraph, R may have a finite value, eventhough k can be equated to zero.   ✶

8.9

✶ **Appendix II: The Particles of Modern Physics**

Modern cosmology is so impregnated, almost colonized, by modern physics and group theory that, even without touching the difficult mathematical aspects of group theory, it is necessary to present some very condensed ideas about *quantum electrodynamics*. We, therefore, include tables and basic data concerning the particles (real or virtual) and interactions of modern physics, which seem to play such a large part in the new interpretations of the (new) big bang.

8.9.1
The Nature of Matter

From the beginning of the history of science, philosophers have disputed extensively about the nature of matter. Many of them (Plato, in the *Timaeus*, already quoted earlier in this book, but still more Democritus and the Stoics, and later Lucretius) advocated the "atomic" hypothesis.

According to their theories, matter is not divisible at infinity. It ultimately reduces to indivisible particles, the "atoms." *Atomism* has its
story, just like the ideas about the Universe, and similarly, there have
been several currents of thought. For reasons linked to the dogma of
transsubstantiation, the Catholic Church of the Counter-Reformation
was against atomism. Some historians claim that the condemnation of
Galileo was due mostly to Galileo's atomism. Later, the battle experienced a change of scale, not of spirit. Some physicists, such as Dalton,
and many others, proposed a real structure of matter made of different
atoms (different of course from Plato's atoms). Other physicists, positi-
vists at the extreme (such as Berthelot) did not believe in the atomic theory except as a practical model, more symbolic than real. Nonetheless,
several discoveries have successively made clear that *matter* is made of
molecules (billions of known molecules!), molecules made of *atoms*
(some hundred different species), atoms made of a *nucleus* and its planetary cloud of *electrons* (up to about one hundred), nuclei of *neutrons* and
protons (up to more than one hundred of each in the heavier nuclei) held
together, and neutrons and protons themselves made of *quarks* (six species), held together by *gluons*. It is the general scheme of this structure
that is represented in Fig. 8.46.

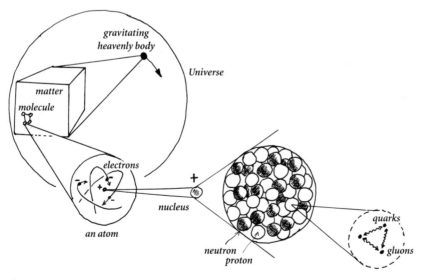

Fig. 8.46

Obviously neither molecules nor atoms are simple. They can be destroyed at moderate energies, by moderate temperatures (the atmosphere of the Sun is too hot to contain more than a very few very simple molecules. The hot centers of evolved stars contain only a very few nuclei still accompanied by one or a few electrons). Molecules and atoms can be reduced to what are traditionally called *elementary particles,* i.e. neutrons and protons, and electrons, and even, in some respects, nuclei of atoms, which can be subjected to high energy experiments without being necessarily broken down into their more elementary components.

Almost all of these elementary particles (proton, neutron, electron) have been observed in the laboratory, and in either cosmic radiation or particle accelerators for years. Isolated quarks have not been observed so far. Neutrons and protons are massive and sometimes called nucleons. They are *baryons,* heavy particles. Electrons are lighter. They are *leptons.* Some quarks have, in the nucleon, a very large "effective" mass. They are not observed in the most powerful accelerators. Their existence and properties have been deduced (like the atoms of Dalton's time) for theoretical reasons.

The structure of matter being thus described, other elementary particles (previously not regarded as necessary to a theory of matter) have been discovered, first in cosmic radiation (*mesons* of all sorts), and finally in the more and more powerful particle accelerators (other mesons, other baryons – *hyperons* –, other leptons – *positrons, neutrinos*). Others have been inferred from theoretical considerations (*gravitons,* such particles such as $W^{\pm}$, Z, etc. (see Table 8.6) and the Higgs particle). We should certainly be careful not to omit the *photon,* an essential particle in many interactions, hypothetically suggested by several scientists of the past, Newton for example, and finally introduced as a particle by Einstein, from his analysis of the photoelectric processes .

For each particle, there is a corresponding *antiparticle,* stable if the particle is stable. The characteristics of it are such that a collision between an antiparticle and a particle leads to the annihilation of both and a production of energy (i.e. photons!). Antiparticles have been discovered mostly in large accelerators. The positron, for example, is an "antielectron." They are designated by the same symbol as the particle, generally topped by a horizontal line (such as "a" for the particle, and "ā" for the antiparticle), except where the two charges are of opposite signs, the annihilation being essentially an annihilation of charge (such as for e^{+} and e^{-}).

Particles are characterized first by their *stability* or *instability*. The more stable a particle is, the more it is a *bona fide* member of the family of particles, the behavior of which can be predicted, making it a possible subject of laboratory experiments. Particles are characterized, therefore, by their *lifetime* τ. In an effort to explain such stability (of atoms, of nuclei, of protons or neutrons, of quarks), and satisfy its theoretical needs, new particles have been introduced, which permit the interaction to maintain the stability, not to dissolve it. A classical example is that of a heavy nucleus. If made only of protons, all positively charged, these protons would repel each other. No nucleus would be stable, hence the need for the introduction of neutrons, later observed as real, and of mesons, also observed as real (see Table 8.5).

In order to give a global description of the large number of particles so found, some physicists, in the sixties, developed a vision of the elementary particles quite similar to that of the elementary nuclei, i.e., a periodic table of elementary particles. The "holes" in the various families so defined ("octets," primarily) therefore predicted some particle to be found experimentally.

Along a simultaneous line of thought, physicists have therefore tried to characterize elementary particles by some obvious qualities, identical to those of "macromatter," such as *mass m* (often expressed in energy units – eV, keV, MeV, GeV, according to Einstein's equation $E = m\,c^2$), *charge q*, and *angular momentum ω*. The existence of discrete energy states of particles (atoms) led to Planck's discovery of the quantum of energy (and the correlative introduction in physics of the fundamental constant h, Planck's constant, after G, c, and the Boltzmann constant k) and the quantification of these physical characteristics. Therefore, the quantum numbers B (baryon number), q (charge), and s (spin, i.e. angular momentum) were introduced. At the same time, there was a need to "organize" the particles in groups, the "hypercharge" Y, the "isotopic spin" vector (i.e. its three components: I_1, I_2, I_3), and even the "strangeness" S, and the "color" of quarks, all quantities coming out of the needs of group theory.

8.9.2

The Principles of Symmetry

Group theory is a basic tool in the physics of elementary particles. It is based on the consideration of physical symmetries, and of conservation laws associated with them.

In various fields of physics, the concept of symmetry has been used. In crystallography for example, the symmetry between left-handed and right-handed twisted molecules leads to important considerations about the effect of these molecules upon polarized light. In the theory of particles, three types of symmetries have been introduced.[38] The C-symmetry (or antisymmetry) introduces a correspondence between the particle and its antiparticle. Antiparticles have a baryonic number equal to that of the corresponding particle, but of an opposite sign. The P-symmetry is an operation which replaces a particle with its mirror image (for example momentums that are equal but of different sign). This is the symmetry of "parity." The T-symmetry is an operation which reverses the sign of time, which inverses the arrow of time. More generally, let us consider two states, S_1 and S_2, defined by a set of properties, for a given system. When by an interchange of properties between S_1 and S_2, the same system is found, we have defined a symmetry of the system. In essence, the symmetries of a physical system permit us to define an operation under which the system is invariant. For each type of symmetry, a specific mathematical tool is called for. Group theory is, for the interactions between elementary particles, that mathematical tool.

The difference in the rules of symmetry and antisymmetry shows itself in the behavior of antiparticles. Particles obeying the rule of antisymmetry annihilate their antiparticles (protons annihilate antiprotons, electrons annihilate positrons…). However, bosons are symmetrical with respect to matter and antimatter. Experience shows that, in some cases (electrons and protons) the antisymmetric rule is enforced. No two antisymmetrical particles can be in the same state. This imposes on them the introduction of spin, which is the quantized expression of their angular momentum. Spin can be measured. Some particles (mesons, of spin 0; photons, of spin 1; gravitons, of spin 2) are ruled by symmetry. These are the *bosons* which act as "mediators" in the interactions. Other

---

[38] It is amusing to note that the idea of *antipodes* and *antichnes* (see Fig. 3.11) is essentially based on a feeling for symmetry.

particles are ruled by antisymmetry (electrons of spin 1/2). These are the *fermions*. Note that the quarks of charge $\pm (1/3)e$ or $\pm (2/3)e$ are governed by a more complex rule of symmetry, similar to that of a triangle, identical to itself by a rotation of $2\pi/3$. Actually, there are in the particular physics context, three main "groups" of symmetries: SU(3), SU(2), and U(1). They describe the symmetries associated with the various interactions as follows: SU(3) with the strong interaction; SU(2), combined with U(1), with the "electro-weak" interaction; which combination, when broken, gives rise to U(1), associated with the electromagnetic interaction. This means that these groups have been operational in defining the quantum numbers that characterize the elementary particles.

The symmetry and antisymmetry considerations led scientists to link the physics of interactions to aspects of mathematical group theory, some special group of transformations being associated with certain of the physical interactions. Group theory may be considered a convenient language for the physics of interactions.

It is possible to conceive of a state of matter where symmetries of all kinds are present. This is *supersymmetry* (or *SUSY*), a hypothetical state of the Universe. However, the moment we admit there are different types of interactions, not reducible each to the other, the medium as a whole does not obey all the rules of symmetry.

One might, therefore, imagine that the Universe was at first completely symmetrical, in all respects, and ruled by a single type of interaction. An interesting aspect of these theories is, therefore, the notion of a possible "breaking of symmetry," a notion introduced in 1967 by Weinberg and Salam. One important parameter is the energy of the medium (the energy density of the "vacuum," so to speak). It governs the energy of the "mediators" of the interactions. At low energy, the four mediators we now invoke (gravitons, photons, W particles $[W^-, W^+, Z^0]$, and gluons) behave differently. They have their individual behaviors. If we leave the gravitons aside, the other three mediators imply the groups of symmetries U(1), SU(2), and SU(3). At high energies, this is not true, and they behave in the same way. At the time of the big bang, energies were high. The T-symmetry is certainly violated, but T and P symmetries can be violated under the present conditions of experimentation in the laboratory. The lowering of the energy of the medium leads, therefore, to a spontaneous breaking of symmetry, possibly in successive jumps, successive breakings of symmetries, possibly almost at the same time as indicated in Fig. 8.33 (p. 491) and 8.47.

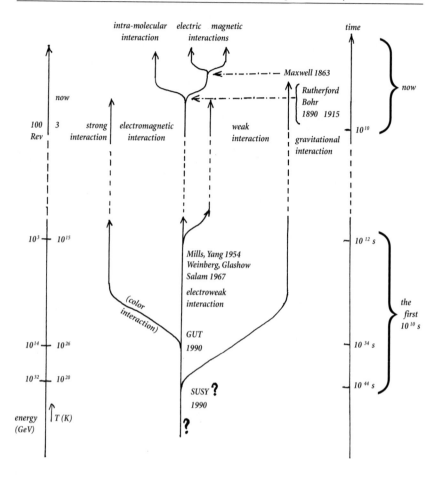

Fig. 8.47

Table 8.3. *Families of quarks and leptons*

	Quarks		Leptons	
Family 1	u	(up)	ν_e	electron neutrino
	d	(down)	e	electron
Family 2	c	(charmed)	ν_μ	muonic neutrino
	s	(strange)	μ	muon
Family 3	t	(truth, or top)	ν_τ	tau-onic neutrino
	b	(beauty)	τ	tau-on

8.9.3
Particles: Their Quantum Numbers and Their Classification

As a result of the application of group theory, various groups of particles may be defined, each characterized by some kind of symmetry. Assuming that all particles are made out of quarks and leptons, (and perhaps implying that they constitute the ultimate bricks of matter, indivisible as in the Greek "atomos" – ἀτόμος meaning that which cannot be divided), the theory shows that three families can be defined, as exemplified in Table 8.3.

Ordinary matter is made up of the quarks and leptons of the first family. Electrons belong in this group. The quarks u and d combined produce the neutron (u,d,d) and the proton (u,u,d). The two particles are

Fig. 8.48   *proton*   *neutron*

Table 8.4. *Pairs of families of particles*

	vs		
Bosons	vs	Fermions	Bosons are "vector," or "carrier," or
photon		proton	"mediator" particles in interactions.
pion		neutron	Fermions are the interacting particle.
kaon		quark	
W			
Hadrons	vs	Leptons	Hadrons are strongly interacting
baryon		electron	particles.
meson		neutrino	Leptons are not!
		muon	
Baryons	vs	Mesons	Both are hadrons; they differ by their
proton		pion	spin. Baryons: s = N (integer) + $\frac{1}{2}$;
neutron		kaon	mesons; s = N (integer).
(all are		(all are	
fermions)		bosons)	

Table 8.5. *Dating the main discoveries*

Particle	Theory		Observation		Discover
photon	Einstein	1925	Lenard	1899–1908	photoelectric effect
electron	Faraday, etc. J.J. Thomson	1880	Millikan	1908	lab measure of e, m_e
proton	nuclei, atoms Rutherford	1911		1915–1919	laboratory
	Bohr	1929		1950	accelerators cosmic rays
antiproton	Dirac	1928	Segrè Chamberlain	1955	accelerators
neutron	Rutherford	1920	Chadwick	1932	laboratory cosmic rays
mesons as "virtual"	Yukawa	1935			
muon			Anderson	1936	cosmic rays
pion			Powell	1947	cosmic rays
antiparticles positron	Dirac	1928	Anderson	1932	cosmic rays
neutrino	Pauli, 1930 Fermi, 1933		Reines, Cowan		
antineutrino				1956	nuclear reactor
Ω^-	Gell-Mann	1961		1963	large
	Gell-Mann, Ne'eman	1964			accelerator
quark	Gell-Mann, Zweig	1964			
meson tauon	Yukawa	1935	Richter, Perl	1975	large accelerator
gauge boson	Glashow, Weinberg, Salam	1961–72	Rubbia, Van der Meer	1983	large accelerator

... to be continued in further years!

often represented symbolically as in Fig. 8.48. Families 2 and 3 are constituent of the many unstable particles that we have already mentioned.

By emphasizing the difference between them, Table 8.4 associates all the types of particles known in pairs.

Table 8.5 gives the main theoretical predictions and actual discoveries of some elementary particles. Table 8.6 groups all the particles we have mentioned, with a very simplified classification, giving their mass and their main quantum numbers.

8.9.4
Interactions

This brings us to a very difficult question. According to classical physics, the interaction between two charged particles, or two massive particles, takes place through the existence of an electromagnetic field, or a gravitational field. The action created in the field by the motion of one of these particles modifies the field locally. This local change propagates, at the velocity of light, and affects the other particle. The carrier of the interaction is, in both cases, *the field*. This means also that the field itself has a content of energy and momentum. The equations describing these actions are the Maxwell-Lorentz equations and the Einstein equations.

This picture has been altered by quantum physics. The fields and particles are no longer so distinct from one another. The *carriers of the interaction are particles of the field*, i.e. photons (the reality of which, as particles, was demonstrated experimentally by Einstein) or light waves, and gravitons or gravitational waves (still an abstract concept). Similarly electrons and protons, or other particles, can be described as the quanta of a field. The individuality of the particles is more or less lost. The interaction takes place between many states of electrons, photons, etc. The vacuum of classical physics is empty, devoid of anything, but the vacuum of quantum physics is full of particles in their state of lower energy. It contains "virtual" pairs of particles (such as electron-positron pairs) which would annihilate each other. It contains several unstable particles which play a part as carriers of interactions. An interaction may involve several of these unstable particles. All this means that the potential energy of the quantum vacuum is not zero. The theory which establishes these interactions in quantum physics is *quantum electrodynamics* (QED), a very rich branch of modern physics. The interaction itself is conveniently symbolized by a *Feynmann diagram* such as the one represented in Fig. 8.49 (the interaction between particles 1 and 2, giving rise to the formation of particles 3 and 4, a certain "carrier" particle, a wavy line, acting between the two).

Of course all interactions obey strict conservation laws, meaning more or less that they work under the rules of groups of symmetries. The elec-

Table 8.6. *Elementary particles of modern physics*

Family	Name	Symbol	Mass (MeV)	Baryon number	Charge	Spin	Strangeness	Lifetime (seconds)
Vectors	graviton	g	0?	0	0	2	0	stable
(bosons)	photon	γ	0?	0	0	1	0	stable
	gluon(s)	g	0?	0				
	W	W^{+-}	$\sim 10^5$	0	+1, −1	1		
	Z	Z^0	$\sim 10^5$	0	0			
Quarks	up	$u\,\bar{u}$	5.1	0	2/3, −2/3	1/2		stable
(fermions)	down	$d\,\bar{d}$	11	0	−1/3, 1/3	1/2		stable
	charm	$c\,\bar{c}$	1270		−1/3, 1/3	1/2		stable
	strange	$s\,\bar{s}$	215		2/3, −2/3	1/2		stable
	beauty	$b\,\bar{b}$	4250		2/3, −2/3	1/2		stable
	top (truth)	$t\,\bar{t}$	174000		−1/3, 1/3	1/2		stable
Leptons	neutrinos:							
(fermions)	electronic	$\nu_e\bar{\nu}_e$	>0?	0	0	1/2	0	stable
	muonic	$\nu_\mu\bar{\nu}_m$	>0?	0	0	1/2	0	stable
	tauonic	$\nu_\tau\bar{\nu}_t$	>0?	0	0	1/2	0	stable
	electron/ positron	e^-e^+	0.511	0	−1, +1	1/2	0	stable
	muon	$\mu^+\mu^-$	105.66	0	+1, −1	1/2	0	stable
	tauon	$\tau^+\tau^-$	1776.96	0	+1, −1	1/2	0	stable
Hadrons mesons (leptons, fermions)	pion	π^{+-}	139.57	0	+1, −1	0		2.6×10^{-8}
		π^0	134.96	0	0	0		0.6×10^{-16}
	kaon	K^+K^-	493.67	0	+1, −1	0	±1	1.2×10^{-8}
		K^0K^0	497.67	0	0	0	±1	0.9×10^{-10}
	eta	η	548.8	0	0	0		2.5×10^{-19}
baryons (fermions)	proton	$p\,\bar{p}$	938.28	1	+1, −1	1/2	0	stable
	neutron	$n\,\bar{n}$	939.57	1	0	1/2	0	898s
hyperons (fermions)	lambda	$\Lambda\bar{\Lambda}$	1115.36	1	0	1/2	±1	2.6×10^{-10}
	sigma	$\Sigma^+\bar{\Sigma}^+$	1189.36	1	+1, −1	1/2	±1	0.8×10^{-10}
		$\Sigma^0\bar{\Sigma}^0$	1192.46	1	0	1/2	±1	5.8×10^{-20}
		$\Sigma^-\bar{\Sigma}^-$	1197.34	1	−1, +1	1/2	±1	1.5×10^{-10}
	xi	$\Xi^0\bar{\Xi}^0$	1314.3	1	0	1/2	±2	2.9×10^{-10}
		$\Xi^-\bar{\Xi}^-$	1321.3	1	−1, +1	1/2	±2	1.6×10^{-10}
	omega	$\Omega^-\bar{\Omega}^-$	1672.5	1	−1, +1	3/2	±3	0.810^{-10}

This list is of course far from exhaustive; it is limited to the particles of which the life-time is larger than 10^{-23} seconds.

Fig. 8.49

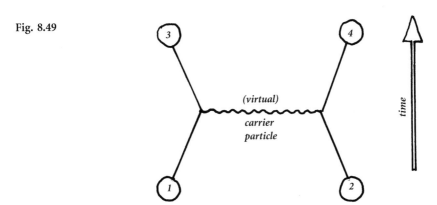

tric charge, the baryon number, the spin, the strangeness have to be conserved during the interactions, but some other numbers are not conserved in some interactions, as we shall see.

As is well known, the four types of interactions are the *gravitational interaction* (exchange of a graviton, or action of the gravitational field); the *electromagnetic interaction* (which holds an atom together by the interaction between the positive nucleus, and the electrons of the atomic structure, with typically the exchange of a photon); the *weak interaction* (which holds the protons and neutrons inside the nucleus, involving typically an exchange of a pion, and also accounts for the decay of fermions); and the *strong interaction* (which holds the proton and the neutron together, and involves quarks and virtual particles, or gluons).

The language of interactions is rather difficult. Two particles are involved, and a third acts as a *carrier*, or a mediator. In each of the four cases of interactions, there exists a *coupling constant* which characterizes the interaction. Let us try to give some idea of the coupling constant. In the case of electromagnetic interactions, we must remember one basic law of electromagnetism, Coulomb's law, according to which the force exerted between two electrons is proportional to the product of their charge, and inversely proportional to the square of the distance which separates them. The product of two charges is therefore equal to the product of $q_1 \, q_2 = Fr^2$ (*F* being Coulomb's force) and/or *(Frt) (r/t)* (angular momentum of the force multiplied by a velocity). If we want to introduce dimensionless constants in a way which is as objective as we can, not depending upon the system of units, we may use the non-dimensional ratios of angular momentum to the Planck constant *h*, and of velocity to the velocity of light *c*.

As charges are quantized, and can be introduced as such, one can use the non-dimensional ratio Q_1 and Q_2 of charges q_1 and q_2 to the electron (or proton) charge e. Then the product of the two charges involved can be expressed as follows, using non-dimensional quantities:

$q_1 q_2 = hc (e^2 /hc) (Q_1 Q_2)$. The ratio e^2/ hc appears there naturally. It is denominated by the symbol α, called the "coupling constant," or the "fine structure constant," because of its intervention in some spectroscopic interpretations. The numerical value of α is 1/137.

Similarly, by introducing dimensionless ratios in all types of interactions, one can define the constants α_0 (for gravitational coupling); α (the electro-weak interaction), were α is a function of the Sum $\alpha_1 + \alpha_2$, where α_1 corresponds to the electromagnetic coupling and α_2 to the weak coupling; and α_3 (strong interactions). The values of these constants have been determined experimentally, but they depend on energy involved in the interaction. Typically, the electro-weak coupling has at low energy the values $\alpha_1 = 1/60$ and $\alpha_2 = 1/30$, and converges towards the same value 1/42 (Fig. 8.33, p. 491) around an energy of 10^{13} GeV. The strong coupling (called the "color interaction," as it involves color interactions between quarks inside the proton) has a value of $\alpha_3 = 1/9$ at low energies (down to 1/40 at about 10^{14} GeV). The gravitational coupling has a constant $\alpha_0 = Gm_p{}^2/hc = 10^{-40}$ (expressed relative to the proton mass, as a unit).

The *weak interaction* implies decays of particles we know as *fermions*. Two typical examples are: (i) The β-decay of a neutron (a stable particle, in a sense, but with a lifetime on the order of 17 minutes): the neutron n decays into a proton p, an electron e⁻, and an antineutrino v; and (ii) the scattering electron-neutrino: electron + neutrino --> electron + neutrino. The electrons and neutrinos are stable.

These are weak interactions in that they imply little energy. Moreover, they are essentially local. In the first formulation, they amount to an exchange of charge between the main actors of the interaction, n and p, or e and v.

A characteristic of weak interaction is the conservation of the *lepton number* for each pair of the type e^-, v_e; μ, v_μ; τ, v_τ, between particles of the same family, all fermions: electrons, muons, tau-mesons (or tauons), charged and non-charged neutrinos of the three types known. In each of the reactions, the lepton number L (number of leptons) is conserved for each pair, the antiparticle having a lepton number equal to $L = -1$, for a corresponding given particle of lepton number 1. The lepton number of a lepton is (strangely enough!) equal to 0.

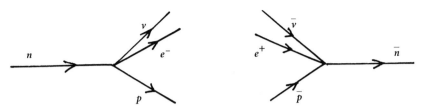

Fig. 8.50 *An illustration of the CPT theorem.* On the left, a neutron n decays into p, ν, and e⁻. On the right, according to the symmetry theorem (the "CPT theorem"), an antiproton $\bar{p}$ combines with a neutrino and a positron to produce an antineutron. The direction of time has been reversed (T-parity), particles have been replaced by their antiparticles (C-parity), and the spatial parity has been reversed (P-parity).

A difference between the electromagnetic interaction and the weak interaction still exists. *Electromagnetic interactions preserve the "parity"* which characterizes reactions for which a symmetrical reaction exists, particles being replaced by antiparticles, arrow of time being reversed, as in Fig. 8.50 (after Narlikar).

Weak interactions do not necessarily preserve parity. Such reactions as are described here above cannot be purely and simply reversed.

In the weak interactions, as in the electromagnetic interactions, some mediator, a *boson*, has to exist. In the electromagnetic and gravitational interactions, the mediators are the photons and gravitons, which travel far and fast, and which have a long life. In the weak interaction, the exchange is conducted through pions, or through heavy neutral particles called W-bosons, of spin 1. These interactions can be represented by Feynmann diagrams, as exemplified in Fig. 8.51. This representation, more symbolic than corresponding to some geometrical reality (but somewhat physical in that a computation of the amplitude of the probability of various processes is described), is very practical, and often used in the practice of elementary interactions. Knowing the quark structure of the neutron and proton, the interaction may often be represented by a Feynmann-like diagram, following each quark.

The particles so far described (protons, neutrons, electrons, mesons) are relatively stable. Their stability is maintained by what is called the strong interaction, which does not act outside these particles. The basic ingredient of the *strong interaction* inside the nucleus, (i.e. involving neutrons and protons) is the *quark*, the basic element of the nucleons, mesons, and other particles. The transmitter (or carrier, or mediator) is called the *gluon*. Six types (six *flavors*) of quarks have been identified

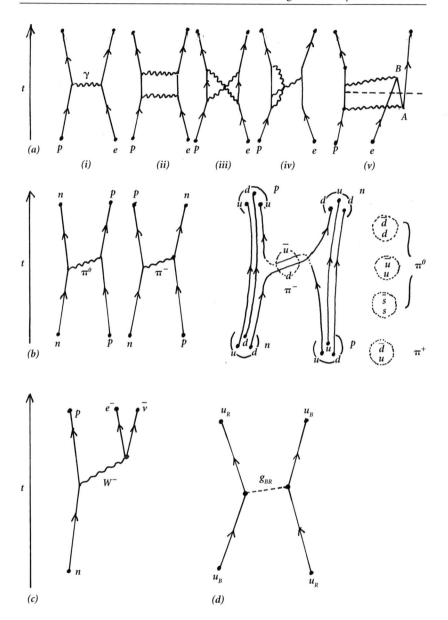

(from theoretical consideration), each having three possible *colors*. Similarly there are eight gluons. A strong interaction can be represented by a Feynmann diagram, such as the one in Fig. 8.49. Again, strong interactions can be described in terms of group theory. The groups involved are more complicated than those intervening in the case of weak interactions. A special characteristic of quarks should be noted. Each of the six types of quarks is also characterized by a certain quality, its color. Each quark can have three colors, red (R), green (G), and blue (B), the sum of which is "white," or "colorless," just as the sum of three quantum numbers may be 0. There are also three anti-colors ($\overline{R}$, $\overline{G}$, $\overline{B}$). Only a combination of three complementary colors can give rise to an observable particle. Inside a nucleon, gluons also have colors, and form a "colorless-white" ring together with the three quarks they unite.

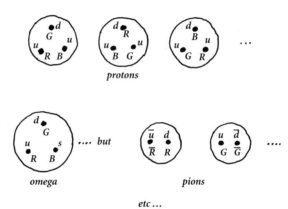

Fig. 8.52 *The color structure in particles: a few examples.* The protons are made of three quarks, of three different colors R, G, B. So is also, for example, the Ω particle. Pions are made of only two quarks, one of any given color, the other one of the corresponding "anti-color."

Fig. 8.51a–d *Feynmann diagrams: several typical cases.* (a) Interaction of a proton and an electron, involving either one or two photons. The diagram on the right shows that at the time indicated by a dotted line, the electron line is seen as three particles: section BA is interpreted as a virtual positron. (b) A weak interaction between a neutron and a proton. A charged pion (second Feynmann diagram) produces an exchange of the two particles. On the right, a detailed version of the same diagram is drawn, and explains how the second diagram may be interpreted as an exchange of two quarks. (c) In a weak interaction a neutron disintegrates into a proton, an electron and a neutrino, a boson vector W⁻ is implied. Note this reaction is the same as in Fig. 8.50, left side. (d) A strong interaction between quarks. A gluon is the virtual vector particle. The "color" is invariant.

This theory (*quantum chromodynamics, or QCD*) is still in its infancy. It is still the theory of groups of symmetry which permits the prediction that there may be eight gluons. Three quarks (of different colors) form a nucleon; two quarks a meson. Similar to the quarks, the gluons are not observable as isolated particles, in spite of their mass (equal to zero, according some theoreticians). They move in groups, in color groups, forming colorless gluon-bundles. We have not introduced QCD, and the quantum numbers (the colors) associated with it in Table 8.4, but in Fig. 8.52, we give a few examples of this behavior. We should perhaps mention that some theorists at the present time are trying to look for particles of a more elementary nature, the "subquarks," which could be the elementary bricks of all particles so far listed. For the time being, this idea is still entirely speculative. There is no "need" for this simplification, but the demon of simplicity (Occam again) might lead to such a development. However, we would make clear once more that for us, there is no reason except of a metaphysical nature to be alarmed by the very great number of "elementary particles" and the complex interactions between them.   ✶

Chapter 9

General Conclusion

At the dawn of the 21st century, we find ourselves in some ways at the same point as we were three millennia ago. The horizons of our instruments have expanded in a tremendous way (see p. 319ff.). The time scales of the phenomena we have explored have dramatically changed from a few millennia to billions of years. The physics of most of the objects known in the Universe is much better understood, and quite consistent with the laws of physics as discovered on Earth. Not only is the Earth no longer the center of the Universe, with Man at its own heart, but the Sun is not even at the center of our Galaxy; and our Galaxy is one among billions, far from the center of our local cluster, which is itself far from the center of our supercluster, not to speak of structures of a higher degree about which we as yet know nothing. Still, the battle rages among three different approaches to the Universe as a whole, much as it did in ancient Greece.

There is little doubt that the extension of techniques, in classical optics, in optics of all wavelength ranges (ultraviolet, infrared, radio, X-rays, gamma-rays), in space instrumentation, and in space exploration *in situ*, has been and will always be the transmitter of basic progress. The ideas have to adapt themselves, progressively, to the new findings. This adaptation is becoming rather rapid. A thousand years ago, the time-lag between a new observed phenomenon and an interpretation was often counted in centuries. Now it is counted in years, often in months. The development of powerful computers, of astute software, of simulation methods, have allowed theorists to compare their theories with existing observations, to develop the potentially observable consequences of their models, and to improve them by successive (very rapid) iterative approximations.

There is no revolution in ideas other than that generated by new observations. Still, throughout the history of science, progress has taken the form of the extension of the known world to the still unknown one rather than the form of any real revolution. The 17th century saw the beginning of observational power. During the 18th and 19th centuries, great advances were made in the exploration of the planetary system, then of the Galaxy, with its stars, its nebulae, its stellar clusters, and its interstellar matter. In the 20th century, we have achieved an almost complete understanding of stellar evolution. A few limited problems still remain to be explored, but the general picture is more or less clear. We have also discovered the extragalactic world, with its many components, galaxies of very diverse shapes and structures, clusters of galaxies, radiosources, and quasars and we have carefully explored them, measuring distances, evaluating time-scales, and discovering some unifying principles.

However, we still know very little, almost nothing, about galactic evolution. Are the AGNs young exploding galaxies? Is this phenomenon general or exceptional? Are galaxies always centered around a black hole? Or does that apply to only a few galaxies? What are quasars, really? Early stages of galactic life? late stages of galactic life? small ejecta of some galaxies by the AGNs? I would be hard pressed to give sure answers to any of these questions, which will be central in 21st century astronomy.

Then what about the Universe? We do not even know whether it is a thermodynamically isolated system, or not! It is certainly evolving locally, but globally, are we looking at an exploding Universe, born out of nothing a few billion years ago? Or has it existed forever, in various successive stages, from some "quantum Universe" to the present day Universe? Or is it a steady Universe, constant and eternal, subject only to local fluctuations in which galaxies are born and die in many successive generations? How has life come about in that Universe? Will we be able to answer these questions before the middle or the end of the next century?

Over and above all of these questions, those solved in the past, those still to be solved, I hope that the readers of this book have become aware of the constant feature of the philosophical attitudes of astronomers and cosmologists. They are roughly divided in three groups (notwithstanding the fact that they overlap more and more one upon the other), which we shall label with quotations marks, "Pythagorians," "Platonists," and "Aristotelians." We use these designations according to the meanings we have given them earlier (p. 40, 41), often different from their usual accepted meanings.

The "Pythagorians" would like to see a beautiful and simple Universe, obeying simple numerical laws, beautiful geometries. They equate unknown quantities to zero (for simplicity). They have recourse to beautiful constructions (such as GUT and SUSY), developed mostly from the magnificent and coherent mathematical construction called group theory. Still, they consider observations necessary tests, but not necessary guides. The young Kepler, centuries ago, and the GUT physicists among contemporary scientists, are typical of this group.

The "Platonists" are constantly searching for the lost spirituality of science. In a naive way, some of them identify the "fiat lux" with the big bang, or the origins of the Universe with the Creation, 6000 or so years ago. Less naively, they "prefer" the big bang models, which seem to imply a "Creator" to some investigators. Sometimes, on the other hand, they are seduced by models that have metaphysically opposite implications, such as eternity and infinity. They are easily seduced by the anthropic models. Newton, the 1917 – Einstein, and the standard big bang cosmologists, are Platonists of various types. So various indeed that it is quite unlikely that Einstein would have appreciated the so-called anthropic principle, for example.

The "Aristotelians" continue, steadily, to attempt to "σώζειν τὰ φαινόμενα," to save the phenomena. Hence, they do not disregard a priori any observation as insignificant, or biased. They request checks of the observations but also tests of the refutations. For them observations come first. Galileo, Einstein in general, and many other astronomers of good will, are Aristotelians.

Of course, some individuals have belonged to either one or the other of these three groups, at different times in their lives. Moreover, the three are not necessarily completely contradictory. The watershed in the attitudes is, therefore, perceived differently by different authors. We are quite conscious that our classification, as rough as it may be, as cautious as we may try to be, may be subject to controversy. For example, some counter to our three currents of thought, with another division based more or less on the concept of the nature of matter, but partly similar to ours. Atomism emerged with the Pythagorians, was developed by Anaxagoras, and much more coherently by the great Democritus, then by the Epicureans, Epicurus, and later, Lucretius and others. Plato and Aristotle, and their followers did not concern themselves too much with this structure of matter (even though the theory of the four elements is a step towards that preoccupation). They described the Universe as embedded in a closed sphere. Aristotle's Universe was finite. Archytas questioned that view by the famous metaphor of the arrow which will never reach the border of the Universe. For the Stoics, followers of Zeno, the Universe was not limited by the boundary, but extended infinitely into a vacuum. This is certainly an interesting system of classification, but we prefer ours, however simplified and subject to criticism it may be.

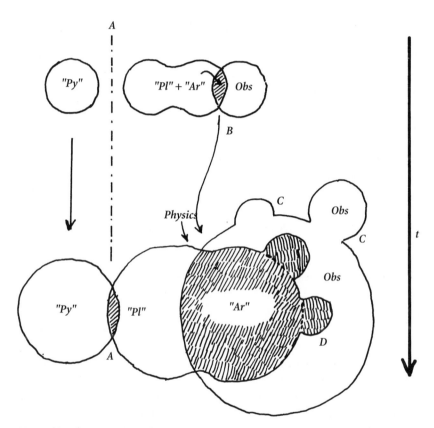

Fig. 9.1 *The domains of scientific investigations.* One note from 500 BC to the modern times, a huge increase of the domains of investigation. The observations and the theories are more and more in agreement and (B) interaction, except from some idealistic constructions which are still extrapolations without basis and pseudopods of the observations (C), which are still out from any theoretical counterpart, not to mention non-idealistic views, necessarily in accordance with the observations (D) which progress away from the "Platonician" domain of idealist theory. The "Pythagorean" views, from time to time, join now (A) the fields of classical physical thought, as exemplified by the "Platonician" – "Aristotelian" domain.

The evolution of cosmology since antiquity has seen a displacement, a change, in the different domains of understanding and knowledge that one can define (Fig. 9.1). One domain is that of observational material. It expands continuously, and at present very rapidly.

The domain of theory, which intersects necessarily with the domain of observation, is also expanding, but more slowly. The intersection of the two domains is much stronger now than it has ever been. However, the progress of observation tends to counteract that tendency, and sometimes leads to the manifestation of pseudopods which depart from the domain of legitimate theory. Within the domain of bona fide theories, the "Platonic" point of view predominates. The "Aristotelian" point of view concentrates at the intersection, avoiding self-justification by observations that are not yet understood, as well as by metaphysical intuition. Apart from both spheres, which have progressed together since antiquity, the "Pythagorian" point of view had, in the past, almost no contact with observations. More recently, the two domain have begun to intersect, but not very much. Kepler was no doubt conscious that his early views about the harmony of the spheres and the polyhedrons did not fit the observations, thus he moved to the "Aristotelian" part of the theoretical domain! Einstein, between 1917 and the fifties, made a similar journey.

It seems to me that the future is clear. New observations will come at an ever faster rate. They will achieve better and better quality. Theoreticians, sometimes seduced by a desire for a necessary simplicity or beauty, sometimes hoping to regain the lost spirituality which seems to them inseparable from the physical world, and sometimes attempting only to precisely account for as many observations as possible, will progress together, but often with ongoing disputes, towards a view of the Universe. Sometimes it will be a unified view; sometimes a view limited to realistic statements; and, sometimes a view completely detached from reality, purely aesthetic and metaphysical. Such is "progress." The more we know, the more we want to know. Intuition, logic, and imagination are the weapons in this forward march. Observations are always there, as a safeguard against foolishness, but will we ever achieve a completely acceptable and coherent view of the Universe?

Further Reading

The author, whose main sources are in French, is greatly indebted to Professor Michael Hoskin for providing a comparable list of sources for the English readers. (Chapters given in parentheses following each entry in the list refer to the chapters to which the literature is relevant.)

(Anthology) (1964) *Les penseurs grecs avant Socrate.* GF Flammarion, Paris, pp 367 (Chap. 1)

Aiton EF (1972) *The vortex theory of planetary motions.* American Elsevier, New York (Chap. 5)

Arrhenius S (1910) *L'évolution des mondes.* Béranger, Paris, pp 246 (French translation from Swedish; apparently no English translation) (Chap. 6)

Blamont JE (1993) *Le chiffre et le songe, histoire politique de la découverte.* Odile Jacob, Paris, pp 941 (Chaps. 3–8)

Born M (1969) *Die Relativitätstheorie Einsteins,* 5th edn. (*Einstein's theory of relativity*; English translation from German, new edition) Dover, New York (Chap. 7)

Caspar M (1995) *Johannes Kepler,* 4th ed. GNT-Verlag, Stuttgart, pp 506 (Chap. 4)

Conger GP (1922) 1950, *Theories of macrocosmos and microcosmos.* Columbia University Press, new edition 1976, pp 146, Russell & Russell, New York (Chap. 3)

De La Cotardière P (1987) *Dictionnaire de l'astronomie.* Larousse, Paris, pp 325 (all Chaps.)

Couderc P (1945) (several new editions; last under the title: *Histoire de l'astronomie classique*) *Les étapes de l'astronomie.* Presses Univ. de France, Coll. Que sais-je? Paris, pp 128 (all Chaps.)

Crombie AC (1952) *Augustine to Galileo: the history of science AD 400–1650*, vol 2. Falcon Press (new edition W. Heinemann) London, pp 588 (Chaps. 3–5)

d'Abro (1950) *The evolution of scientific thought*, 2nd edn. Dover, New York, pp 481 (Chaps. 5, 7, 8)

Descartes R (date not available) *Le monde ou le traité de la lumière*. In: Oeuvres de Descartes, vol 11. Leopold Cerf, Paris, pp 118 (Chap. 5)

Descartes R (1912) *Discourse on method*; *Meditations on the first philosophy*; *Principles of philosophy* (new edition 1986). Dent, London, pp 254 (English translation by John Veicht) (Chap. 5)

Dreyer JLE (1906) *A history of planetary systems from Thales to Kepler*. Cambridge University Press, 1953, reproduced under the title: A history of astronomy from Thales to Kepler, Dover, New York, pp 438 (Chaps. 1–4)

Duhem (1959) *Le système du monde*, 10 volumes, Hermann, Paris, 1987, English translation: selection from the 10 volumes: *Medieval cosmology: Theories of infinity, place, time, void, and the plurality of worlds*. (Translated by R. Ariew.) University of Chicago Press, pp xxxi and 601 (Chaps. 1–4)

Eddington AS (1920) *Space, time and gravitation*. New edition 1970. Cambridge University Press, pp 218 (Chaps. 7, 8)

Feynman R (1965) *The character of physical laws*. British Broadcasting Corporation, London, pp 297. New edition 1992, Penguin books, London, pp 173 (Chap. 8)

Fournier d'Albe EE (1907) *Two new worlds*. Longmans Gren & Co, London, pp 157 (Chap. 6)

Gingerich O (1992) *The great Copernicus chase and other adventures in astronomical history*. Cambridge University Press, pp xii + 304 (Chaps. 4–6)

Gingerich O (ed) (1984) *The general history of astronomy*, vol 4 A. Cambridge University Press, pp 200 (Chaps. 7, 8)

Grenet M (1994) *La passion des astres au XVII-ème siècle*. Hachette, Paris, pp 295 (Coll. La vie quotidienne. No English translation) (Chap. 5)

Gribbin J (1986) *In search of the Big Bang*. Helmann, London, pp xviii + 413 (Chap. 8)

Hawking SW (1988) *A brief history of time*. Bantam Press, New York (Chap. 8)

Hoskin M (ed) (1997) *The Cambridge illustrated history of astronomy*. Cambridge University Press, pp viii + 392 (all Chaps.)

Hoskin M (ed) (1999) *The Cambridge concise history of astronomy*. Cambridge University Press, pp xiv + 362 (all Chaps.)

Hoskin M (1982) *Stellar astronomy: historical essays*. Science History Publ., pp iv + 197 (Chap. 6)

Kafatos M, Kondo Y (1996) *Examining the Big Bang and diffuse background radiation*. IAU Symposium No. 168. Kluwer, Dordrecht, pp 586 (Chap. 8)

Kant I (1754) *Universal natural history*. New edition 1969, The University of Michigan Press, pp 180 (English translation from German) (Chap. 6)

Kourganoff V (date not available) *Initiation à la théorie de la relativité*. Presses Univ. France, pp 180 (Coll. Science vivante) (Chap. 7)

Koyré A (1973) *The astronomical revolution: Copernicus, Kepler, Borelli*. Methuen, London, pp 530 (English translation from French by Relj Maddison) (Chap. 4)

Lambert JH (1976) *Cosmological letters on the arrangement of the world-edifice*. Scottish Acad. Press, Edinburgh, pp viii + 245 (English translation from French and German) (Chap. 6)

Laplace PS de (1796) *L'exposition du système du monde*. Fayard, Paris, pp 675 (new edition 1835, reprint 1984; apparently no English translation available) (Chap. 6)

Mach E (1904) *Die Mechanik. La mécanique*. Hermann, Paris, pp 498 (French translated from German; apparently no English translation) (Chaps. 7, 8)

Michel PH (1973) *The cosmology of Giordano Bruno*. Hermann, Paris, pp 306 (Chap. 4)

Misner CW, Thorne KS, Wheeler JA (1973) *Gravitation*. Freeman & Co, San Francisco, pp 1279 (Chap. 8)

Neugebauer O (1957) *The exact sciences in antiquity*, 2nd edn. Brown University Press, Providencer, R. I, pp xvi + 240 (Chaps. 1–3)

Neugebauer O (1983) *Astronomy and history, selected essays*. Springer, New York, pp 146 (Chaps. 1–3)

Newton Isaac (1934) *Principia* (English translation from Latin by Motte), 1992 revised by F. Cajori F, 1999 English translation by Cohen, I. Bernard & Whitman, Anne, Univ. of California press, Los Angeles (Chap. 5)

North J (1994) *The Fontana history of astronomy and cosmology*. Fontana, London, pp xxvii + 697 (all Chaps.)

North JD (1965) *The measure of the Universe: a history of modern cosmology.* Clarendon Press, Oxford, pp xxvii + 436 (Chaps. 6–8)

Nottale L (1993) *Fractal space-time and microphysics.* World Science Publ., Singapore, pp 333 (Chap. 8)

Pannekoek (1961) *A history of astronomy.* G. Allen & Unwin, London, repr., pp 521, Dover, New York (all Chaps.)

Poincaré H (1911) *Leçons sur les hypothèses cosmogoniques.* Hermann, Paris, pp 294 (apparently no English translation) (Chap. 7)

Stephenson B (1987) *Kepler's physical astronomy.* Springer, Berlin Heidelberg New York, pp viii + 217 (Chap. 4)

Stephenson B (1994) *The music of the heavens: Kepler's harmonic astronomy.* Princeton University Press, Princeton, pp xii + 260 (Chap. 4)

Tannery P (1930) *Pour l'histoire de la science hellène.* Gauthiers-Villars, Paris, re-printed 1990, J. Gabay, Paris (Chaps. 1, 2)

Taub LC (1993) *Ptolemy's universe.* Open court, Chicago, pp xvi + 188 (Chaps. 3, 4)

Tolman R (1934) *Thermodynamics and cosmology.* Clarendon Press, Oxford, pp 501 (Chaps. 7, 8)

Treder HJ, Von Borzeszkowski HH, Van der Merwe A, Yourgrau W (1980) *Fundamental principles of general relativity.* Plenum Press, New York, pp 216 (Chap. 8)

Walker C (ed) (1996) *Astronomy before the telescope.* British Museum Press, London, pp 352 (Chaps. 1–4)

Weyl H (1950) (1952) *Space, time, matter.* Dover, USA, pp 330 (English translated from German) (Chap. 8)

Figure Acknowledgements

Fig. 3.6a   Museum of the History of Science, Oxford, UK; Paul Freestone, phot.
Fig. 3.6b   (Commission of Sundials of the) Astronomical Society, Paris
Fig. 3.8   Forschungs- und Landesbibliothek Gotha; Germany. Memb.II 141,
               Bl. 21r
Fig. 3.14a   Tycho Brahe, *Astronomiae instauratae mechanica*
               (Wandesburgi 1598). Heidelberg University Library, Germany
Fig. 3.14b+c   Cliché Bibliothèque nationale de France, Paris
Fig. 3.15   Werner & Apianus, P., *Introductio geographica Petri Apiani*,
               Ingolstadii 1533. (Observatoire de Paris Library, Paris)

Fig. 4.5a+b,
and Fig. 4.21 Johannes Kepler, *Mysterium Cosmographicum*
               (Astronomisches Rechen-Institut, Heidelberg)
Fig. 4.14   Tycho Brahe, *De Stella Nova* (Institut de France Library, Paris)
Fig. 4.16a,b   Tycho Brahe, *Astronomiae instauratae mechanica.*
               (Wandesburgi 1598).
               (Reproduced from: *Cambridge Illustrated History*, M. Hoskin (ed.),
               Cambridge University Press 1997)
Fig. 4.23   Johannes Kepler, *Harmoniae Mundi.*
               (Institut de France Library, Paris)
Fig. 4.29   Reproduced from: S. Toumin, J. Goodfield, *The Fabric of the Heavens.*
               Harper & Brothers 1961
Fig. 4.34–35   Johannes Kepler, *Astronomia nova.* (Institut de France Library, Paris)
Fig. 4.36   Thomas Digges, *Perfect Description of the Celestial Orbes.*
               From Typ Bl. B76JC (University of St. Andrews, Fife, UK)
Fig. 4.38a–c   Raymond d'Hollander, *L'Astrolabe*, Ed: Musée Paul Dupuy et Ass. fr.
               Topographie, Toulouse 1993, pp. 151.
Fig. 4.39   Petrus Apianus, *Cosmographia.* Antwerpen 1584
               (Observatoire de Paris Library, Paris)
Fig. 4.40, 4.43 Observatoire de Paris Library, Paris
Fig. 4.42,   Galileo Galilei, *Sidereus Nuncius.* Venice 1610
               (Reproduced from edition, published by Archival Facsimiles Ltd.
               1987; Astronomisches Rechen-Institut, Heidelberg)
Fig. 4.44   Galileo Galilei, *Il Saggiatore.* Firenze 1655
               (Observatoire de Paris Library, Paris)
Fig. 4.46   Galileo Galilei, *Sidereus Nuncius.* Venice 1610
               (Institut de France Library, Paris)

Fig. 5.5a   Descartes, *Le monde ou le traité de la lumière.*
                   (Institut de France Library, Paris)
Fig. 5.5b,
5.6–7   Descartes, *Principes de la Philosophie.* 1647
                   (Institut de France Library, Paris)
Figs. 5.8, 5.10,
5.12–16 I.   Newton, *Principia.* Cajori edition
                   (Institut de France Library, Paris)
Fig. 5.18–19   C. Huygens, *Systema Saturnium.* Hagae-Comitis, A. Vlacq 1659
                   (Astronomisches Rechen-Institut, Heidelberg)
Fig. 5.20   O. Römer, *Journal des Savants,* 1676. (Institut de France Library,
                   Paris)
Fig. 5.27   A. Pannekoek, *A History of Astronomy,* G. Allen & Unwin, London,
                   repr. Dover, New York 1961
Fig. 6.5   M. Harwit, *Cosmic Discovery – The Search. Scope and Heritage of
                   Astronomy.* Basic Books, New York 1981. Courtesy of Martin Harwit
Fig. 6.6   Erasmus, *L'Éloge de la Folie,* taken from the vol. 1 of the French Edi-
                   tion of *Oeuvres Morales et Philosophiques d'Érasme de Rotterdam,* in
                   10 vols., Publ. Constantin Castéran, Paris 1937.
                   Illustration by H. Holbein (Institut de France Library, Paris)
Fig. 6.7   B. v. Fontenelle, *Entretien sur la pluralité des mondes.*
                   Amsterdam 1719.
                   (Reproduced from German edition, Berlin 1780, p. 440;
                   Astronomisches Rechen-Institut, Heidelberg).
Fig. 6.8   Thomas Wright of Durham, an *Original Theory or New Hypothesis
                   of the Universe.* 1750. Plate XXXI. (Reproduced from facsimile re-
                   print, Astronomisches Rechen-Institut, Heidelberg)
Fig. 7.18   G. Gamow, *Mr. Tompkins in Wonderland.* Cambridge 1978.
                   Courtesy of Cambridge University Press
Fig. 8.11a   *A Source Book in Astronomy and Astrophysics, 1900–1975* (K.R. Lang,
                   O. Gingerich, eds.). Harvard University Press 1979, p. 728.
Fig. 8.11b   E. Hubble, M.L. Humason, *Astroph. J.* **74**, 43 (1931)
Fig. 8.11c   *Galaxies and the Universe* (A. Sandage, M. Sandage, J. Kristian, eds.).
                   The University of Chicago Press 1975, p. 782
Fig. 8.15a–c   Courtesy of NASA, USA
Fig. 8.16   Courtesy of V. de Lapparent, IAP, Paris, France
Fig. 8.19a–c   Courtesy of H. Arp, Garching (München), Germany
Fig. 8.21   Courtesy of A. Stockton, Hawaii, USA
Fig. 8.22a,b   Courtesy of H. Arp, Garching (München), Germany
Fig. 8.23   Courtesy of W. Napier and B. Guthrie, UK
Fig. 8.24a,b   Courtesy of J.-M. Souriau, (Aix-en-Provence) France
Fig. 8.25   G.G. Pooley, M. Ryle, *Mon. Not. R. Astr. Soc.* **139**, 515–528 (1968)
Fig. 8.38a,b   I.E. Segal, *Mathematical Cosmology and Extragalactic Astronomy.*
                   Academic Press, New York 1976, pp. 16–17.
                   Courtesy of I.E. Segal, USA.
Fig. 8.39   Courtesy of H. Arp, Garching (München), Germany

Name Index*

---

* The author has tried, by means of cross references, to give all orthographies of various names; in addition, apologies are expressed for the absence of several initials
of given names, dates of birth or other dates, as well as for unavoidable inaccura-
cies.

Subject Index

The first "subject index" to look at should indeed be the **Table of Contents** (page VII of this book). One should use the present index either to find a definition of a term or notion, or to "navigate" through the book. Note that *Greek words* and *Greek-initialled words* appear in this index at the beginning of the index in alphabetical order; *acronyms* appear in strict alphabetical order, followed by *"see"* and the complete term. *Star abbreviations* appear in alphabetical order, notably in the Greek list at the beginning. The same applies to the *particles* of modern physics. Entries that begin with a *number* immediately follow the entries with Greek letters.

Note also that page numbers written in **bold face** refer to an important passage concerning the entry, or to a definition of this entry. Note also that page numbers in *italics* refer to a figure or table where this entry appears, either in the text, in the figure or table, or in its caption. Page numbers, however, are generally not indicated when the figure or the table is included in the bold face numbered passage of the text concerning this particular entry.

The author apologizes for any possible, and almost unavoidable, errors in this long index.